# MULTICOMPONENT CHROMATOGRAPHY

## *Theory of Interference*

# CHROMATOGRAPHIC SCIENCE

*A Series of Monographs*





OTHER VOLUMES IN PREPARATION

# MULTICOMPONENT CHROMATOGRAPHY

## *Theory of Interference*

FRIEDRICH HELFFERICH
SHELL DEVELOPMENT COMPANY
EMERYVILLE RESEARCH CENTER
EMERYVILLE, CALIFORNIA

AND

GERHARD KLEIN
SEA WATER CONVERSION LABORATORY
UNIVERSITY OF CALIFORNIA
RICHMOND, CALIFORNIA

MARCEL DEKKER, INC., New York 1970

MARCEL DEKKER, INC.
*95 Madison Avenue, New York, New York 10016*

LIBRARY OF CONGRESS CATALOG CARD NUMBER 79-106898

PRINTED IN THE UNITED STATES OF AMERICA

# PREFACE

This book presents original work of the authors; it is not a monograph in the usual sense, in that it calls on the state of the art only where needed for proper perspective. To publish such a work as a book rather than as a series of papers may be unusual, but we have preferred to do so because the unity and complex interrelations of an essentially self-contained theoretical framework would have been seriously impaired and disrupted by publication in separate parts. We trust the concepts developed here will prove of sufficient permanent value to warrant publication in this form, and we are grateful to the editors and publisher for their willingness to share this belief and offer their Chromatographic Science Series as a forum.

We have perhaps not been entirely successful in our quest for a descriptive title. "Chromatography" as commonly understood is both too broad and too narrow a term to characterize the subject: too broad in that we have concentrated on only a particular aspect of chromatographic behavior, that of interference of the sorbable species, an aspect that is not even of greatest concern to the practical user of analytical chromatography; and too narrow in that the principles apply beyond chromatography to any type of system with multicomponent flow through a porous or dispersed sorbent and may be of interest to, say, the chemical engineer or geologist as much as to the analytical chemist. At the core are new concepts relating even more generally to dynamic systems with interference, as found in many other fields of physics and engineering. Yet, for lack of established terms for the more general aspects, we have let the title refer only to chromatography, as the process most directly related to our subject.

The origin of this work dates back to a day in 1963, when the authors shared a flight to a research conference. One of us (G. K.) was at that time preparing work on numerical calculations for multicomponent fixed-bed systems, and the other, as an interested bystander, ventured intuitive predictions of some results. Intriguingly, these predictions later proved correct, challenging their proponent to rationalize his notions and seek proofs. The conceptual framework of the theory was complete in 1964 and was supplemented by the $h$ transformation within a year. From that time on, the second author, having brought the first phase of his separate studies of multicomponent systems to a logical conclusion, participated actively in bringing the treatment into its present form. His direct contributions have been the development of the computer programs and the calculation of the more complex examples, activities in which his associates at the Sea Water Conversion Laboratory of the University of California have shared and which will be reported in more detail in separate publications. The ultimate responsibility for design and implementation lies with the first author, on whom all errors should be blamed.

It is a pleasure to acknowledge the assistance of many colleagues and friends, without whom this work could never have reached its present state. We are grateful to Theodore Vermeulen for valuable counsel and for encouragement offered in his capacity as leader of the Sea Water Conversion Laboratory's research group on ion-exchange separations; to Don L. Peterson, Reinhard Schlögl, John Beek, Olvi Mangasarian, Daniel Tondeur, Lester P. Van Brocklin, and Paul C. Mangelsdorf, Jr., for constructive criticism and searching questions that helped to shape the theory in its early stages; to Jeffrey A. Barnett and Manucher Nassiri for their participation in developing calculation methods; to Suzanne Anderson, Louise Crooks, and Carlotta Anderson for their tireless work on the manuscript; and to various members of Shell Development Company's graphic arts and library staff for assistance in producing the illustrations and checking literature references. We are furthermore indebted to the Office of Saline Water of the U.S. Department of the Interior for sponsoring a major part of the contributions made by members of the Sea Water Conversion Laboratory.

*Berkeley, California*
*December, 1969*

FRIEDRICH HELFFERICH
GERHARD KLEIN

# CONTENTS

# MULTICOMPONENT CHROMATOGRAPHY

*Theory of Interference*

# 1

# INTRODUCTION

This book deals with but a small portion of the wide expanse of chromatographic theory, and a few remarks are called for to clarify its scope and its relation to other theoretical treatments.

The classical theories of chromatography, from the plate model of Martin and Synge [1] to Giddings' statistical treatment [2], operate with the basic premise that the various sorbable species in the column do not interact. This assumption conveniently permits each multicomponent case to be treated as a superposition of independent single-component cases, which can be calculated separately. The assumption is entirely adequate for conventional analytical chromatography at low concentrations and low degrees of sorbent loading, but becomes untenable when high concentrations in the mobile or stationary phase lead to interference of the sorbable species with one another, for example, through their competition for the limited number of available sorption sites. The species then can no longer be considered separately, and each multicomponent case must be treated as a whole. This seemingly minor complication has serious consequences for the theoretical treatment and may be said to add a new dimension to chromatographic behavior. As a result of the growing interest in preparative chromatographic separations and other applications involving high concentrations, multicomponent systems with interference are now encountered more and more often in practice. This book is the first attempt to provide a comprehensive, if rudimentary, treatment of such systems. It thus is intended to supplement, rather than to compete with, the many existing excellent treatises on chromatography.

Interference of the species may arise from various types of interactions: The species may affect one another's kinetic properties (such as sorption rates), they may affect one another's equilibrium distribution between the mobile and stationary phases, and they may be transformed into one another by chemical reactions. In this book, only interference caused by the interdependence of the equilibrium distributions will concern us; variations of kinetic properties merely produce second-order effects in altering the extent of disturbances, and chemical reactions lead beyond what is normally considered the scope of chromatographic theory.*

The implications of interference in chromatography have long been recognized. The basic problem was correctly assessed and stated, in terms of differential equations, as early as 1943 by DeVault [4]. Solutions for various special cases have been given; in particular, the present approach owes much to the pioneer work of Glueckauf [5] and Sillén [6]. These and all later solutions, however, build on a tacit assumption confining them a priori to a certain type of simple behavior—to what we shall later call "coherence"—and thus can cover but a fraction of the broad spectrum of potential phenomena. The treatment presented here was designed with the more ambitious aim of providing general rules not subject to such restrictions and applicable, at least in principle, to any problem of multicomponent flow through porous or dispersed sorbent media, from conventional chromatography to, say, leaching of soils by rain water. Ideally, such a treatment will provide a unifying concept comprising theories for any particular types of chromatographic and related processes as special cases, the classical theories of chromatography included, which will appear as the limiting case of vanishing interference.

Most experts will agree that a "complete" theory of chromatography accounting for all possible effects, if it could ever be developed, would be of unmanageable complexity. Therefore, the "grand design" of a general, unifying concept can only be envisaged if detail is sacrificed. This the present authors have done abundantly: Disregarding all the various

* The elements of a general theory of chromatography of reacting substances have recently been advanced by Collins and Deans [3]. So far, however, this development has remained restricted to small perturbations in systems otherwise at equilibrium.

band- and peak-broadening disturbances usually held to be the central topic of chromatographic theory, they have, at least for the moment, returned to the premises of the so-called equilibrium theories* of the early forties [4, 7, 8]. The problem of how to extend the treatment to account for disturbances is briefly discussed in Chapter 5, but, here, much work still remains to be done.

Granted the simple premises of the equilibrium theories, a general treatment of chromatography with interference may, at first glance, seem to require no more than a straightforward integration of differential equations already stated by DeVault. Inherent difficulties must be overcome, however, which have intrigued DeVault and some of his most perceptive followers (see, for example, the critical assessment by Baylé and Klinkenberg [9]). To obtain mathematical solutions one must specify the form of the equilibrium relations governing the species distributions between the mobile and stationary phases, and one cannot expect any rules derived from such solutions for particular equilibrium properties to be valid in general. Moreover, even for specified equilibrium properties, DeVault's equations have infinite multiplicities of solutions except under certain very simple conditions, to which no theory aspiring to be general can let itself be confined. To avoid multiple solutions is merely a matter of mathematical technique, but to derive general rules applicable to arbitrary equilibrium relations is a problem that cannot be solved by mathematics alone; here, a conceptual understanding of the phenomena and their physical causes must first be gained, and this is the foremost goal the present treatment has set out to accomplish.

Still, the question is bound to be asked: Instead of developing a primitive equilibrium theory of interference that requires later corrections for disturbances, why not incorporate corrections for interference into the modern, highly refined nonequilibrium theories that already account for disturbances? Unfortunately, this is impossible in principle. Although chromatography is intrinsically a dynamic process, its essential features are controlled by the equilibrium distributions of the species between the mobile and stationary phases. Equilibrium properties determine the sequence, types, approximate forms, and propagation velocities of the composition variations arising under any given operating conditions, while kinetic and dynamic effects merely render all these variations somewhat more diffuse. One might say that equilibrium determines the first, and

* Named for their premise of local equilibrium between the mobile and the stationary phase. Ideal plug flow and absence of axial diffusion are also presumed.

kinetics the second and higher statistical moments. Therefore, an equilibrium theory can be refined by corrections for nonequilibrium effects, but a greater complexity of equilibrium properties produces basically new effects that cannot be accounted for by corrections or minor modifications. This is true for the self-sharpening effects and so-called constant and proportionate patterns produced by the concentration dependence of the distribution coefficient ("nonlinear" chromatography), and even more so for the effects produced by interference. To give but one illustration: That a concentration variation of *one* sorbable component at the inlet of a column in equilibrium with an $n$-component influent produces in general not one, but $n-1$ signals which travel through the column at different velocities and each involve concentration variations of *all* components, is an interference effect that, obviously, mere corrections in a classical theory cannot adequately describe.

The treatment in this book does not introduce new premises or hypotheses. Its mathematical basis is the same as that of the classical equilibrium theories (except that interference requires simultaneous differential equations, one for each species, to be solved instead of a single such equation). Therefore, its basic soundness is as unquestionable as that of the equilibrium theories, and its shortcomings in quantitative details are as glaring, at least as far as they are not allayed by the outlined possible extensions. Arguable and awaiting verification is merely the extent to which the treatment has succeeded in capturing the intrinsic features of interference and is capable of reliable predictions.

While nothing new has been added to the mathematical basis, unconventional mathematical techniques are employed and new concepts introduced to derive general rules, solutions, and solution procedures. A quantitative and a conceptual approach are used in combination. The quantitative treatment, leading to convenient explicit criteria and solutions for many cases of practical interest, operates with a nonlinear transformation of the concentration variables by means of the so-called $H$ function. While it may well be possible to generalize this transformation for more complex systems, its use is so far confined to certain simple types of equilibrium relations (constant separation factors, Langmuir-type isotherms). The conceptual approach, which provides a general understanding of the phenomena and their physical causes, is free of such restrictions. Its basis is a material-balance argument in terms of the velocities at which molecules and given concentration values of the species travel in the column. The key concept is that of "coherence," a condition in which the concentrations of the different species advance "in phase," that is, do not shift relative to one another. The value of this concept stems from the fact that all phenomena in the column can be viewed as a development toward the stable condition of coherence. For coherence to be established from the outset, the operating conditions must be of a certain simple type, and only such cases have been covered by other theories for interfering solutes. In the present treatment, the universal trend toward coherence, once proved, permits the behavior to be understood and predicted under any arbitrary noncoherent conditions as well.

It is historically interesting to note that all previous theories for interfering solutes, far from questioning the attainment and stability of so restrictive a condition as coherence, have taken an entirely coherent behavior for granted under the simple conditions to which they had confined themselves. Apparently, only Baylé and Klinkenberg [9] recognized that an unproved postulate was involved. In a sense, the introduction of the concept, and the proof of attainment, of coherence as provided by the present treatment is needed to remove an element of speculation from the basis of the earlier theories.

To go with the $H$-function formalism, coherence, and related concepts, a jargon has been coined which permits the phenomena to be described economically. Apologies are offered to the casual reader who may be appalled when opening the book at random to find himself confronted with unfamiliar language. He will find all new concepts and expressions defined and discussed where introduced (in Chapter 3, Sections IV and V).

True to its title, this book is an entirely theoretical treatise and does not present experimental data or results. It merely contains verification that the treatment, when applied to well-known chromatographic operations such as frontal analysis, elution development, and displacement development, is in complete agreement with experience—were this not so, the book would not have been written. Beyond such cases with known behavior, the treatment yields a wealth of predictions for systems so far not studied experimentally. Such as yet completely unsupported predictions must be regarded as speculative, but are nevertheless offered because their experimental test will provide the best opportunities to confirm the theory or reveal its shortcomings.

Seen from a more general point of view, three dominant elements of the present treatment can be distinguished and deserve comment. The first is the mathematical development in terms of the $H$ function. The significance of this function is not confined to the problem at hand, for which it provides concise and easy-to-use solutions and solution procedures for systems with simple equilibrium properties, but extends to a variety of other dynamic multicomponent problems with interference, such as in electrophoresis, ultracentrifugation, distillation, and steady-state electrodiffusion across membranes. For some of these, operations equivalent to the present transformation with the *H* function have been used previously in various guises by different authors, but have not been characterized and developed as completely as here [10–13]. In fact, it is curious to discover how often the transformation with the *H* function or one of its equivalents has been independently reinvented and rederived, including the present development, which its authors at first mistakenly believed to be entirely original. To arrive at a concise definition of the conditions under which this transformation provides the key to simple solutions of interference problems in any area of physics, physical chemistry, and engineering would be a fascinating objective for further study.

The second general element, an entirely original one, is the conceptual development starting out from an analysis of the velocities of molecules and concentrations and culminating in the principle of coherence. To think in terms of velocities is, of course, not new to chromatographers, but the implications of the fundamental difference between the rates of advance of individual molecules and of given concentration values of a species [14] have gone almost completely unnoticed, and the failure to distinguish between these two velocities has led to various paradoxes and contradictions in chromatographic theory [14, 15]. In the present treatment, a recasting of the fundamental material-balance considerations of chromatography in terms of these velocities provides the key to a general, phenomenological understanding of physical causes and effects, including the principle of development toward coherence from arbitrary noncoherent initial conditions. Here, too, analogies with other areas of physics become apparent, most notably the almost complete formal analogy of the self-sharpening and nonsharpening boundaries of nonlinear chromatography with the shock fronts and rarefaction waves of fluid mechanics. The concept of coherence is, of course, well-known from optics, and a glamorous one at that in this day of the laser. It is tempting to speculate how many more analogies will be found, and to what extent the principles and concepts developed here can be profitably applied to problems in other fields of science.

The third element of this treatment is one of applications and coverage and is closely linked to the other two. Whereas previous theories of chromatography or fixed-bed operation with interfering solutes have aimed at giving solutions for particular cases—that is, for specified equilibrium properties and operating conditions—the present treatment is essentially oriented toward providing a conceptual understanding of, and predictive tools applicable to, any column phenomena regardless of their origin and of the system's equilibrium properties. Various examples of particular types of operation, with specified equilibrium properties, are analyzed in considerable detail, but these are mainly intended to illustrate the application of the general principles to practical cases. Thus, the sets of rules deduced for responses to various types of influent composition changes cover but a small fraction of the effective range of the tools and techniques provided here. A more complete coverage is ruled out by the need to keep the book's length within reason, if not by the authors' limited imagination.

## REFERENCES

1. A. J. P. Martin and R. L. M. Synge, *Biochem. J.*, **35**, 1358 (1941).
2. J. C. Giddings, *Dynamics of Chromatography*, Part I, Marcel Dekker, New York, 1965.
3. C. G. Collins and H. A. Deans, *A. I. Ch. E. J.*, **14**, 25 (1968).
4. D. DeVault, *J. Am. Chem. Soc.*, **65**, 532 (1943).
5. E. Glückauf, *Proc. Roy. Soc.* (*London*), **A 186**, 35 (1946); *Discussions Faraday Soc.*, 7, 12 (1949).
6. L. G. Sillén, *Arkiv Kemi*, **2**, 477 (1950).
7. J. N. Wilson, *J. Am. Chem. Soc.*, **62**, 1583 (1940).
8. J. Weiss, *J. Chem. Soc.*, **1943**, 297.
9. G. G. Baylé and A. Klinkenberg, *Rec. Trav. Chim.*, **73**, 1037 (1954).
10. H. Pleijel, *Z. Physik. Chem.*, **72**, 1 (1910).
11. V. P. Dole, *J. Am. Chem. Soc.*, **67**, 1119 (1945).
12. A. J. V. Underwood, *Chem. Eng. Progr.*, **44**, 603 (1948).
13. R. Schlögl, *Z. Physik. Chem.* (*Frankfurt*), **1**, 305 (1954).
14. F. Helfferich, *J. Chem. Educ.*, **41**, 410 (1964).
15. M. J. E. Golay, in *Gas Chromatography 1964* (A. Goldup, ed.), Institute of Petroleum, London, 1965, p. 143.

# 2

# DEFINITIONS AND BASIC CONCEPTS

In this chapter, the basic concepts, terms, units, and descriptive tools used in the present approach will be defined and explained. Although some of the variables and diagrammatic representations—such as species velocities, simplex and distance-time diagrams—are not commonly used in chromatographic theory, none of the concepts is basically new, and all are readily translated into forms with which the reader may be better acquainted.

Not covered in this chapter are a family of new concepts, of which coherence is the most fundamental. These will emerge from the theoretical treatment in the next chapter and cannot profitably be discussed until various fundamental properties of systems with interfering species have been clarified.

## I. Basic Terms

*Mobile and Stationary Phase.* The theoretical approach in this book deals, in the broadest sense, with the general problem of percolation of a multicomponent *mobile phase* through a dispersed, or porous, multicomponent *stationary phase*, with transfer of species between the two phases. (The term stationary, chosen to conform with common chromatographic usage, should not be taken to exclude adaptation of the theory to countercurrent systems, in which both phases move.) The mobile phase may be a liquid or gas. The stationary phase usually is a solid or gel, but may also be a liquid supported on a solid carrier, as, for example, in gas–liquid chromatography.

*Geometry.* Although the theoretical treatment is not confined to chromatography and fixed-bed operation, the standard terminology for such operations will be employed. The system considered thus is called a *column*, packed with a sorbent constituting (or supporting) the stationary phase. The fluid filling the void between the sorbent particles constitutes the mobile phase.

Fluid enters the column at the *inlet* as *influent*, and emerges at the *exit* as *effluent*. Inlet and exit are understood to be the locations at which the fluid first and last contacts the stationary phase. That is, no distinction is made between "column" and "bed," and any liquid head above the bed

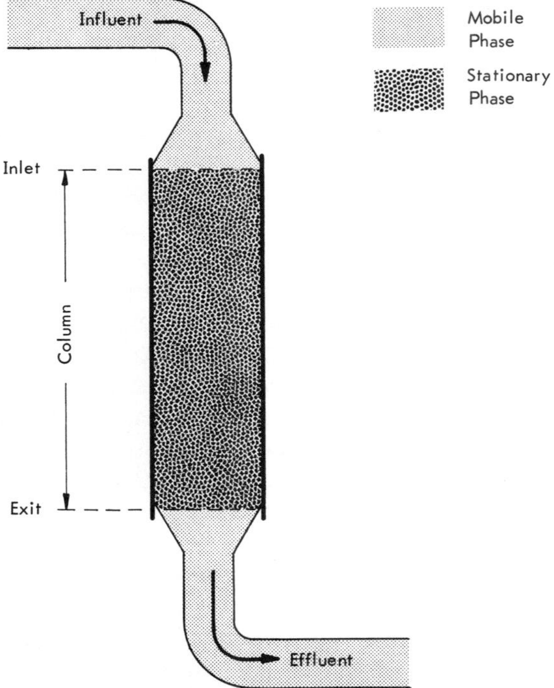


**Fig. 2.1.** Schematic picture of column operation.

thus is counted as influent (or, in upflow operation, as effluent) rather than as contents of the column (see Fig. 2.1).

Within the column, *upstream* and *downstream* denote relative positions with respect to the direction of mobile-phase flow. As far as practicable, terms such as top, bottom, above, and below are avoided because of their ambiguity when the direction of flow is not specified.

*Sorption, Desorption, and Stoichiometric Exchange. Sorption* is commonly understood to be uptake of molecular or ionic species by one phase from another, in general by a more condensed phase (*sorbent*) from a less condensed one. *Desorption* is the reverse of sorption. Neither term implies a particular mechanism, such as retention by a surface (adsorption) or dissolution within a bulk phase (absorption). In the present context, the stationary phase is the more condensed one, and sorption thus refers to uptake of species by this phase.

Often, but not necessarily, sorption of one species is accompanied by desorption of another. This is called *exchange sorption.* In certain special cases of exchange sorption, the sorbent has a fixed, constant capacity for sorbable species and is at all times loaded to this capacity. Exchange of species between the phases then is necessarily *stoichiometric*, i.e., the amounts sorbed and desorbed equal one another. The special case of stoichiometric exchange is approximated, for example, in ion exchange with dilute solutions because, here, conservation of electroneutrality requires the fixed charges of the ion exchanger to be compensated at all times by an equivalent amount of exchangeable counterions in the ion exchanger. (Of course, the exchange is stoichiometric in terms of counterion equivalents rather than moles.)

*Sorbable Species.* The theoretical treatment in this book is chiefly concerned with the behavior of molecular or ionic species that can be sorbed and desorbed. In chromatographic usage, such species are called *solutes*, *sorbable components*, or *sorbable species.* Here, the short term *species*, unless qualified, will be understood to designate sorbable species.

The sorbable species are not the exclusive constituents of the system. The mobile phase may or may not comprise additional, inert constituents, e.g., an inert solvent or carrier gas. The stationary phase usually has a constituent that is insoluble in the mobile phase; the theory also applies to vapor–liquid systems without such a constituent, but the stationary liquid phase then is supported on an inert solid carrier. For the sake of simplicity, only the sorbable species are counted as components. Thus, an *n-component system* is taken to be a system with $n$ sorbable species. For example, a system in which ion exchange of $Na^+$, $H^+$, and $Ag^+$ is the relevant process is called a three-component system; its species are the three exchangeable cations; the count of species or components thus does not include the co-ions, the solvent, or the matrix of the ion exchanger.

*Trace Species and Trace-Component Systems.* For the purpose of the present approach, a *trace species* is defined as a species whose concentrations in both the mobile and stationary phases are infinitesimal compared to those of at least one other species. Species having finite concentrations will be designated as *bulk species*, where distinction from trace species is called for.

Of particular interest in certain parts of the present treatment are systems with one bulk species and all other species at trace levels. Such systems will, for short, be called *trace-component systems.* Thus, this term implies that there is no more than one bulk species.

## II. Composition Variables

*Concentrations.* The concentration of a species is usually defined as the amount (e.g., number of moles) of the species per unit volume or weight of the respective phase. In chromatography, however, equations of simpler forms are obtained if the concentrations in the mobile as well as the stationary phase are given *per unit volume of the column* rather than of the respective phase. This definition will be used here for the mobile- and stationary-phase concentrations $C_i$ and $\bar{C}_i$ of species *i.*

The relations between the concentrations so defined and those most frequently used by other authors are

$$C_i = \varepsilon c_i \qquad \text{and} \qquad \bar{C}_i = \rho q_i \tag{2.1}$$

where $c_i$ is the mobile-phase concentration per unit volume of the mobile phase, $q_i$ is the stationary-phase concentration per unit weight of the stationary phase, $\varepsilon$ is the fractional interparticle void volume of the bed, and $\rho$ is the weight of the stationary phase per unit volume of the column.

It will be understood that, unless otherwise noted, concentrations $C_i$ and $\bar{C}_i$ are in moles per cubic centimeter. However, other units can be substituted (e.g., equivalents for moles), provided this is done consistently.

For deriving certain relations it will prove expedient to introduce the *overall concentration* $\tilde{C}_i$ of species *i*, defined as the combined amount of this species *in both phases* per unit volume of the column. Thus,

$$\tilde{C}_i \equiv \bar{C}_i + C_i \tag{2.2}$$

(It is understood that the amount of species $i$ is averaged over a volume large enough that the phase-volume ratio is at its average for the bed.)

The *total concentrations* $C$ and $\bar{C}$, in the mobile and the stationary phase, respectively, are defined as the sums of the concentrations of all sorbable species in the respective phase:

$$C \equiv \sum_i C_i \qquad \text{and} \qquad \bar{C} \equiv \sum_i \bar{C}_i \tag{2.3}$$

(Nonsorbable species such as inert solvent or carrier gas, co-ions and solvent in ion exchange, etc., are not included in the summation.)

For systems with constant total concentrations $C$ and $\bar{C}$, extensive use will be made of *normalized concentrations* $x_i$ and $y_i$, defined as

$$x_i \equiv C_i/C \qquad \text{and} \qquad y_i \equiv \bar{C}_i/\bar{C} \tag{2.4}$$

In view of the definitions (2.3),

$$\sum_i x_i = 1 \qquad \text{and} \qquad \sum_i y_i = 1 \tag{2.5}$$

*Compositions.* A complete set of species concentrations in the mobile or the stationary phase constitutes a *composition*. For example, $\{C_1, \ldots, C_n\}$ or $\{x_1, \ldots, x_n\}$ is a mobile-phase composition. For convenience, a shorthand vector notation, with bold symbols, will be used for such sets. Thus, for example,

$$\mathbf{x} \equiv \{x_1, \ldots, x_n\} \qquad \text{and} \qquad \mathbf{y} \equiv \{y_1, \ldots, y_n\} \tag{2.6}$$

*Composition Space and Simplex.* Any composition of an $n$-component mobile or stationary phase can be represented by a point in the so-called *composition space*, i.e., in a coordinate system with the $n$ species concentrations as coordinates. Generally, orthogonal carthesian coordinates are used, and the composition space then is $n$-dimensional.

If the total concentration is a given constant, then only $n-1$ independent concentrations define a composition, the $n$th being fixed by Eq. (2.3) or (2.5). The number of dimensions of the composition space can accordingly be reduced to $n-1$. A convenient way of graphical representation in such cases is by means of a so-called *simplex*, as illustrated in Figs. 2.2 and 2.3. The restriction to a constant total concentration confines the possible compositions of an $n$-component phase in the conventional $n$-dimensional space with orthogonal coordinates to an $(n-1)$-dimensional hyperplane (a plane in a three-component system, a straight line in a two-component system). This is shown for two- and three-component systems in Fig. 2.2, left. The $(n-1)$-dimensional line, plane, or hyperplane to which the compositions are confined is the "simplex" of the system, as shown in Fig. 2.2, right. The simplex of a two-component system thus is a straight line, that of a three-component system is an equilateral triangle, that of a four-component system is a regular tetrahedron (see Fig. 2.3), and, in general, that of an $n$-component system is an $(n-1)$-dimensional, $n$-cornered, regular polyhedron.

Since the reduction to $n-1$ dimensions simplifies diagrams as well as topological arguments, the simplex representation will be used extensively

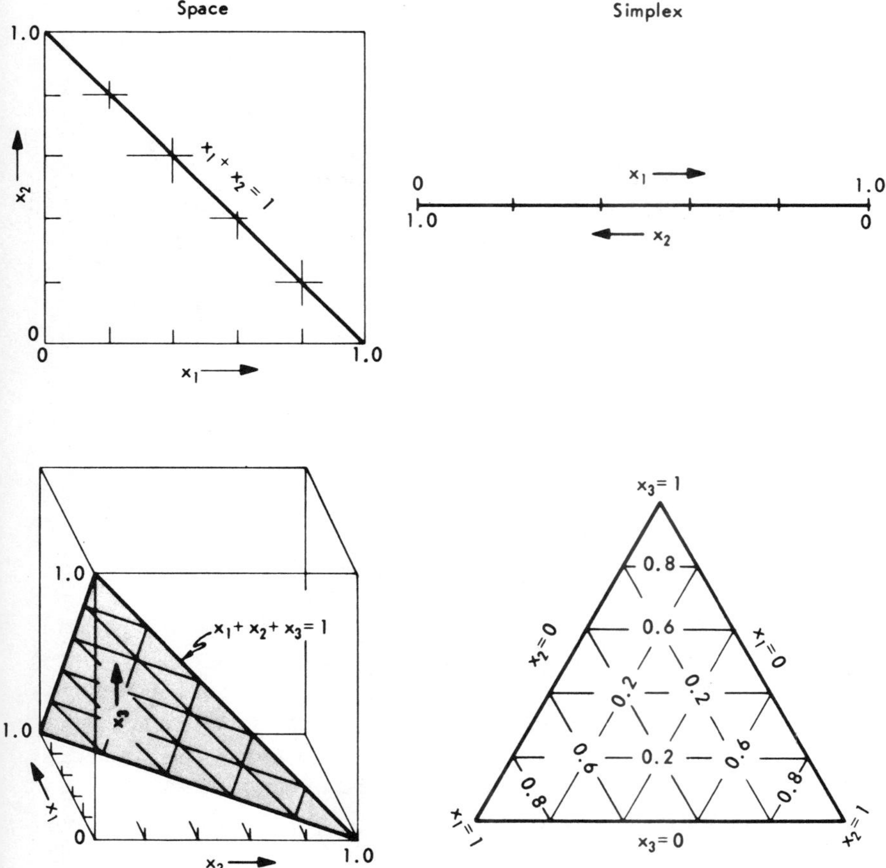


**Fig. 2.2.** Composition space with orthogonal concentration coordinates (left) and composition simplex (right), for two- and three-component systems with constraint of constant total concentration.

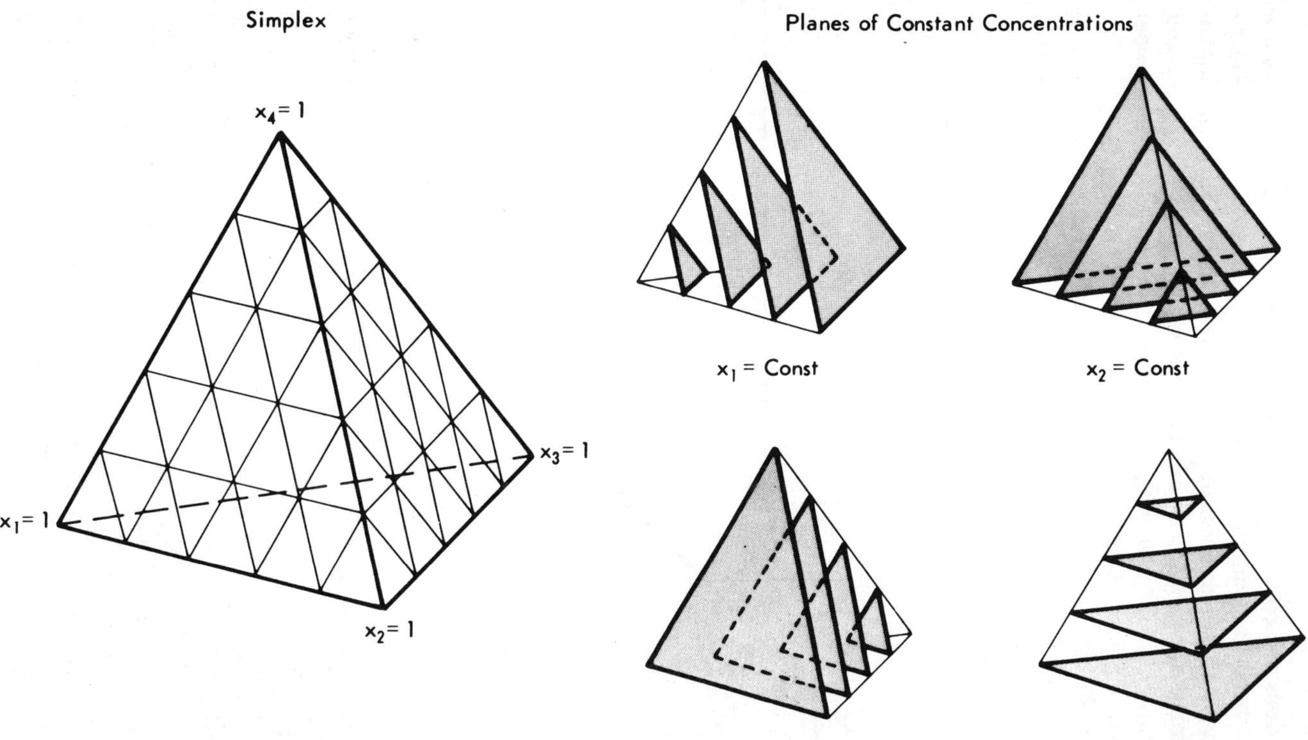


**Fig. 2.3.** Composition simplex of four-component system (left), and loci of constant concentrations of components (right).

in the next chapters. Of course, the use of a simplex instead of an $n$-dimensional orthogonal coordinate system is entirely a matter of convenient representation and visualization and has no bearing on mathematical derivations and equations.

## III. Representation of Equilibrium Properties

The following terms and concepts will be used for characterizing and representing the equilibrium distribution of species between the mobile and stationary phases.

*Distribution Ratios and Separation Factors.* The *distribution ratio* $\bar{C}_i/C_i$ of a species $i$ is defined as the ratio of the concentrations, both per unit volume of the column, of the species in the stationary and the mobile phase. The distribution ratio is not identical with the commonly used distribution (or partition) coefficient, since the latter is defined in terms of concentrations per unit volume or weight of the respective phases.

For systems with constant total concentrations it will be convenient to use the *normalized distribution ratio* $y_i/x_i$ in conjunction with the normalized concentrations; the two ratios differ merely by a constant factor:

$$y_i/x_i = (C/\bar{C})(\bar{C}_i/C_i) \tag{2.7}$$

The binary *separation factor* of two species $i$ and $j$ is defined as

$$\alpha_{ij} \equiv \frac{\bar{C}_i C_j}{C_i \bar{C}_j} \quad \text{or} \quad \alpha_{ij} \equiv \frac{y_i x_j}{x_i y_j} \quad \text{(in equilibrium)} \tag{2.8}$$

[In view of Eqs. (2.4) and (2.5), these two definitions are equivalent.] The separation factor $\alpha_{ij}$ thus is the ratio of the distribution ratios of the two species, $i$ and $j$. The definition (2.8) for the binary separation factor applies regardless of the presence of additional species. Note that the definition implies

$$\alpha_{ji} = 1/\alpha_{ij}, \qquad \alpha_{ij}\alpha_{jk} = \alpha_{ik}, \qquad \text{and} \qquad \alpha_{ii} = 1 \tag{2.9}$$

*Affinity and Affinity Sequence.* The term *affinity*—or, explicitly, affinity for the stationary phase—will be used as a relative measure of the tendency of a species to be sorbed by the stationary phase. Thus, species $i$ is said to have higher affinity than species $j$ if its distribution ratio is larger than that of $j$, i.e., if

$$\bar{C}_i/C_i > \bar{C}_j/C_j \qquad \text{(in equilibrium)} \tag{2.10}$$

In view of Eq. (2.8), condition (2.10) is equivalent to $\alpha_{ij} > 1$.

The *affinity sequence* is defined as the sequence of sorbable species, arranged in the order of decreasing affinity. That is, for any member of the sequence, the distribution ratio is larger than those of all subsequent members. The affinity sequence may be invariant or may vary with composition. In systems with invariant affinity sequence, the species will be numbered in the order of this sequence, i.e., so that any species $i$ has a higher affinity than species $i+1$.

$$\bar{C}_i/C_i > \bar{C}_{i+1}/C_{i+1} \qquad \text{for all } i = 1, \ldots, n-1 \text{ (in equilibrium)} \tag{2.11}$$

*Interfering Species.* Species will be called *interfering* if they affect one another's equilibrium distribution between the mobile and stationary phases, and *noninterfering* if they do not.

In conventional theories of chromatography, it is usually postulated that the stationary-phase concentration of any species $i$, in equilibrium, is a function of the mobile-phase concentration of that species alone:

$$\bar{C}_i = f(C_i) \qquad \text{(in equilibrium)} \tag{2.12}$$

regardless of the presence of other species. By definition, then, the species in such systems are noninterfering. In contrast, the present approach deals specifically with interfering species, i.e., with systems in which the stationary-phase concentration of any species $i$ depends not only on the mobile-phase concentration of that species, but on those of all other species as well:

$$\bar{C}_i = f(C_1, \ldots, C_n) \qquad \text{(in equilibrium)} \tag{2.13}$$

*Competitive and Synergistic Sorption.* It will be expedient to distinguish between different types of interference. Commonly, interference arises from the competition of the molecules of the various species for the limited capacity or limited number of sorption "sites" of the sorbent. In this case, any addition of molecules of an arbitrary species $k$ to the mobile phase (at otherwise constant mobile-phase composition), providing increased competition, will reduce the stationary-phase concentrations of all other species $j$:

$$\left(\frac{\partial \bar{C}_j}{\partial C_k}\right)_{C_i(i \neq k)} < 0 \qquad \text{for all } j \text{ and } k \text{ with } j \neq k \tag{2.14}$$

(Here and below, only isothermal equilibrium systems are considered.) The higher the affinity of the added molecules, the stronger is the competition they provide, and a stronger competition usually results in a greater reduction of the stationary-phase concentrations of the other species:

$$\left(\frac{\partial \bar{C}_j}{\partial C_k}\right)_{C_i(i\neq k)} < \left(\frac{\partial \bar{C}_j}{\partial C_l}\right)_{C_i(i\neq l)} < 0 \quad \text{if } \alpha_{kl} > 1 \qquad \begin{array}{l}\text{for all } j, k, \text{and } l\\ \text{with } j \neq k, l \text{ and } k \neq l\end{array} \tag{2.15}$$

Condition (2.15), which comprises the weaker condition (2.14), is typical for interference arising from competition. Sorption behavior obeying condition (2.15) will be called *competitive sorption.*

It is also possible that addition of molecules of a species $k$ to the mobile phase increases rather than reduces the stationary-phase concentrations of one or several other species $j$:

$$\left(\frac{\partial \bar{C}_j}{\partial C_k}\right)_{C_i(i\neq k)} > 0 \qquad (j \neq k) \tag{2.16}$$

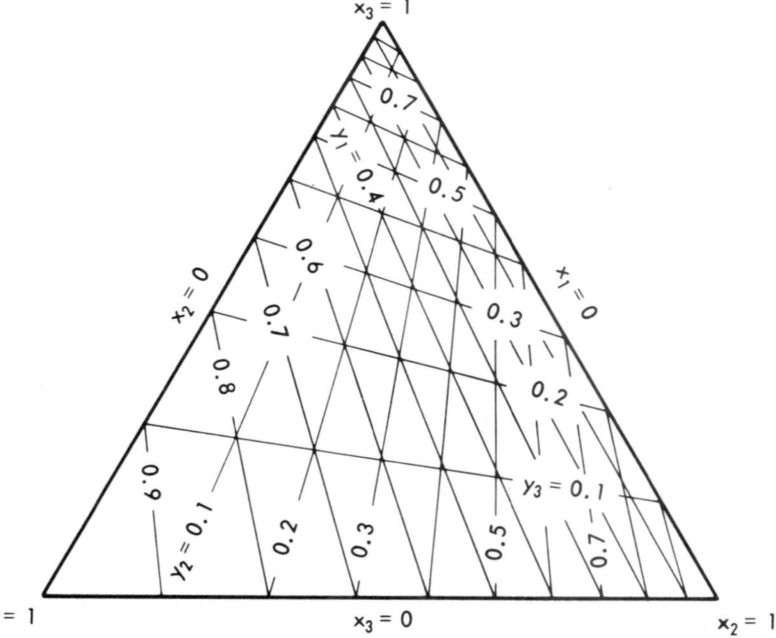


**Fig. 2.4.** Loci of constant stationary-phase concentrations in mobile-phase composition simplex of three-component system (for constant separation factors $\alpha_{12} = 2$, $\alpha_{13} = 4$).

Such enhancement of sorption of one species by another may arise, for example, from a tendency of the molecules of the two species to form aggregates or complexes with one another in the stationary phase. Sorption behavior obeying condition (2.16) will be called *synergistic sorption*.

*Sorption Isotherms.* The *sorption isotherm* is a functional relation or diagram giving the stationary-phase composition as a function of the mobile-phase composition, in equilibrium and at constant temperature.

In systems with only one sorbable species, the isotherm is usually shown as a plot of the stationary-phase versus the mobile-phase concentration (e.g., see Fig. 3.4). In $n$-component systems without interference, there are $n$ independent isotherms of this kind, one for each species. Single-species isotherms may be *linear* or *nonlinear*. A nonlinear isotherm, as in Fig. 3.4, implies that the distribution ratio varies with composition.

In multicomponent systems with interference, the stationary-phase concentration of each species is, by definition, a function of all mobile-phase concentrations [see Eq. (2.13)], and the isotherm thus is multidimensional. As an example, the equilibrium distribution in a three-component system with stoichiometric exchange and constant separation factors is illustrated in Fig. 2.4, which shows loci of constant stationary-phase concentrations of the various species in the mobile-phase composition simplex.

## IV. Representation of Column Behavior

The terms and formats used for the description and diagrammatic representation of column behavior will be summarized in this section. Most of the descriptive tools are those commonly used in chromatography. However, a few terms and concepts have been newly coined or designed for the specific needs of the present theoretical approach.

### A. General Properties

*Profiles and Histories.* It is convenient to describe column behavior in terms of concentration profiles or histories. A *concentration profile* shows the concentration of the respective species as a function of the distance from the column inlet at a fixed, specified time. A *concentration history* shows the concentration as a function of time at a fixed, specified distance from the column inlet.

Profiles and histories may show mobile-phase, stationary-phase, or overall concentrations. Unless otherwise stated, however, they will be taken as referring to the mobile phase.

A set of synchronous concentration profiles of all species constitutes a *composition profile*. Similarly, a set of syntopic concentration histories of all species constitutes a *composition history*.

A profile referring to zero time is called *initial profile*. Histories referring to the column inlet and exit are called *influent history* and *effluent history*,* respectively.

*Plateau Zones, Boundaries, Steps, and Pulses*. The following terms will be used to describe characteristic features of profiles and histories. A *plateau zone* is a finite region of constant composition, and a *boundary*† is a composition variation. A boundary is said to be *diffuse* if the variation is gradual, and *sharp* if the variation is abrupt (i.e., discontinuous). An abrupt composition variation is also called a *composition step*. For coherent boundaries and steps, see Chapter 3, Section IV.B.

The terms plateau zone and boundary, unless qualified, will be understood to refer to composition, whereas the term step is also applied to other variables (e.g., *concentration step*). Of course, discontinuous variations are idealizations merely approximated in real systems.

A concentration variation involving a maximum or minimum will be called a *concentration pulse*. The "peaks" in conventional chromatography are concentration pulses by this definition; however, the term pulse is not confined to variations with zero or equal concentrations on the two sides (e.g., see Fig. 2.5). The two portions of a pulse, on the upstream and downstream sides of the extremum, are its *rear* and *front flanks*. Pulses with one, two, and more than two extrema of the respective variable will be called *single*, *double*, and *multiple pulses*.

A set of concurrent concentration pulses of different species constitutes a *composition pulse*. For instance, sample introduction into the carrier gas in conventional gas chromatography gives rise to an *influent composition pulse* (or, for short, *influent pulse*) at the column inlet. For coherent composition pulses, see Chapter 3, Section IV.G.

---

* The term "breakthrough curve", often found in chemical engineering literature, describes the effluent history in particular types of operation.

† *Boundary* is an accepted, although perhaps not very fortunately chosen, term in chromatography and electrophoresis. In other theories of chromatography, boundaries have been variously called fronts, concentration changes, waves, and transitions.

*Composition Routes.* The sequence of compositions in a composition profile or history can be mapped in the composition space or simplex. The resulting curve is called a *composition route.* There are *composition profile routes* and *composition history routes,* and either may be in terms of mobile-phase, stationary-phase, or overall concentrations. The brief term composition route (or, simply, route), unless qualified, will be taken to denote a profile route in terms of mobile-phase concentrations, since other routes will be used only in exceptional cases.

Composition routes are not synonymous with composition paths, which will be introduced and defined in Chapter 3, Section IV.C.

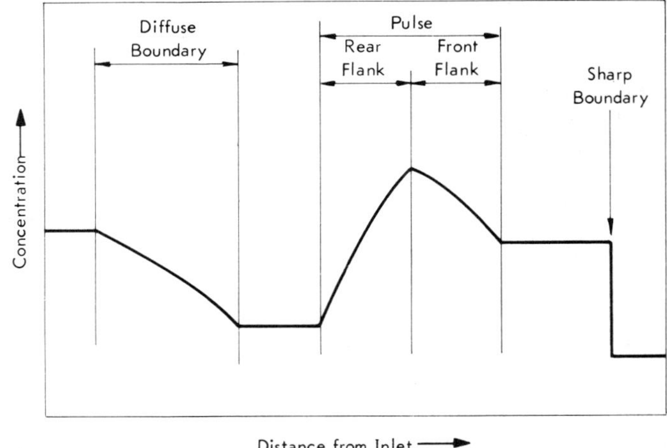


**Fig. 2.5.** Concentration profile with diffuse boundary, pulse, and sharp boundary.

*Response and Response Patterns.* The term *response* (or *column response*) expresses the behavior of the species in the column as produced by specified initial and influent conditions. The term is generally used in the quantitative sense, denoting the complete solution $\mathbf{C}(z, t)$, $\overline{\mathbf{C}}(z, t)$ under given conditions.

The response at any given stage of development (time, or distance from inlet) can be characterized by its *response pattern* (or, for short, *pattern*), which indicates the sequence and general properties of the plateau zones, boundaries, and pulses in the column. The pattern does not, however, give quantitative information as to concentrations, widths of plateau zones and boundaries, etc.

The response pattern may change as development progresses. However, in many types of operation of practical interest, a final pattern is eventually attained. In conventional chromatography, for example, a final pattern has been attained when the various peaks are completely resolved.

## B. Velocities

The conceptual approach in the next chapter is largely based on an examination of the velocities at which the mobile phase, molecules of sorbable species, concentrations, steps, etc., advance through the column.

The use of velocities is contingent on that of distance and time as the independent variables. This choice of variables seems natural for describing column behavior. Yet, most theories of chromatography operate with other independent variables, for example, with throughput volume [1–10] instead of time, and with amount of sorbent [5, 11] or bed volume [7–10] upstream of a location instead of its distance from the inlet, in order to make the treatment directly applicable to systems with variable as well as constant flow rate and cross-sectional area of the column (see Chapter 5, Section I). In all other respects, however, the alternative variables are essentially equivalent to distance and time. The present approach operates with distance, time, and velocities not for mathematical expedience, but because these more readily visualized quantities allow the conceptual parts of the theory to be stated in simpler terms.

With the distinction between the various velocities, concepts not previously used in chromatographic theory are introduced. The definitions will therefore be examined and interpreted in detail.

*True Velocities.* Unless converted to the adjusted-velocity scale (see p. 24), all velocities will be taken as *linear velocities* $\partial z/\partial t$, where $z$ is the linear distance from the column inlet, and $t$ is time. Units of $z$ and $t$ are understood to be centimeters and seconds, but other units can be substituted if this is done consistently. Where a distinction from adjusted velocities is called for, the linear velocities are named *true velocities.*

By definition, true velocities are positive for motion in the direction of mobile-phase flow, and negative for motion in the opposite direction.

*Types of Velocities.* Velocities of three different types will appear, (1) that of bulk mobile-phase flow, (2) those of molecules (or ions) of given species, and (3) those of given values or steps of a variable such as a concentration or composition.

1. The *mobile-phase velocity,* $u_0$, is a convective velocity stating the flow rate of a volume element of the bulk mobile phase. It can be measured, for

example, by observing the advance of a marker which remains entirely in the bulk mobile phase and is not retarded by interactions with the stationary phase (e.g., by shear, friction, or adhesion).

2. The *species velocity*, $u_i$, of a given species $i$ is defined as the average rate of advance, in the direction of flow, of the molecules or ions of this species. It can be measured by observing the advance of a tracer (e.g., radioisotope, distinguishable stable isotope, or molecule tagged with either) of the respective species.

The species velocities are lower than the mobile-phase velocity because the molecules of the species spend part of their time in the stationary phase and thereby are retarded relative to the mobile-phase flow. In general, the species velocities are positive, since an average molecular motion against the direction of flow can occur only under very exceptional circumstances.

3. A typical example for a velocity of the third type is the *concentration velocity* of a species, defined as the rate of advance of a given concentration value of the species in the direction of flow. For example, the concentration velocity $u_{C_i}$ of a given concentration value $C_i$ is defined as

$$u_{C_i} \equiv \left(\frac{\partial z}{\partial t}\right)_{C_i} \tag{2.17}$$

A concentration velocity may refer to the mobile-phase, stationary-phase, or overall concentration of a species. These three concentrations, of the same species and at the same location and time, do not necessarily advance at the same rate, so that their velocities $u_{C_i}$, $u_{\bar{C}_i}$, and $u_{\tilde{C}_i}$ must be distinguished. Velocities of given values of other variables are defined in an analogous manner (for *composition velocity* and *root velocity*, see Chapter 3, Sections IV.A and V.A, respectively). Similarly, the *step velocity* is defined as the rate of advance of a step of a given variable.

The rate of advance of a given value of a variable is well defined only if this value occupies a distinct location $z$ at any moment in the time interval of interest. Therefore, the velocity of this value is physically significant only where the variable has a gradient, not where it has a plateau. By the same token, a step velocity can be defined only where the respective variable has a step which remains sharp while advancing.

The concentration velocity (or step velocity) refers to propagation of a variation or disturbance, as distinct from the species velocity, which refers to travel of molecules. Describing different physical phenomena, the two velocities generally differ from one another [12]. The distinction is not commonly made in chromatography, but is familiar from fluid mechanics (wave velocity and particle velocity; see also Chapter 3, Section III.B). Crude analogies are the differences between the propagation of a flood wave racing down a river and the flow of the river water, and between the propagation of a traffic jam on a highway and the progress of the individual cars [12]. The flood wave reaches the estuary well in advance of the rain-water molecules that caused the flood upstream. A traffic jam is even propagated *against* the direction of traffic flow, as arriving cars are forced to slow down or stop *behind* preceding ones while these resume speed (see Fig. 2.6). It will be seen that, in chromatographic systems with interference, concentration values may travel at higher rates than, or in the opposite direction as, the molecules of the respective species (see Chapter 3, Section III.B).

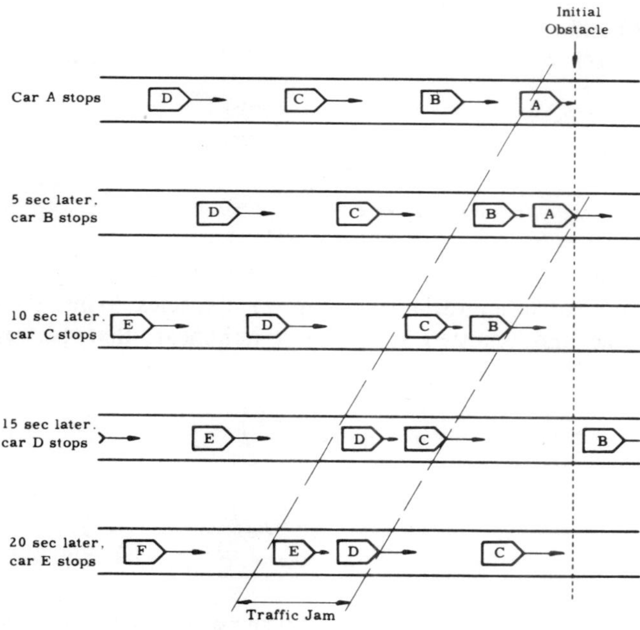


**Fig. 2.6.** Propagation of a traffic jam (from F. Helfferich [12], reproduced from *Journal of Chemical Education* with permission of the American Chemical Society).

*Adjusted Time and Adjusted Velocities.* In the quantitative treatment of chromatographic phenomena, a significant reduction in mathematical complexity can be achieved if so-called adjusted, instead of true, time and velocities are used. For systems with constant total concentrations and constant mobile-phase velocity, with which we shall mainly be concerned, the use of the adjusted time $\tau$ defined as

$$\tau \equiv \frac{Cu_0}{\bar{C}}\left(t - \frac{z}{u_0}\right) \tag{2.18}$$

will prove convenient. *Adjusted velocities* $\mathscr{u}_i$, $\mathscr{u}_{Ci}$, etc., are defined in terms of adjusted instead of true time; for example, the adjusted velocity $\mathscr{u}_{C_i}$ of a given concentration value $C_i$ is, by definition,

$$\mathscr{u}_{C_i} \equiv (\partial z/\partial \tau)_{C_i} \tag{2.19}$$

The adjusted time has formally the dimension of a length,* while the adjusted velocities are dimensionless, as is readily established with Eqs. (2.18) and (2.19).

The relations between the true and adjusted velocities are

$$u = \frac{u_0}{1 + \bar{C}/C\mathscr{u}} \tag{2.20}$$

$$\mathscr{u} = \frac{\bar{C}u}{C(u_0 - u)} \tag{2.21}$$

where $u$ and $\mathscr{u}$ stand for any pair of true and adjusted velocities of a concentration or other variable, a step, or a species. While the relation between true and adjusted velocities is nonlinear, that between their reciprocals is linear:

$$\frac{1}{u} = \frac{1}{u_0}\left(\frac{\bar{C}}{C\mathscr{u}} + 1\right) \tag{2.22}$$

$$\frac{1}{\mathscr{u}} = \frac{Cu_0}{\bar{C}}\left(\frac{1}{u} - \frac{1}{u_0}\right) \tag{2.23}$$

---

* A time variable with the dimension of a length may meet with some objections on aesthetic grounds and can for most practical purposes be avoided through the use of dimensionless variables $z/z_0$ and $\tau/z_0$ for distance and time, where $z_0$ is some characteristic length, e.g., of the column or of an initial band of species. However, this procedure is not practicable in the present general approach, much of which deals with phenomena at arbitrary locations in columns of indefinite length under unspecified initial conditions.

Equations (2.20) to (2.23) can be derived as follows. Partial differentiation of Eq. (2.18) with respect to $z$ at, say, constant $C_i$ yields

$$\left(\frac{\partial \tau}{\partial z}\right)_{C_i} = \frac{Cu_0}{\bar{C}}\left[\left(\frac{\partial t}{\partial z}\right)_{C_i} - \frac{1}{u_0}\right] \tag{2.24}$$

Replacing the derivatives by means of Eqs. (2.17) and (2.19) one obtains

$$\frac{1}{u_{C_i}} = \frac{Cu_0}{\bar{C}}\left(\frac{1}{u_{C_i}} - \frac{1}{u_0}\right) \tag{2.25}$$

This relation is of the form of Eq. (2.23). The derivation for step velocities, species velocities, etc., is analogous. Equations (2.20) to (2.22) are readily obtained from (2.23) by rearrangement.

The transformation of true into adjusted time in Eq. (2.18) involves an "adjustment" by the term $-z/u_0$ and a normalization by the factor $Cu_0/\bar{C}$. The adjustment is essential for simplifying the mathematical treatment, whereas the normalization merely eliminates constant factors from a number of equations. In various guises, the adjustment appears in most theories of chromatography. An adjusted time $t-z/u_0$ with or without normalization factor has been used by many previous authors [13–17]. In gas chromatography, the adjustment is equivalent to the transition from true to adjusted ("apparent") retention times or volumes by subtraction of the retention time or volume of the "air peak" [18–21]; this becomes apparent when one considers that $z/u_0$ is the retention time of an inert marker, traveling at the velocity $u_0$, in a column of length $z$. The use of an adjusted time is furthermore equivalent to that of a throughput volume or other throughput variable defined so as not to include the fluid initially in the void space of the bed ([4, 6, 11]; see also Chapter 5, Section I). The variable $\psi$ introduced by Sillén [6] and named "throughput parameter" **T** by Vermeulen and co-workers ([7–10]; symbol $t/s$ or **Z** used in the earlier publications) is an adjusted and normalized variable and is related to the adjusted time $\tau$, defined in Eq. (2.18), by*

$$\psi = \mathbf{T} = \tau/z \tag{2.26}$$

Because of the adjustment by $-z/u_0$, the adjusted time at a fixed true time varies with the distance $z$ from the column inlet. The longer this distance, the larger is the shift of the adjusted relative to the true time

---

* This variable involving both $\tau$ and $z$ (or their equivalents) was designed for, and is essentially restricted to, systems in which it can serve as the sole independent variable, obviating the need for separate distance and time variables; see also Chapter 3, Section III.B.1, and Chapter 4, Section I.H.

scale. The physical significance of the adjustment is that for any volume element of the bulk mobile phase (or inert marker in that phase) traveling through the column the adjusted time "stands still," i.e., the volume element passes through each column layer at the same adjusted time. The adjusted time at any column layer begins to count not from $t = 0$, but from the moment at which the layer is reached by the mobile-phase volume elements that had entered the column at $t = 0$. (Occasionally, these volume elements have been called the "solvent front" or "carrier-gas front.") The distance-time diagram, to be discussed later in this section, will further illustrate the nature of the adjusted time.

As stated earlier, the adjusted velocities are defined in terms of the adjusted instead of the true time. On the adjusted scale, the velocity of the mobile phase is infinite since, as stated above, the adjusted time is defined so as to "stand still" for the traveling volume elements of this phase. Entities (e.g., concentrations) traveling faster than the mobile phase have negative adjusted velocities since, on the adjusted time scale, they arrive "earlier" at locations downstream than they had departed from upstream. (Compare the negative retention time that would be calculated in gas chromatography for a peak emerging ahead of the air peak.) For $\tau$ as defined in Eq. (2.18), it is readily shown with Eq. (2.21) that

$$
\begin{array}{llll}
-\bar{C}/C < \upsilon < 0 & \text{if} \quad u < 0, & \upsilon = 0 & \text{if} \quad u = 0, \\
\upsilon > 0 & \text{if}\ 0 < u < u_0, & \upsilon \to \pm\infty & \text{if} \quad u = u_0, \\
\upsilon < -\bar{C}/C & \text{if} \quad u > u_0 & &
\end{array}
\tag{2.27}
$$

[As in Eqs. (2.20) to (2.23), $u$ and $\upsilon$ stand for any pair of related true and adjusted velocities.] With the normalization as in Eq. (2.18), the adjusted velocity of an entity becomes the ratio of two quantities: the amount of stationary phase traversed by the entity per unit time, and the amount of mobile phase overtaking the entity per unit time, both amounts being expressed in moles (or equivalents, if that concentration scale is chosen) of sorbable species in the respective phase. Also, with this normalization, the average species velocity $\Sigma_i y_i \upsilon_i$ at any composition is unity.

The nature of the adjusted velocity may be illustrated by the analogy to a sailboat driven by a following wind. Let the velocity of the boat through the water (presumed stagnant) be $u$ knots, and that of the wind $u_0$ knots. The apparent wind velocity, as observed from the boat, then is $u_0 - u$, and the velocity of the boat as a fraction of the apparent wind velocity is $u/(u_0 - u)$. In a sense, this fraction is a measure of the effectiveness of wind utilization, while its reciprocal is a measure of the effectiveness of friction with the water. The fractional velocity $u/(u_0-u)$ has the characteristics of an adjusted velocity, $u$ being the true velocity of the object, and $u_0$ that of the flow acting as driving force (in chromatography, the mobile-phase flow). As Eq. (2.21) shows, the adjusted velocity $\mathscr{u}$ used here differs from $u/(u_0-u)$ merely by a constant normalization factor.

The extension of the concept of adjusted velocities to systems with variable flow rate and total concentrations will be discussed in Chapter 5, Section I.

## C. Distance-Time Diagrams

A particularly valuable tool for depicting column responses is the distance-time diagram,* illustrated in Figs. 2.7 and 2.8. In such a diagram, composition variables (e.g., concentrations) are shown as contour lines in a coordinate system with time or adjusted time as the abscissa and distance from the column inlet, plotted downwards, as the ordinate. The advantage over a representation in terms of profiles or histories is that the entire response can be shown in a single diagram.

Given its distance-time diagram, the course of a column operation is readily visualized. Suppose that pictures of the column, showing the distribution of the species, were taken at regular time intervals and arranged side by side, as illustrated in Fig. 2.7 for the separation of a binary mixture. Connecting the points of equal concentrations (or of concentration steps) in such a side-by-side arrangement, one obtains what amounts to a distance-time diagram. The figure, in which the distance-time diagram is superimposed on the column pictures, illustrates the ease with which such a diagram is mentally translated into a time sequence of events in the column. It is to facilitate this translation that the distance is plotted downwards, to correspond to the usual downflow operation of columns.

For the sake of clarity, an ideal separation was chosen as the example in Fig. 2.7, i.e., the pulses of the two species were presumed to retain their initial square-wave shapes. Accordingly, the distance-time diagram in this figure merely displays the charted courses of the concentration steps at the front and rear ends of the two pulses. A more realistic distance-time

---

* The distance-time diagram was designed for the present treatment. Essentially equivalent representations with different coordinate variables, interchanged co-ordinates, and opposite direction of the distance scale were used previously by Sillén [6], whose pioneer work has, however, gone largely unnoticed.

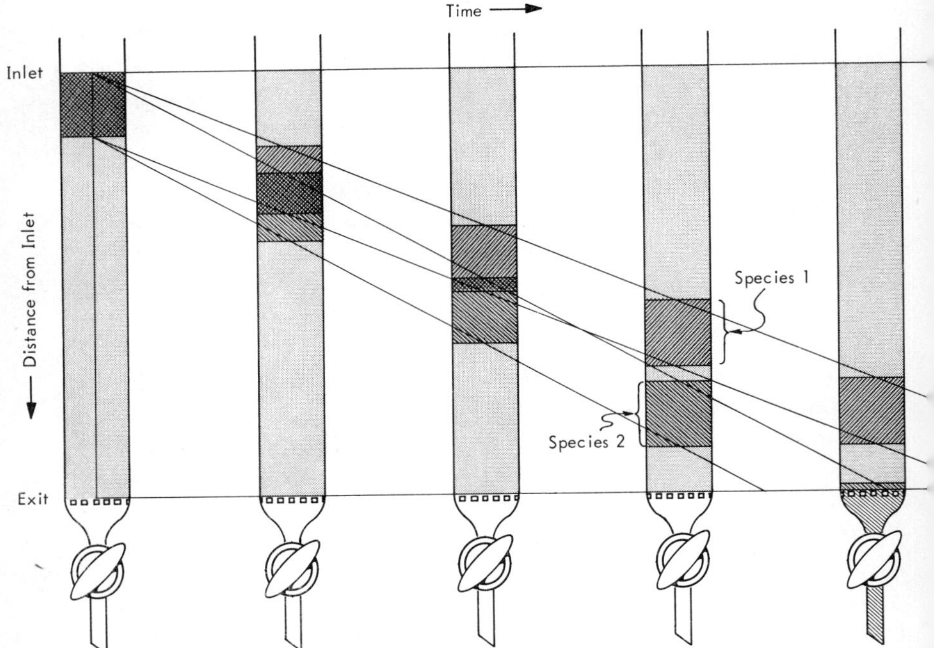


**Fig. 2.7.** Distance-time diagram superimposed on column pictures, for ideal separation of binary mixture (schematic).

diagram, that of elution of an attenuating pulse, is shown in Fig. 2.8. Here, the contour lines are the courses of arbitrarily chosen concentration values.

Whether traced by steps or fixed values of a composition variable, the courses in distance-time diagrams will be called *trajectories*. As in Fig. 2.8, regions of gradual concentration variations will be shown as shaded.

Vertical cuts through the distance-time diagram are loci of constant time (or adjusted time, see below), and horizontal cuts are loci of constant distance from the inlet. Variations of composition variables along vertical and horizontal cuts thus constitute profiles and histories, respectively. In Fig. 2.8, concentration histories at three indicated distances from the inlet are shown alongside the distance-time diagram.

Distance-time diagrams may be drawn with either true or adjusted time as the abscissa. Normally, true time will be used in schematic or

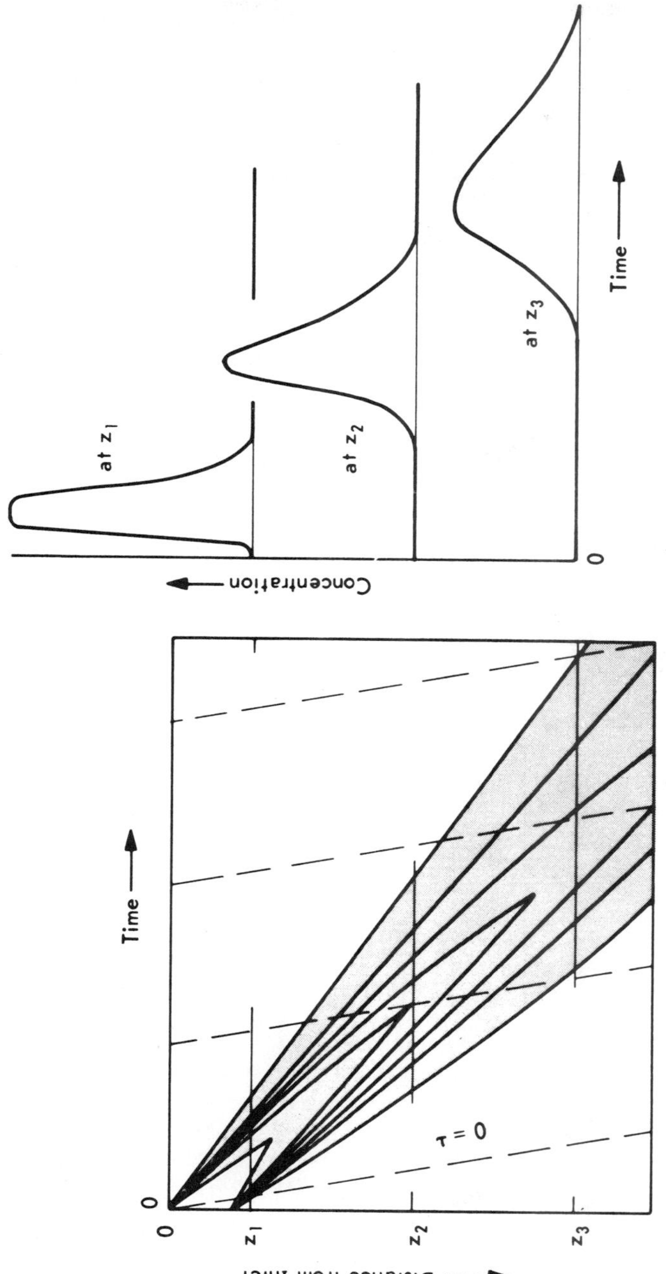


**Fig. 2.8.** Distance-time diagram (left) and concentration profiles (right) for elution of a concentration pulse under realistic conditions (schematic).

qualitative diagrams, and adjusted time in quantitative diagrams illustrating mathematical relations derived in terms of adjusted quantities. The qualitative appearance and topology of the diagram is not affected by the transition from true to adjusted time. In Fig. 2.8, drawn with true time as the abscissa, loci of constant adjusted time are shown as thin broken lines; the lines are parallel and inclined at such an angle that the distance between $t = 0$ and $\tau = 0$ at any distance $z$ from the inlet is $z/u_0$ [see Eq. (2.18)], i.e., the slope is $u_0$. Transformation to adjusted time deforms the diagram in such a way that the lines of constant $\tau$ become vertical.

The slope $\partial z/\partial t$ or $\partial z/\partial \tau$ of a trajectory in the distance-time diagram with true or adjusted time is given by the true or adjusted velocity of the respective entity [see Eqs. (2.17) and (2.19)]. The trajectories of the mobile-phase volume elements run along lines of constant $\tau$ and, corresponding to the true velocity $u_0$ and infinite adjusted velocity, have the slope $u_0$ in $(z, t)$ diagrams and are vertical in $(z, \tau)$ diagrams.

The transformation of true into adjusted time and the corresponding transformation of the distance-time diagram involve the mobile-phase velocity $u_0$ and the concentration ratio $C/\bar{C}$. However, these two parameters also affect the true velocities of all entities and thus the trajectory slopes in the $(z, t)$ diagram. In this way it comes about that the $(z, \tau)$ rather than the $(z, t)$ diagram is insensitive and, except for disturbances, even invariant to changes of the parameter values $u_0$ and $C/\bar{C}$, provided the other parameters remain unchanged. This invariance, which makes the adjusted diagram more general and therefore more useful, is not accidental but is an intended result of the time adjustment.

## REFERENCES


1. J. N. Wilson, *J. Am. Chem. Soc.*, **62**, 1583 (1940).
2. D. DeVault, *J. Am. Chem. Soc.*, **65**, 532 (1943).
3. J. Weiss, *J. Chem. Soc.*, **1943**, 297.
4. J. E. Walter, *J. Chem. Phys.*, **13**, 229 (1945).
5. E. Glückauf, *Nature*, **156**, 748 (1945); *Proc. Roy. Soc.* (*London*), **A 186**, 35 (1946).
6. L. G. Sillén, *Nature*, **166**, 722 (1950); *Arkiv Kemi*, **2**, 477 (1950).
7. N. K. Hiester and T. Vermeulen, *Chem. Eng. Progr.*, **48**, 505 (1952).
8. T. Vermeulen, in *Advances in Chemical Engineering* (T. B. Drew and J. W. Hoopes, Jr., eds.), Vol. 2, Academic Press, New York, 1958, p. 147.
9. N. K. Hiester, T. Vermeulen, and G. Klein, in *Chemical Engineers' Handbook* (R. H. Perry, C. H. Chilton, and S. D. Kirkpatrick, eds.), 4th ed., McGraw-Hill, New York, 1963, Sec. 16.
10. G. Klein, D. Tondeur, and T. Vermeulen, *Ind. Eng. Chem. Fundamentals*, **6**, 339 (1967).

11. J. A. Faucher, Jr., R. W. Southworth, and H. C. Thomas, *J. Chem. Phys.*, **20**, 157 (1952).
12. F. Helfferich, *J. Chem. Educ.*, **41**, 410 (1964).
13. R. H. Beaton and C. C. Furnas, *Ind. Eng. Chem.*, **33**, 1500 (1941).
14. H. C. Thomas, *J. Am. Chem. Soc.*, **66**, 1664 (1944) [Eq. (4) defining the adjusted time contains a misprint; $t$ should be substituted for $x$]; *Ann. N. Y. Acad. Sci.*, **49**, 161 (1948).
15. N. R. Amundson, *J. Phys. Chem.*, **52**, 1153 (1948); **54**, 812 (1950).
16. J. B. Rosen, *J. Chem. Phys.*, **20**, 387 (1952).
17. S. Goldstein, *Proc. Roy. Soc. (London)*, **A 219**, 151 and 171 (1953).
18. International Union of Pure and Applied Chemistry, "Preliminary Recommendations on Nomenclature and Presentation of Data in Gas Chromatography," in *Gas Chromatography 1960* (R. P. W. Scott, ed.), Butterworths, London, 1960, p. 423; *Pure and Appl. Chem.*, **1**, 177 (1960).
19. A. I. M. Keulemans, *Gas Chromatography*, 2nd ed., Reinhold, New York, 1959, p. 16.
20. A. B. Littlewood, *Gas Chromatography*, Academic Press, New York, 1962, p. 27.
21. H. Purnell, *Gas Chromatography*, Wiley, New York, 1962, p. 93.

# 3

# THEORETICAL BASIS

The theoretical treatment in this chapter is focused on the effects of mutual interference of the sorbable species in chromatographic columns. To bring out clearly the consequences arising exclusively from interference and to separate them from complications also found in other systems and known from earlier theories, drastically simplifying assumptions about column properties and equilibria are made at the outset. The relaxation of these premises will be discussed in a later chapter. On the other hand, the treatment is not restricted to particular types of operation, but rather, is applicable to any conceivable operating conditions: Any number of species in arbitrary concentrations may be present in the influent and the column, the initial concentration profiles in the column may be uniform or nonuniform, and the influent composition may be varied arbitrarily, i.e., abruptly, gradually, and repeatedly. The presentation thus provides a unifying, if idealized, description in which particular types of operation covered by earlier theories, such as frontal analysis, elution development, displacement development, and vacancy chromatography, appear as special cases.

The main purpose of this chapter is to provide a general understanding of column behavior. The immense variety of chromatographic patterns, confusing indeed when viewed empirically, arises from only a few basic phenomena occurring in the column. Once these are well understood, column responses even under complex operating conditions can be predicted with relative ease. The attempt is therefore made to strike at the root by discussing, in general terms and with the aid of new concepts, the underlying physical interactions. Examples illustrating the deduction of column responses are little more than incidental and mainly serve to show that known patterns are indeed obtained as special cases. A host of other patterns can be derived, and complete coverage has not been attempted.

A secondary purpose of this chapter is to set forth the rudiments of a quantitative theory, amenable to refinement by inclusion of nonequilibrium effects neglected here. A nonlinear transformation of variables by means of the so-called *H* function provides the key to simple solutions and solution procedures applicable to any operating conditions. While the deduction of rules relies for rigor on the quantitative mathematical treatment, enough conceptual detail and interpretations are given so that mathematical derivations may be skipped without loss in continuity. Lengthy mathematical proofs, required in a few instances, are relegated to the Appendix. Beyond these, very little mathematics will be needed even in the quantitative treatment.

Unfortunately, the above-mentioned transformation of variables and the solutions obtained with it are contingent on certain simple equilibrium properties of the interfering species and are not readily generalized. However, the key parts of the theory, stated in the form of inequalities, conditions, and qualitative rules, can be derived for arbitrary equilibrium properties from conceptual arguments alone and need to be qualified only in exceptional cases, as will be discussed briefly in Chapter 5.

The line of attack taken here differs from usual practice, which is to derive directly equations for effluent histories under given conditions. Here, instead, the first step is to derive rules to which boundaries and plateau zones are subject regardless of their origin and which, therefore, are valid under any operating conditions. These rules then serve as building blocks which can be combined in different ways to account for column responses under various conditions. This approach may tax the reader's patience in the early stages, but appears best suited to provide the intended broad coverage.

## I. Premises

The premises stated below will allow us to concentrate first on the simplest possible systems with interfering species and to exclude, for the time being, complications which can be taken into account at a later stage.

The column is assumed to be uniformly packed, to have uniform cross section, and to exhibit ideal plug flow. The flow rate is considered to be constant with time and uniform throughout the column. The species in the column are assumed to be transferred exclusively by mobile-phase flow, i.e., molecular diffusion in axial direction is disregarded. Also, it is assumed that the rate of equilibration of the stationary with the mobile phase is high enough for local equilibrium between the phases to exist everywhere in the column. These premises are the same as in the classical "equilibrium theories" of chromatography [1–6]. The effects caused by deviations are well known from experience and theory [7–10] and will later prove to be no different in the systems studied here than in any others (see Chapter 5, Section V).

In addition, the following conventions and assumptions regarding the sorbable species are introduced. An arbitrary number $n$ of sorbable species are allowed to be present. The species are numbered 1, ..., $n$ in the sequence of decreasing affinity for the stationary phase. With this convention, the distribution ratios and separation factors obey the inequalities

$$\bar{C}_i/C_i > \bar{C}_j/C_j \qquad (i < j) \tag{3.1}$$

$$1 < \alpha_{ij} < \alpha_{ik} \qquad (i < j < k) \tag{3.2}$$

for all $i, j$, and $k$ at any given composition. The binary separation factors of all pairs of species are assumed to be constant, i.e., independent of composition:

$$\alpha_{ij} \equiv \frac{\bar{C}_i C_j}{C_i \bar{C}_j} = \text{const} \qquad \text{for all } i \text{ and } j \tag{3.3}$$

It is furthermore assumed that transfer of species between the mobile and the stationary phase takes place exclusively by stoichiometric exchange. The total concentration $\bar{C}$ in the stationary phase then remains constant, independent of time and location:

$$\bar{C} \equiv \sum_i \bar{C}_i = \text{const} \tag{3.4}$$

The total concentration of the influent will also be held constant. In view of the earlier assumptions of stoichiometric exchange and species transport exclusively by mobile-phase flow, the total concentration $C$ of the mobile phase in the column then is also independent of time and location:

$$C \equiv \sum_i C_i = \text{const} \tag{3.5}$$

The assumptions (3 ·o (3.5), although commonly made in simple theories of ion-exchang. ·.olumns [4, 7–9, 11–13], are highly restrictive. In particular, the postula e of stoichiometric exchange is obeyed in few systems other than ior exchange with very dilute solutions. [For exchange of ions of different valences, of course, concentrations must be expressed in terms of equivalents rather than moles for the premise (3.4) to apply.] Fortunately, the treatment is readily extended to systems with non-stoichiometric sorption by introduction of a dummy variable, with which a formal equivalence of stoichiometric systems with $n+1$ components and nonstoichiometric systems with $n$ components can be established (see Chapter 5, Section III.A). The other assumptions can also be relaxed at least partially without serious changes in the overall qualitative picture, as will be discussed in Chapter 5.

## II. Equilibria

Any chromatographic process is based on differences in the equilibrium distributions of the sorbable species between the mobile and the stationary phase. Systems with interfering species are no exception. The particular properties distinguishing these systems are a direct consequence of their equilibrium behavior, namely, of the interdependence of the equilibrium distributions of the various species. Interference implies that the stationary-phase concentration of any given species depends not only on the mobile-phase concentration of that species, as in systems without interference, but also on the concentrations of all other species [see Eq. (2.13)]. The premises in the previous section specify particular equilibrium properties, which will now be examined.

From Eqs. (2.8) and (2.3) to (2.5), defining the separation factors $\alpha_{ij}$, total concentrations $C$ and $\bar{C}$, and normalized concentrations $x_i$ and $y_i$, one obtains for the stationary-phase concentration of the arbitrary species $j$ as a function of the mobile-phase concentrations of all species

$$\bar{C}_j = \frac{\bar{C}C_j}{\sum_i(\alpha_{ij}C_i)} \quad \text{or} \quad y_j = \frac{x_j}{\sum_i(\alpha_{ij}x_i)} \quad \text{for all } j \tag{3.6}$$

and for the distribution ratio of the arbitrary species $j$

$$\frac{\bar{C}_j}{C_j} = \frac{\bar{C}}{\sum_i(\alpha_{ij}C_i)} \quad \text{or} \quad \frac{y_j}{x_j} = \frac{1}{\sum_i(\alpha_{ij}x_i)} \quad \text{for all } j \tag{3.7}$$

These relations are general since they are derived exclusively from definitions. With the premises of constant total concentrations in both phases and Eq. (3.6), the compositions of both phases are completely determined by $n-1$ concentrations in either phase if $C$, $\bar{C}$, and the separation factors are given. With the premise of constant separation factors, the distribution ratios $\bar{C}_i/C_i$ or $y_i/x_i$ of all species are in constant ratios to one another [see Eq. (3.3)] and thus *increase or decrease jointly* when the composition is varied.

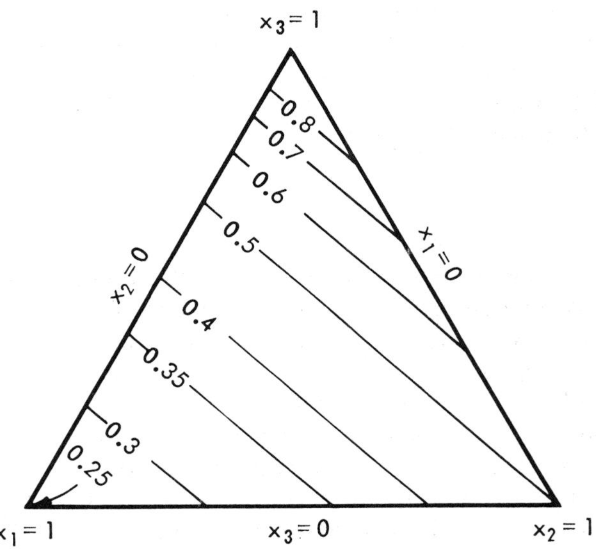


**Fig. 3.1.** Contour lines of constant distribution ratios in mobile-phase composition diagram of three-component system with $\alpha_{12}=2$, $\alpha_{13}=4$. Numbers on contour lines state factors by which distribution ratios of all species are smaller than at composition $x_3 \rightarrow 1$.

The effect of interference on the species distributions is reflected by Eqs. (3.7). The distribution ratio $\bar{C}_j/C_j$ or $y_j/x_j$ of any given species increases when the species with low affinities for the stationary phase (having low values of $\alpha_{ij}$) become more predominant. That a predominance of low-affinity species should lead to higher distribution ratios may seem contrary to intuition, but is as one should expect if one considers that all molecules compete for the limited capacity of the stationary phase and that the molecules of any given species have less competition from low-affinity than from high-affinity companions. Figure 3.1 illustrates the composition dependence for a typical three-component system. Here, the loci of constant distribution ratios are parallel straight lines. In the general case of an $n$-component system, they are parallel $(n-2)$-dimensional hyperplanes.

The limiting case of *trace-component systems*, i.e., with all but one species at trace levels, is of particular interest. With $k$ as the only bulk species,

$$C_k \to C, \quad C_i \ll C \qquad \text{or} \qquad x_k \to 1, \quad x_i \ll 1 \qquad \text{for all } i \neq k \quad (3.8)$$

equations (3.6) reduce to

$$\lim \bar{C}_i = \frac{\bar{C}}{C}\alpha_{ik}C_i \qquad \text{or} \qquad \lim_{x_k \to 1} y_i = \alpha_{ik}x_i \qquad \text{for all } i \quad (3.9)$$

(Note that $1/\alpha_{ki} = \alpha_{ik}$.) The distribution ratios $\bar{C}_i/C_i$ or $y_i/x_i$ then depend only on the respective separation factors $\alpha_{ik}$. In particular, the distribution of any given trace species is independent of the presence or absence of other trace species and of their concentrations and separation factors. The large excess of $k$ thus "uncouples" the trace species, i.e., suppresses their interferences with one another.

For ease of reference, the equilibrium behavior can be summarized by the following rules:

1. All distribution ratios increase or decrease together.
2. The distribution ratios increase when concentrations of species of lower affinities are increased at the expense of those of higher affinities.
3. Trace species do not affect one another (provided another species is present in large excess).

## III. Velocities and Continuity

The conceptual approach in this chapter is largely based on an analysis of species velocities and concentration velocities and their interrelations (for definitions, see Chapter 2, Section IV.B). In this analysis, we consider a column in which some arbitrary nonuniform concentration profiles of the various species have been created (e.g., by influent composition changes). The exact nature of the profiles and their origin are not as yet of concern.

### A. Species Velocities

We examine first the species velocity of an arbitrary species $i$ at a location (distance from column inlet) where its mobile- and stationary-phase concentrations are $C_i$ and $\bar{C}_i$, respectively. The molecules make headway, at the rate of mobile-phase flow $u_0$, only while they are in the moving fluid, but are at rest while they are in the stationary phase. Of the molecules of species $i$ in a volume element of the column, only the fraction $C_i/(\bar{C}_i+C_i)$ is in the mobile phase and contributes to the species velocity. Accordingly, this velocity is*

$$u_i = u_0 \frac{C_i}{\bar{C}_i+C_i} = \frac{u_0}{1+\bar{C}_i/C_i} \qquad \text{for all } i \tag{3.10}$$

The species velocities at a given flow rate $u_0$ thus are exclusively determined by the distribution ratios $\bar{C}_i/C_i$, which are composition-dependent equilibrium properties. The velocities at any given composition and flow rate therefore are determined by equilibrium properties alone and are independent of the operating conditions.

At any given composition, the species with higher affinities have the higher distribution ratios and, accordingly, the lower species velocities:

$$u_i < u_j \quad \text{if} \quad i < j \qquad \text{for all } i \text{ and } j \tag{3.11}$$

as follows from Eqs. (3.10) and (3.1).

The composition dependence of the species velocities can be derived from that of the distribution ratios. It was shown in the previous section that all distribution ratios, in response to composition changes, remain constant or increase or decrease together. According to Eq. (3.10), the species velocities then do the same:

$$\begin{matrix} u_i' < u_i'' & \text{if} \quad u_j' < u_j'' \\ u_i' = u_i'' & \text{if} \quad u_j' = u_j'' \end{matrix} \qquad \text{for all } i \text{ and } j \tag{3.12}$$

(where primes and double primes refer to two different compositions). Where the species with lower affinities are present in higher proportions, all distribution ratios are higher and all species velocities thus are lower. Figure 3.2 illustrates the composition dependence for a typical three-component system, where the loci of constant species velocities are parallel straight lines. In the general case of an $n$-component system, they are parallel $(n-2)$-dimensional hyperplanes. In accordance with Eq. (3.10), loci of constant species velocities coincide with those of constant distribution ratios (compare Figs. 3.2 and 3.1). Condition (3.12) is generally valid for systems with competitive sorption.

---

* This velocity expression was first derived by LeRosen [14], but was somewhat misleadingly labeled "velocity of a zone."

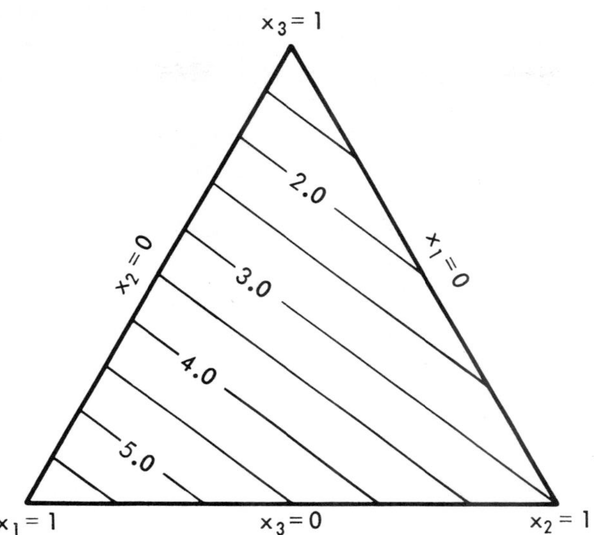


**Fig. 3.2.** Contour lines of constant species velocities in mobile-phase composition diagram of three-component system with $\alpha_{12} = 3$, $\alpha_{13} = 6$, and $C \ll \bar{C}$. Numbers on contour lines state factors by which all species velocities are higher than at composition $x_3 \to 1$.

One aspect of this composition dependence calls for a comment. If one considers, at any given time, two zones in different parts of the column and with different compositions, one generally finds *all* species velocities to be higher in one than in the other. At first glance, this may seem to be in conflict with the premise of uniform total concentration throughout the column, which requires the total species flow (number of molecules of all species passing through a fixed cross section per unit time) to be the same everywhere in the column [i.e., $\Sigma_i(u_i\tilde{C}_i) = \text{const}$]. However, as illustrated schematically in Fig. 3.3, uniform total flow requires, rather than excludes, species-velocity variations. From zone 1 to zone 2, the species velocities must decrease to compensate for the composition shift in favor of the faster species 2 so as to maintain the same total flow. Similarly, from zone 2 to zone 3, the species velocities must increase to compensate for the composition shift in favor of the slower species 1.

The rates of advance of zones or compositions as such cannot be deduced from these considerations, since they are determined by concentration velocities rather than by species velocities or total flow.

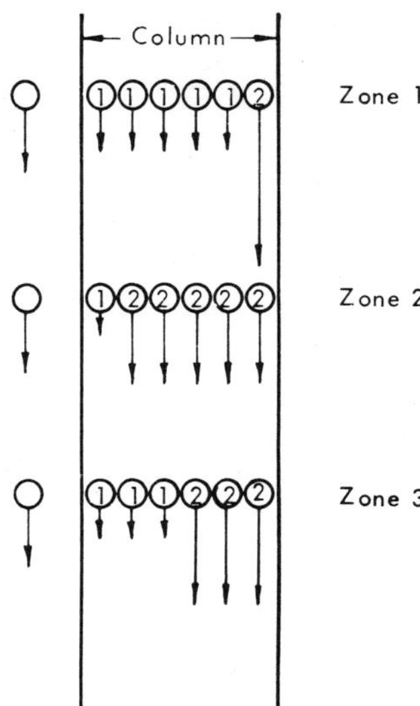


**Fig. 3.3.** Species velocities in three zones with different compositions in a two-component system (schematic). Lengths of arrows are in proportion to species velocities. Average species velocity in each zone is shown on the left.

The conclusions can be summarized as follows:

1. At any given composition, the species with higher affinities have the lower species velocities.
2. With changes in composition, all species velocities remain constant or increase or decrease together. They increase if the composition shifts in favor of species of higher affinities.

## B. Concentration Velocities and Continuity Requirements

The concentration velocity of a species is defined as the rate of advance of a specified concentration of that species. As pointed out earlier, species velocities and concentration velocities refer to two entirely different physical phenomena, namely, to motion of molecules and to propagation of concentration changes (see Chapter 2, Section IV.B). Moreover, concentration velocities, unlike species velocities, cannot be defined everywhere in the column. The rate of advance of a concentration can be unambiguously stated only if the concentration is at a unique location at any given time, not if it extends uniformly over a zone of finite length between boundaries moving with different speeds. Concentration velocities can therefore be defined only where concentration gradients exist. The special case of sharp concentration steps will be discussed separately. All derivations are for a single arbitrary species $i$ and are independent of the behavior of the other species.

Concentration velocities can be defined for mobile-phase, stationary-phase, and overall concentrations. (The overall concentration $\tilde{C}_i \equiv C_i + \bar{C}_i$ is the combined amount of $i$ in both phases per unit volume of the column.) These velocities are not necessarily equal. The quantity commonly used is the mobile-phase concentration velocity. However, the overall concentration velocity will also be examined because its relation to the species velocity provides the basis for deriving general rules in Sections IV.B and V.C.

### 1. *Mobile-Phase Concentration Velocities*

Although the term "concentration velocity" has rarely been used in the past, the concept dates back to the earliest pioneer work in chromatography [1–6, 11, 15–17] and constitutes the core of practically all contemporary theories. In terms of the variables used in the present approach, the velocity of a given mobile-phase concentration $C_i$ is

$$u_{C_i} \equiv \left(\frac{\partial z}{\partial t}\right)_{C_i} = \frac{u_0}{1+(\partial \bar{C}_i/\partial C_i)_z} \qquad \text{for all } i \tag{3.13}$$

The derivation is based on a material-balance argument, namely, that the change in content of a volume element must equal the difference between the amounts entering and leaving.

*Derivation.* The concentrations $C_i$ and $\bar{C}_i$ can be expressed as functions

$$C_i(z, t, \ldots) \qquad \text{and} \qquad \bar{C}_i(z, t, \ldots) \tag{3.14}$$

The additional arguments not listed—operating conditions (initial concentrations of all species as functions of location, influent concentrations of all species as functions of time, flow rate as a function of time) and equilibrium parameters—are fixed for any given operation and therefore need not be carried in the subsequent derivation. The functions (3.14) must exist, since otherwise the column behavior under completely specified conditions would be indeterminate.

With the premise of species transfer exclusively by mobile-phase flow, the flux of a species $i$ in the direction of flow is

$$J_i = u_0 C_i \tag{3.15}$$

Continuity requires

$$\left(\frac{\partial(\bar{C}_i + C_i)}{\partial t}\right)_z = -\operatorname{div} J_i = -u_0\left(\frac{\partial C_i}{\partial z}\right)_t \tag{3.16}$$

This is the fundamental differential material-balance equation (also called conservation equation) on which most theories of chromatography are based. Replacing $(\partial C_i/\partial z)_t$ in Eq. (3.16) by applying the general rule for partial derivatives of three variables $q$, $r$, and $s$:

$$(\partial q/\partial r)_s = -(\partial q/\partial s)_r(\partial s/\partial r)_q \tag{3.17}$$

and then solving for $(\partial z/\partial t)_{C_i}$ and writing $(\partial \bar{C}_i/\partial C_i)_z$ for $(\partial \bar{C}_i/\partial t)_z/(\partial C_i/\partial t)_z$, one obtains Eq. (3.13).

The derivation is not restricted to local equilibrium in the column or to particular forms of the equilibrium relations. Equivalent derivations were given by DeVault [2], Glueckauf [18, 19], Sillén [20], Baylé and Klinkenberg [21], and many later authors.

According to Eq. (3.13), the mobile-phase concentration velocity depends on the ratio $(\partial \bar{C}_i/\partial C_i)_z$ of the concentration changes in the stationary and mobile phases rather than on the ratio $\bar{C}_i/C_i$ of the concentrations, as does the species velocity. The physical cause becomes apparent if one considers, for example, the advance of a "front" of higher mobile-phase concentration of species $i$ through a column layer initially containing $i$ at a lower concentration. As the higher concentration enters the layer, it brings in an amount of $i$ in excess over that initially present. If the stationary-phase concentration were to remain unchanged, this excess amount would have to stay entirely in the mobile phase, and the front would thus advance with the rate of mobile-phase flow. Indeed, Eq. (3.13) gives $u_{C_i} = u_0$ for $(\partial \bar{C}_i/\partial t)_z = 0$. It is true that many of the individual molecules of species $i$ entering the layer with the advancing front will go into the stationary phase and will therefore be retarded. However, this loss is compensated by an equal number of molecules entering the mobile phase from the stationary phase if the concentration in the latter remains constant. Thus, retardation of individual molecules does not necessarily delay the advance of the front.

The more usual situation is that the concentration in the stationary phase increases with that in the mobile phase. In this case, part of the excess amount entering the layer is drawn off from the mobile phase to raise the concentration in the stationary phase, and the advance of the front is correspondingly retarded. The retardation is in proportion to the concentration increase achieved in the stationary phase, not in proportion to the absolute concentration in that phase. Furthermore, a given concentration increase in the stationary phase is achieved by a smaller fraction of the excess amount, and therefore with less delay of the front, if the concentration increase in the mobile phase is large. In this way, the dependence of the concentration velocity on the ratio of the concentration changes comes about.

It is also possible in principle that the stationary-phase concentration decreases while that of the mobile phase increases. The amount of species $i$ released by the stationary phase may then add to the advance of the front, or may even force the front to move against the direction of flow. Thus, in accordance with Eq. (3.13), the mobile-phase concentration velocity exceeds the velocity of mobile-phase flow, or is negative, if the concentrations in the two phases change in opposite directions. (Propagation of a disturbance against the direction of flow is not uncommon. An example from everyday life is a traffic jam, which moves against traffic flow; see Fig. 2.6.)

The calculation of concentration velocities with Eq. (3.13) is straightforward for systems with only one sorbable species (or with several noninterfering species) in an inert carrier gas or solvent. Here, granted the premise of local equilibrium, $\bar{C}_i$ is a unique function of $C_i$, and the partial derivative $(\partial \bar{C}_i/\partial C_i)_z$ can therefore be replaced by the total derivative $d\bar{C}_i/dC_i$. The same is true for two-component systems with stoichiometric exchange [premises (3.4) and (3.5)], which also have only one independent concentration variable because the concentration of one species is fixed by that of the other. In these classes of systems, the velocity of any given mobile-phase concentration depends exclusively on the functional relation between $\bar{C}_i$ and $C_i$, that is, on equilibrium properties alone and is independent of the operating conditions. Even in these systems, however, the species velocity and concentration velocity at the same composition generally differ from one another, as a comparison of Eqs. (3.10) and (3.13) shows and Fig. 3.4 illustrates: The concentration velocity is deter-

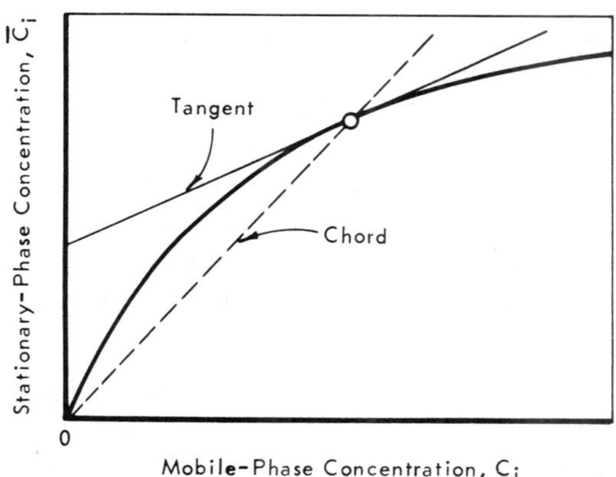


**Fig. 3.4.** Sorption isotherm of a single sorbable species (schematic) with tangent and chord for given composition point.

mined by the slope $d\bar{C}_i/dC_i$ of the tangent to the isotherm at the given composition point, and the species velocity by the slope $\bar{C}_i/C_i$ of the chord from the origin to the composition point.

In multicomponent systems with interfering species, the evaluation of Eq. (3.13) is more difficult. Here, $\bar{C}_i$ depends not on $C_i$ alone, but also on the concentrations of all other species. The quotient $(\partial\bar{C}_i/\partial C_i)_z$ merely represents the ratio of the concomitant concentration changes with time at the fixed location $z$ and leaves the behavior of the other species unspecified. Once the restriction of a functional dependence of $\bar{C}_i$ on $C_i$ alone is removed, these two concentrations can in principle change in the same or opposite directions and either of them can remain constant while the other changes, depending on the behavior of the other species. The ratio of the concentration changes can thus assume any positive or negative value or be zero, and Eq. (3.13) by itself leaves the concentration velocity undetermined. Even for given concentrations of all species, the concentration velocities are not determined by equilibrium properties alone but depend on the time derivatives of the concentrations and thus, in a complex manner, on the operating conditions.

This basic problem was first recognized by DeVault [2] and has led Baylé and Klinkenberg [21] to conclude that a solution cannot be found unless additional postulates and restrictions are introduced. The usual approach, taken by Glueckauf [5, 18, 19], Offord and Weiss [22], Sillén [20], Baylé and Klinkenberg [21], and several later authors [23–27] is to restrict the treatment to very simple operating conditions and to presume that, under these conditions, all concentrations are functions of only one variable (itself a function of $z$ and $t$) rather than of $z$ and $t$ separately. Through the mutual dependence on a single variable, a functional relation between $\bar{C}_i$ and $C_i$ (involving also the operating conditions) can be reestablished. This amounts to postulating, and restricting the treatment to, what will later be called "coherence" (see Section IV). The present approach uses coherence as a key concept, but is not restricted to it.

*Trace-component systems*, with all but one species at trace levels, must be singled out as an exception. As shown in Section II, the large excess of one species, $k$, suppresses interferences of the trace species with one another, and Eq. (3.9) then provides a direct functional relation between the stationary- and mobile-phase concentrations of any trace species. Since $\bar{C}_k \to \bar{C}$ when $C_k \to C$, Eq. (3.13) with Eqs. (3.9), (3.10), and (2.8) reduces to

$$\lim_{C_k \to C} u_{C_i} = \frac{u_0}{1+\alpha_{ik}\bar{C}/C} = u_i \qquad \text{for all } i \neq k \tag{3.18}$$

Thus, for each trace species, the concentration velocity is given by equilibrium properties alone, is independent of the other trace species, and equals the species velocity.

The special case of an ideally sharp *concentration step* is not covered by Eq. (3.13). Here, the definition of the concentration velocity is not applicable and the quotient $(\partial \bar{C}_i/\partial C_i)_z$ does not exist. The "step velocity," i.e., the rate of advance of such a step between fixed concentrations of species $i$, is given by

$$u_{\Delta C_i} = \frac{u_0}{1+\Delta \bar{C}_i/\Delta C_i} \qquad \text{for all } i \tag{3.19}$$

where $\Delta \bar{C}_i$ and $\Delta C_i$ are the concentration differences between the upstream and downstream sides of the step. This relation is readily obtained from a material balance with finite instead of infinitesimal concentration differences. Like the concentration velocity and for the same reason, the step velocity is not given by equilibrium properties alone. Calculation of the velocity with Eq. (3.19) requires prior knowledge of the concentration differences in both phases across the step.

The problem of establishing practicable functional relations between the velocities of values (or steps) of composition variables and the equilibrium properties and operating conditions cannot be solved satisfactorily at this stage and will be taken up after the concept of coherence and the $h$ transformation have been introduced (see Sections IV.D and V.A).

## 2. *Stationary-Phase Concentration Velocities*

The velocities of the mobile-phase and stationary-phase concentrations of the same species and at the same location and time are not necessarily equal. Although not needed for later deductions, a relation for the stationary-phase concentration velocity is given for comparison and completeness:

$$u_{\bar{C}_i} \equiv \left(\frac{\partial z}{\partial t}\right)_{\bar{C}_i} = \frac{u_0 - u_{C_i}}{(\partial \bar{C}_i/\partial C_i)_t} \tag{3.20}$$

This relation is readily derived from the general material-balance equation (3.16) with the rule (3.17) and the definition (2.17) of the mobile-phase concentration velocity.

With Eqs. (3.20) and (3.13) one finds that

$$u_{C_i} = u_{\bar{C}_i} \qquad \text{if} \quad (\partial \bar{C}_i/\partial C_i)_z = (\partial \bar{C}_i/\partial C_i)_t \tag{3.21}$$

The partial derivatives appearing in condition (3.21) may or may not be equal. The particular behavior associated with their equality will be discussed later [see Section IV.A, Eq. (3.43)].

## 3. *Overall Concentration Velocities and Continuity*

The overall concentration

$$\tilde{C}_i \equiv \bar{C}_i + C_i \tag{2.2}$$

of a species is the total amount of the species (in the stationary and mobile phases) per unit volume of the column. This concentration would be found by an observer who has no means of distinguishing between the two phases. A simple relation, derived below, exists between the velocity of a given overall concentration and the species velocity*:

$$u_{\tilde{C}_i} \equiv \left(\frac{\partial z}{\partial t}\right)_{\tilde{C}_i} = u_i + \tilde{C}_i \left(\frac{\partial u_i}{\partial \tilde{C}_i}\right)_t \qquad \text{for all } i \tag{3.22}$$

This expression is analogous to one of the fundamental equations of fluid mechanics, relating the velocity $v_\rho$ of propagation of a given density in a compressible fluid to the particle velocity $u$ and density $\rho$ [28]:

$$v_\rho \equiv (\partial z/\partial t)_\rho = u + \rho(\partial u/\partial \rho)_t \tag{3.23}$$

It is therefore not surprising that the propagation of waves (density variations) and of concentration variations in chromatographic columns have much in common. The similarity, apparently first invoked in a tentative manner by Golay [29], has been discussed in more detail by Mangelsdorf [25]. We shall return to this analogy at a later stage when examining the sharpening characteristics of chromatographic boundaries (see Section IV.E).

*Derivation.* Equation (3.22) is readily derived as follows. The flux of species $i$ in the direction of flow, written in terms of $\tilde{C}_i$, is

$$J_i = u_i \tilde{C}_i \tag{3.24}$$

[see Eqs. (3.15), (3.10), and (2.2)]. The requirement of continuity, as in Eq. (3.16), then becomes

$$\left(\frac{\partial \tilde{C}_i}{\partial t}\right)_z = -\operatorname{div} J_i = -u_i\left(\frac{\partial \tilde{C}_i}{\partial z}\right)_t - \tilde{C}_i\left(\frac{\partial u_i}{\partial z}\right)_t \tag{3.25}$$

Replacing $(\partial \tilde{C}_i/\partial t)_z$ by applying the rule (3.17) and then solving for $(\partial z/\partial t)_{\tilde{C}_i}$ one obtains Eq. (3.22).

The derivation is not restricted to local equilibrium between the mobile and stationary phases or to particular forms of the equilibrium relations.

Equation (3.22) is an alternative formulation, in terms of velocities, of the fundamental differential material balance normally written as in Eq. (3.16). The relation merely states a material-balance restriction which a migrating concentration must obey, but is not by itself sufficient to

---

* The author is indebted to Dr. P. C. Mangelsdorf for suggesting the use of this compact relation and pointing out its analogy to Eq. (3.23), used in fluid mechanics.

determine the concentration velocity. Nevertheless, this relation will provide the conceptual basis for deriving the properties of coherent boundaries and for proving the attainment of coherence from arbitrary initial conditions (see Sections IV.B and V.C).

The physical significance of Eq. (3.22) is as follows: In general, the species velocity and the velocity of the overall concentration of the species differ. Accordingly, the individual molecules of the species, advancing at a higher or lower rate than the concentration profile, move from higher to lower concentrations or vice versa. When doing so, they are required by continuity to increase or decrease their velocity relative to the profile, in much the same way as water flowing in a pipe must accelerate where the pipe narrows, and decelerate where it widens. Consider, for

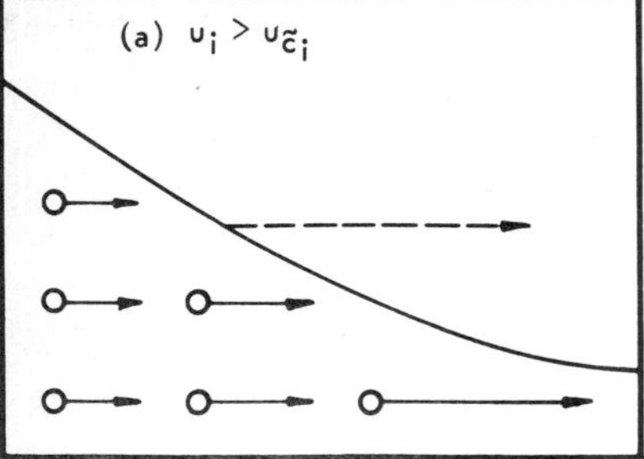

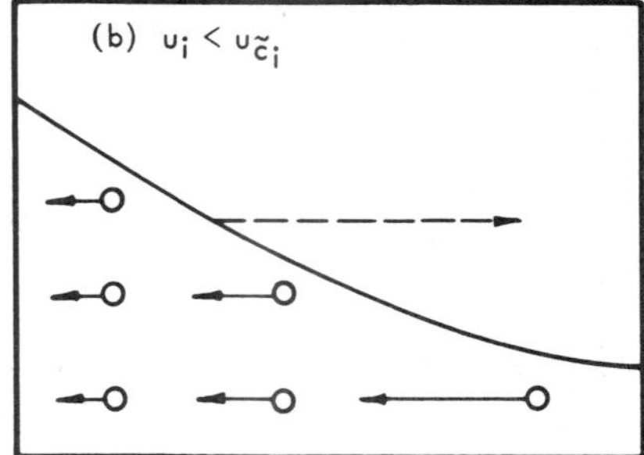


Distance From Column Inlet

**Fig. 3.5.** Velocity variations of molecules moving faster or slower than an advancing front (schematic). Species velocities relative to front are shown as solid arrows, front velocities as dashed arrows.

example, an advancing front of higher concentration. If the species velocity is higher than the concentration velocity, the molecules overtake the front, thus move from higher to lower concentrations, and increase their velocity relative to the front as well as their absolute velocity. The absolute species velocity then is higher on the downstream than on the upstream side of the front (see Fig. 3.5a). On the other hand, if the species velocity is lower than the concentration velocity, the molecules fall behind from lower to higher concentrations and decrease their velocity relative to the front (see Fig. 3.5b). The *absolute* species velocity (i.e., with the bed as the frame of reference), however, then is higher on the upstream side, where the molecules advance more nearly as fast as does the front.

For later deductions, the following consequences of Eq. (3.22) will be important:

$$\begin{aligned} &(\partial u_i/\partial \tilde{C}_i)_t > 0 && \text{if} \quad u_i < u_{\tilde{C}_i} \\ &(\partial u_i/\partial \tilde{C}_i)_t < 0 && \text{if} \quad u_i > u_{\tilde{C}_i} \\ &\left.\begin{aligned} &(\partial u_i/\partial \tilde{C}_i)_t = 0 \\ \text{or} \quad &\tilde{C}_i \to 0 \end{aligned}\right\} && \text{if} \quad u_i = u_{\tilde{C}_i} \end{aligned} \tag{3.26}$$

The first two conditions state that, along the concentration profile at a given time, the species velocity and the overall concentration increase or decrease together if the species are slower than the overall concentration, and vary in opposite directions if the species are faster than this concentration. The special cases illustrated in Fig. 3.5 and discussed above are covered by these two conditions. The third condition states that, for the species velocity and overall concentration velocity to be equal, either the velocities cannot vary with concentration along the profile, or the concentration must tend to zero.

Within finite zones of uniform concentration $\tilde{C}_i$, a concentration velocity cannot be defined. However, in the absence of a concentration gradient,

$$(\partial \tilde{C}_i/\partial z)_t = 0 \tag{3.27}$$

the continuity condition (3.25) requires, for finite $\tilde{C}_i$,

$$\begin{aligned} &(\partial \tilde{C}_i/\partial t)_z > 0 && \text{if} \quad (\partial u_i/\partial z)_t < 0 \\ &(\partial \tilde{C}_i/\partial t)_z < 0 && \text{if} \quad (\partial u_i/\partial z)_t > 0 \\ &(\partial \tilde{C}_i/\partial t)_z = 0 && \text{if} \quad (\partial u_i/\partial z)_t = 0 \end{aligned} \tag{3.28}$$

A variation of the species velocity within the zone thus leads to local accumulation or depletion, making the concentration profile "bulge out" or "cave in," as is illustrated in Fig. 3.6. The zone maintains its uniform concentration only if the species velocity within the zone is also uniform. [For a trace concentration $\tilde{C}_i$ in the zone, the concentration change in the case of varying species velocity is infinitesimal on an absolute, but is still significant on a relative basis since $(1/\tilde{C}_i)(\partial \tilde{C}_i/\partial t)_z$ remains finite.]

The behavior of a concentration step which is and remains ideally sharp is not covered by Eq. (3.22). However, the qualititive rules for the species-velocity changes of molecules passing through the advancing concentration change in either direction, as expressed in conditions (3.26) and illustrated in Fig. 3.5, are valid for concentration steps also. A material

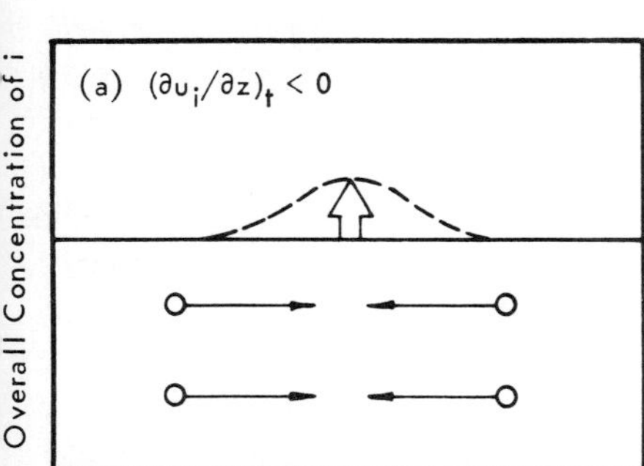


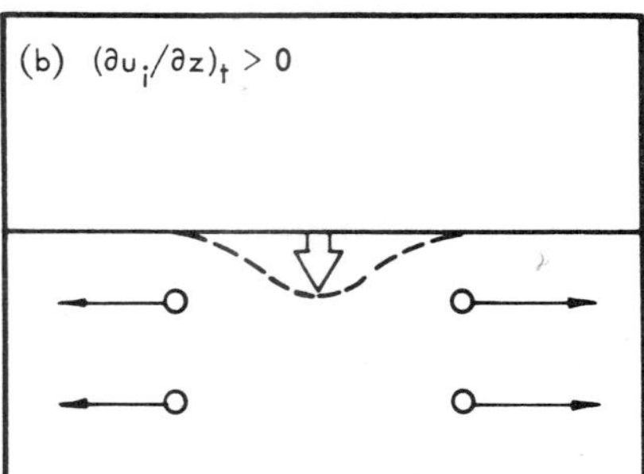


**Fig. 3.6.** Formation of concentration maxima and minima in plateau zones with nonuniform species velocities (schematic). Arrows indicate species velocities relative to that in center of zone. Profile deformations resulting from accumulation or depletion within zone are shown as broken lines.

balance with finite instead of infinitesimal concentration and species-velocity differences gives for the velocity of a step between fixed concentrations of species $i$

$$u_{\Delta\tilde{C}_i} = \frac{u_i'\tilde{C}_i' - u_i''\tilde{C}_i''}{\tilde{C}_i' - \tilde{C}_i''} \tag{3.29}$$

where primes and double primes refer to the upstream and downstream sides, respectively, of the step. Consequences of Eq. (3.29) are

$$\begin{aligned} u_{\Delta\tilde{C}_i} &= u_i'' && \text{if } \tilde{C}_i' = 0 \\ u_{\Delta\tilde{C}_i} &= u_i' && \text{if } \tilde{C}_i'' = 0 \\ u_{\Delta\tilde{C}_i} &= u_i' = u_i'' && \text{if } u_i' = u_i'' \\ \left.\begin{aligned} &u_{\Delta\tilde{C}_i} < u_i', u_i'' \text{ or} \\ &u_{\Delta\tilde{C}_i} > u_i', u_i'' \end{aligned}\right\} & && \text{if } u_i' \neq u_i'' \end{aligned} \tag{3.30}$$

To derive the last of these conditions, one rearranges Eq. (3.29):

$$\frac{u_{\Delta\tilde{C}_i} - u_i''}{u_i' - u_i''} = \frac{\tilde{C}_i'}{\tilde{C}_i' - \tilde{C}_i''} \tag{3.31}$$

since both $\tilde{C}_i'$ and $\tilde{C}_i''$ are positive, the right-hand side of Eq. (3.31) is either negative or larger than unity. Since both $u_i'$ and $u_i''$ are positive, Eq. (3.31) then requires $u_{\Delta\tilde{C}_i}$ to be larger or smaller than both $u_i'$ and $u_i''$.

The first two conditions (3.30) state that, if the species is absent from one side of the step, the step velocity equals the species velocity on the other side of the step. Here, molecules of $i$ cannot pass through the step, which therefore must advance at the same rate as the molecules. The third condition (3.30) results from the fact that molecules must change their velocity when passing from a region of one concentration into that of another, and therefore cannot pass through the step if the species velocity is the same on both sides. The fourth condition (3.30) states that the step velocity cannot be intermediate between the species velocities on the two sides of the step. If it were, the step could not migrate as such, because accumulation or depletion of molecules at the step would occur, in the same way as shown for a zone in Fig. 3.6.

#### 4. *Summary of Rules*

The main conclusions are summarized below.

1. Concentration velocities in systems with more than two interfering species are not determined by equilibrium properties alone, except for trace species.
2. For trace species, the overall concentration velocity equals the species velocity.
3. Along the concentration profile at a given time, the species velocity and the overall concentration increase or decrease jointly if the species is slower than the overall concentration, and vary in opposite directions if the species is faster than the overall concentration.
4. The velocity of a sharp step in overall concentration equals the species velocity on one side if the species is absent from the other side.
5. The velocity of a sharp step cannot be intermediate between the species velocities on the two sides.
6. A zone of initially uniform overall concentration develops a concentration maximum if the species velocity is higher in the upstream than in the downstream part, and a concentration minimum if the reverse is true. Uniform concentration is maintained if the species velocity within the zone is also uniform.

### C. Adjusted Velocities

In later quantitative derivations it will be more convenient to use "adjusted" (and normalized) velocities rather than true velocities. As pointed out in Chapter 2, Section IV.B, the transition from true to adjusted velocity is equivalent to those from true to adjusted retention volume or time in gas chromatography and from true to reduced time, throughput volume, or related variables in other theories.

With the adjusted time $\tau$ as defined by Eq. (2.18) and with normalized concentrations $x_i$ and $y_i$, the fundamental differential material-balance equation (3.16) reduces to

$$(\partial y_i/\partial \tau)_z + (\partial x_i/\partial z)_\tau = 0 \tag{3.32}$$

Replacing either term by means of the general rule (3.17) and solving for $\partial z/\partial \tau$, one obtains for the adjusted velocities of the mobile-phase and stationary-phase concentrations

$$u_{x_i} \equiv (\partial z/\partial \tau)_{x_i} = (\partial x_i/\partial y_i)_z \tag{3.33}$$

$$u_{y_i} \equiv (\partial z/\partial \tau)_{y_i} = (\partial x_i/\partial y_i)_\tau \tag{3.34}$$

[In view of the premises (3.4) and (3.5), there is no difference between the velocities of $C_i$ and $x_i$ or between those of $\bar{C}_i$ and $y_i$.] Similarly, Eqs. (3.10) and (3.19) for the species velocity and step velocity give, after conversion with Eq. (2.21), the respective adjusted velocities

$$u_i = x_i/y_i \tag{3.35}$$

$$u_{\Delta x_i} = \Delta x_i/\Delta y_i \tag{3.36}$$

The adjusted velocities exhibit an interesting symmetry property which allows shortcuts to be taken in various calculations and derivations. The adjusted velocities $\partial z/\partial \tau$ of species, concentrations, and steps are all of the form $x_i/y_i$, $\partial x_i/\partial y_i$, or $\Delta x_i/\Delta y_i$, so that an interchange of $x_i$ and $y_i$ converts them into their reciprocals, $\partial \tau/\partial z$, and thus leads to an interchange of $z$ and $\tau$. The symmetry is most completely displayed by the pair of equations (3.33) and (3.34): One is obtained from the other by interchange of $x_i$ and $y_i$, $z$ and $\tau$, and $1/u$ and $u$. An illustration of how a lengthy derivation can be obviated through the application of this and another symmetry rule will be found in Chapter 4, Section IV.B.5 [see substitutions (4.106)].

The principal conclusions in the earlier parts of this section are not affected by the transition from true to adjusted velocities.

### D. Review and Perspective

In this section, expressions for the species velocity and the various concentration velocities have been derived and analyzed. The species velocity depends only on equilibrium properties, composition, and flow rate, while the concentration velocities also depend on gradients. One may say that the species velocity is a "point property" whereas the concentration velocities are not; that is, at any given location and time, the species velocity is uniquely determined by local, momentary values of the variables (concentrations, flow rate), whereas the concentration velocities, through their dependence on gradients, are also affected by the values of the variables at adjacent points. This dependence on gradients which, for a given composition, can in principle assume any values, poses a fundamental difficulty in the treatment of multicomponent systems. The problem of finding practicable functional relations between the velocities of composition variables and the equilibrium properties and operating conditions has been relegated to later sections.

The various concentration velocities—of the mobile-phase, stationary-phase, and overall concentrations—of a species at a given location and time are not necessarily equal. A criterion for their equality has been stated, but its implications remain to be examined in the next section.

Aside from the introduction and discussion of the various velocities, the principal development in this section is a recasting of the fundamental material-balance equation of chromatography in terms of species and concentration velocities. The new relation, Eq. (3.22), brings out analogies with fluid mechanics. Its implications and consequences, analyzed in detail, will provide the conceptual basis for later deductions, particularly in connection with coherence and its attainment from arbitrary initial conditions.

The considerations and derivations in this section have essentially been confined to a single, arbitrary species in a system with interference. The concerted behavior of multiple species will be examined in the next two sections.

## IV. Coherence

A key concept in the present approach is that of "coherence." In the general case of arbitrary composition profiles or histories in the column, the concentrations of the various species usually shift relative to one another as the species migrate. Particular compositions $\{C_1, \ldots, C_n\}$ in chromatographic boundaries then exist only momentarily since a given concentration of one species, while traveling, does not remain accompanied by fixed concentrations of the other species. Such behavior will be called *noncoherent*. However, under conditions to be discussed in this section, compositions in boundaries may be preserved while migrating. A given concentration of one species then remains accompanied by the same set of concentrations of all other species, i.e., all concentrations are "in phase" with one another. Such behavior will be called *coherent*. A coherent boundary may thus be defined as consisting of traveling loci of constant compositions.

At first glance, coherence appears to be a highly unusual behavior which one could expect to find only under special and very simple operating conditions. This is not so, because coherence is intrinsically a "stable" state, and noncoherence an "unstable" one. Once a boundary has become coherent it will remain so until disturbed. It may sharpen or spread, but will retain coherence since the concentration profiles of the species have ceased to shift relative to one another. In a noncoherent boundary, in contrast, a stable composition profile does not exist because the species concentrations shift relative to one another. It will be seen that they do so in such a way that a stable, coherent composition profile is eventually attained. Far from being restricted to particular operating conditions, coherence thus is a state which, given enough time and column length for undisturbed development, is attained from any arbitrary initial conditions.*

The present section deals with coherent boundaries, plateau zones between such boundaries, and coherent composition pulses, regardless of their origin and location in the column. Since only individual boundaries, zones, and pulses are considered, it will not be necessary to assume that the concentration profiles in other parts of the column are also coherent. How and why coherence develops from arbitrary initial conditions will be shown in Section V.

## A. Coherence Conditions

By definition, coherence requires the conservation of traveling compositions. In a diffuse boundary, this requirement is met only if, at any given time and location, all species concentrations advance at the same rate:

$$u_{C_i} = u_{C_j} = u \qquad \text{for all } i \text{ and } j \tag{3.37}$$

where $u$ is the *composition velocity* common to all species concentrations. With Eq. (3.13) one finds

$$(\partial \bar{C}_i/\partial C_i)_z = (\partial \bar{C}_j/\partial C_j)_z \qquad \text{for all } i \text{ and } j \tag{3.38}$$

---

* For exceptional situations, see Section V.F.2 and Chapter 5, Sections III.B and IV.B.

This relation, obeyed locally in diffuse coherent boundaries, will be called the *differential coherence condition.* Its derivation is not restricted to local equilibrium or to particular forms of the equilibrium relations.

For sharp composition steps, the coherence requirement is analogous, namely, the concentration steps of all species must advance at the same rate:

$$u_{\Delta C_i} = u_{\Delta C_j} = u_{\Delta} \qquad \text{for all } i \text{ and } j \tag{3.39}$$

and thus, with Eq. (3.19),

$$\Delta\bar{C}_i/\Delta C_i = \Delta\bar{C}_j/\Delta C_j \qquad \text{for all } i \text{ and } j \tag{3.40}$$

This relation will be called the *integral coherence condition.*

Although the premises (3.4) and (3.5) allow only $n-1$ concentrations to be varied independently, all $n$ species can obey conditions (3.38) and (3.40). It is readily shown that, regardless of the form of the equilibrium relations, the $n$th species automatically obeys these conditions if all other species do.

The premises (3.4) and (3.5) entail

$$(\partial\bar{C}_k/\partial C_j)_z = -\sum_{i\neq k} (\partial\bar{C}_i/\partial C_j)_z$$

$$(\partial C_k/\partial C_j)_z = -\sum_{i\neq k} (\partial C_i/\partial C_j)_z$$

and thus

$$\left(\frac{\partial\bar{C}_k}{\partial C_k}\right)_z = \frac{\sum\limits_{i\neq k} (\partial\bar{C}_i/\partial C_j)_z}{\sum\limits_{i\neq k} (\partial C_i/\partial C_j)_z} = \frac{(\partial\bar{C}_j/\partial C_j)_z + \sum\limits_{i\neq j,k} (\partial\bar{C}_i/\partial C_i)_z(\partial C_i/\partial C_j)_z}{1 + \sum\limits_{i\neq j,k} (\partial C_i/\partial C_j)_z} \tag{3.41}$$

where $j$ and $k$ are different arbitrary species. If all species other than $k$ obey condition (3.38), then all $(\partial\bar{C}_i/\partial C_i)_z$ in Eq. (3.41) can be replaced by $(\partial\bar{C}_j/\partial C_j)_z$, and this equation reduces to

$$(\partial\bar{C}_k/\partial C_k)_z = (\partial\bar{C}_j/\partial C_j)_z$$

Accordingly, species $k$ obeys condition (3.38) if all other species do. The derivation for condition (3.40) is analogous.

Although written in terms of concentrations in the mobile phase only, condition (3.37) or (3.39) ensures coherence with respect to stationary-phase and overall concentrations as well, provided there is local equilibrium to guarantee that an invariant mobile-phase composition will be accompanied by an invariant stationary-phase composition. Conditions (3.37) and (3.39) thus entail

$$\begin{aligned} u_{C_i} &= u_{\bar{C}_i} = u_{\bar{\bar{C}}_i} = u \\ u_{\Delta C_i} &= u_{\Delta\bar{C}_i} = u_{\Delta\bar{\bar{C}}_i} = u_{\Delta} \end{aligned} \qquad \text{for all } i \tag{3.42}$$

In the event of deviations from local equilibrium, however, conditions (3.42) must be specified as an additional coherence requirement.

In view of condition (3.21), the equality of the mobile-phase and stationary-phase concentration velocities, required by coherence, implies

$$(\partial \bar{C}_i/\partial C_i)_z = (\partial \bar{C}_i/\partial C_i)_t \qquad \text{for all } i \tag{3.43}$$

All concentrations can then be expressed as functions of a single variable (itself a function of $z$ and $t$) rather than as functions of $z$ and $t$ separately. The postulate of dependence on a single variable has served as the basis of earlier theories, which are restricted to coherent systems (see also Section III.B.1).

The coherence conditions (3.38) and (3.40) impose restrictions on concentration changes, not on the concentrations themselves. Any composition can thus be encountered in, or on either side of, a coherent boundary. In two-component systems with local equilibrium, the premises (3.4) and (3.5) of stoichiometric exchange are sufficient to ensure that the coherence conditions are met. Such systems are therefore always coherent. In systems with more than two interfering species, however, the coherence conditions are more restrictive than the premise of stoichiometry. The latter by itself allows the ratios $\partial \bar{C}_i/\partial C_i$ or $\Delta \bar{C}_i/\Delta C_i$ of the concentration variations, and thus the concentration velocities and step velocities, to assume any positive or negative values (see Section III.B.1), but only certain discrete values are compatible with coherence. The solution of this eigenvalue problem will be given in Section IV.D. Without calculation one can recognize that coherence requires the stationary- and mobile-phase concentrations of any species to increase or decrease together:

$$(\partial \bar{C}_i/\partial C_i)_z > 0 \qquad \text{and} \qquad \Delta \bar{C}_i/\Delta C_i > 0 \qquad \text{for all } i \tag{3.44}$$

If the concentrations of one species were to change in opposite directions, condition (3.38) or (3.40) would require those of all other species to do the same. Such equilibrium behavior is thermodynamically impossible, since it would permit a cyclic process with decrease of entropy to be performed.

## B. Coherent Boundaries and Affinity Cut

Basic rules for the behavior of coherent boundaries can be derived from the coherence conditions and the continuity considerations in Section III.B.3. In these deductions, a distinction between mobile-phase and overall concentration velocities (or step velocities) and between mobile-phase and overall concentration changes need not be made since, in coherent boundaries, both velocities are equal and both concentrations change in the same direction [see Eqs. (3.42) and conditions (3.44)]. All rules deduced here can also be obtained from the mathematical derivation in Section IV.D. This derivation, however, is restricted to systems with composition-independent separation factors, whereas for the deductions below the weaker assumptions of an invariant selectivity sequence and of "competitive" sorption (see Chapter 2, Section III) are sufficient.

We have seen that the species velocities at any given composition differ from one another, but that coherence requires the velocities of concentrations or concentration steps of all species to be equal. Therefore, in coherent boundaries in general, the species velocities differ from the composition or step velocity, and the various species pass through the advancing boundary in or against the direction of flow. This behavior has the following implications:

First, coherence requires *all* species velocities to change across the boundary. With changing composition, a species velocity can remain constant only if it equals the concentration velocity and if all other species velocities also remain constant [see conditions (3.26) and (3.12)]. This is not possible in a coherent boundary because the composition or step velocity, common to all species, cannot equal all species velocities simultaneously.

Second, coherence requires *all* concentrations to change across a boundary. The concentration of a species can remain uniform only if the species velocity is also uniform [see conditions (3.28) and Fig. 3.6]. This is not possible in a coherent boundary because, as shown above, all species velocities change. (The exceptional behavior of trace species is discussed on p. 60.)

Third, by far the most important constraint imposed by coherence is the existence, at any point in a coherent boundary, of what will be called an *affinity cut*. This cut divides all species $1, \ldots, j, k, \ldots, n$ (ordered in the sequence of their affinities) into a high-affinity group $1, \ldots, j$ and a low-affinity group $k, \ldots, n$. The cut is between the species with velocities lower and higher than that of the composition (or step), and the concentrations of the species of the two groups vary in opposite directions. For example, in a seven-component system, the cut may be between, say, species 4 and 5; species 1 to 4 then constitute the high-affinity group and are slower than the composition or step, while species 5 to 7 constitute the low-affinity group and are faster, and the concentrations of the former species increase while those of the latter decrease, or vice versa.

The existence of an affinity cut with these properties is a consequence of the following three restrictions derived earlier: (1) All species velocities increase or decrease jointly in the direction of flow [conditions (3.12)]; (2) since all species velocities vary in the same direction, continuity requires the concentrations of species with velocities higher and lower than the composition (or step) velocity to vary in opposite directions [conditions (3.26)]; (3) the species with higher affinities have the lower velocities [condition (3.11)], so that the species having velocities higher and lower than that of the composition (or step) can be identified as having low and high affinities, respectively.

The high- and low-affinity groups must each contain at least one species because the premises (3.4) and (3.5) of constant total concentrations do not allow all concentrations to increase or decrease simultaneously. Accordingly, there are $n-1$ possible positions of the affinity cut, namely, $1 \mid 2, 2 \mid 3, \ldots, n-1 \mid n$, for $n$ species present. (The notation $j \mid k$ is adopted to denote the affinity cut between species $j$ and $k$.) Cuts between species which are not adjacent in the affinity sequence can occur only if the intermediate species are absent from the respective boundary. For example, in the absence of species 3 and 4, the cut may be $2 \mid 5$. Various concentration profiles, compatible and incompatible with the affinity-cut rule, are shown in Fig. 3.7 for a three-component system.

Since the affinity cut is between species having velocities higher and lower than the composition or step velocity, the latter is intermediate between the velocities of the two species marking the cut. Thus, for a sharp coherent boundary with $j \mid k$ cut,

$$u_j', u_j'' \leqq u_\Delta \leqq u_k', u_k'' \tag{3.45}$$

where primes and double primes refer to the upstream and downstream sides of the boundary. [Note that conditions (3.30) prevent $u_\Delta$ from being intermediate between $u_j'$ and $u_j''$ or between $u_k'$ and $u_k''$.] The inequalities apply if $j$ and $k$ are present on both sides of the boundary. An equality appears, according to conditions (3.30), if $j$ or $k$ is present on one side only. Similarly, in a diffuse coherent boundary with $j \mid k$ cut, the local composition velocity is intermediate between the local species velocities of $j$ and $k$:

$$u_j \leqq u \leqq u_k \tag{3.46}$$

The inequalities apply where both $j$ and $k$ are present; an equality appears, according to conditions (3.26), only where the concentration of $j$ or $k$ becomes zero.

The arguments presented so far pertain to sharp coherent steps or to any arbitrary points in diffuse coherent boundaries. An entire coherent boundary may, in principle, consist of portions having different affinity

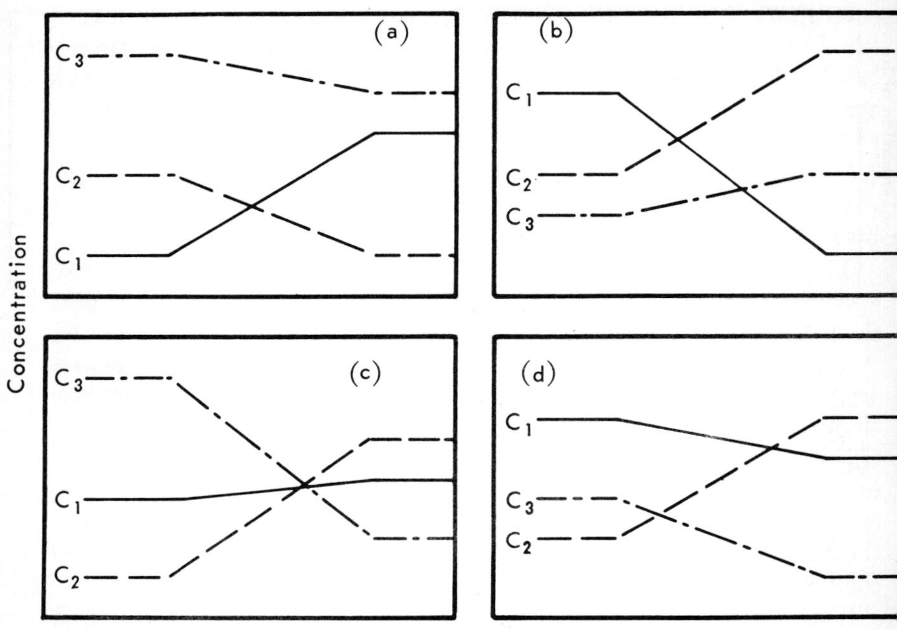


**Fig. 3.7.** Concentration profiles compatible and incompatible with affinity-cut rule (schematic). Profiles a and b (1 | 2 cut) and c (2 | 3 cut) obey rule, profile d does not.

cuts, but the composition velocity then changes discontinuously where the affinity cut changes. If the cuts are, say, $j-1 \mid j$ and $j \mid j+1$ in the upstream and downstream portions of the boundary, respectively, the discontinuous change of the composition velocity from $u < u_j$ to $u > u_j$ leads to the appearance of a plateau zone between the portions with different cuts (see Fig. 3.8a). On the other hand, if the sequence of the cuts is the reverse, the discontinuous change of the composition velocity causes the "faster" upstream and "slower" downstream compositions to interfere with one another (see Fig. 3.8b). A noncoherent profile then arises, as will be discussed in Section VI. It is therefore more expedient to consider portions having different affinity cuts as different coherent boundaries. With this convention, any coherent boundary has a unique affinity cut; it may sharpen or spread when migrating but will neither split into separate boundaries nor develop noncoherence unless it encounters other boundaries.

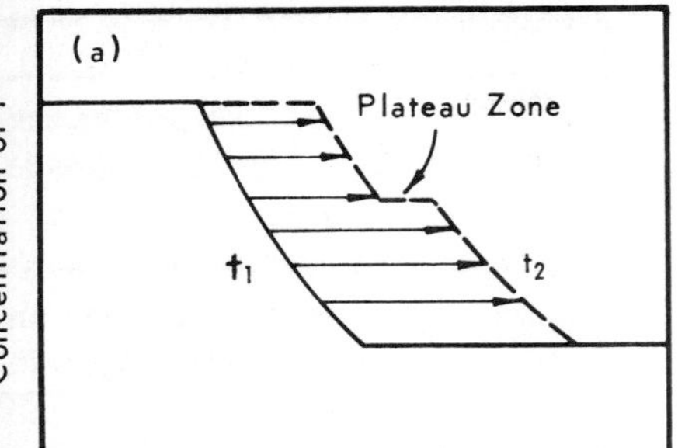

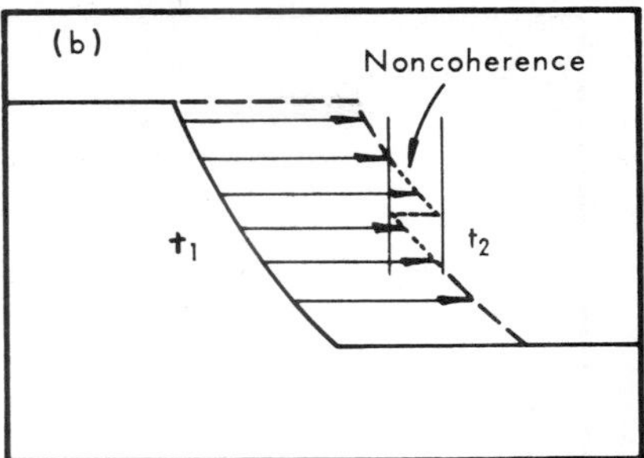


Distance From Column Inlet

**Fig. 3.8.** Development of boundary with different affinity cuts in front and rear portions (schematic). Concentration profile of one species is shown at subsequent times $t_1$ (solid line) and $t_2$ (broken line). (a) Front portion has cut between species of lower affinity than rear portion, plateau zone appears. (b) Front portion has cut between species of higher affinity than rear portion, noncoherence develops.

A coherent boundary with unique affinity cut may consist of sharp and diffuse portions, but no concentrations can change abruptly where others change gradually. In view of the affinity-cut rule, an abrupt concentration change of some species entails an abrupt increase of the proportions of either high- or low-affinity species and thus an abrupt change of all species velocities. The continuity condition (3.25) shows that any species present in finite concentrations undergo abrupt concentration changes where their species velocities change discontinuously.

Since the affinity cut is between the groups of species whose concentrations increase and decrease in the direction of flow, no single species can have a concentration maximum or minimum within a coherent boundary having a unique affinity cut. It is merely possible that *all* species have concentration maxima and minima at the same locus within the boundary. Profiles with such maxima and minima will be called "composition pulses" and will be examined more closely in Section IV.G. The considerations in the present section apply to coherent composition pulses also.

One of the most important implications of the affinity-cut rule is that only a species marking the cut may be absent from one side of the boundary while being present on the other side. All other species must either be present on both sides and within boundary, or must be entirely absent. This follows from the conditions that the composition (or step) velocity is intermediate between the species velocities of the species marking the cut [conditions (3.45) and (3.46)] and also equals the species velocity of a species whose concentration becomes zero [conditions (3.26) and (3.30)]. The two species marking the cut cannot, however, both be absent from the same side of the boundary, because their concentrations vary in opposite directions.

Exceptional behavior is found in *trace-component systems* (i.e., with all but one species at trace levels). Here, any composition variation on an absolute basis is negligible, and the species velocities therefore remain virtually constant. This allows any trace concentration to vary while all others (except that of the bulk species) remain constant, even on a relative basis. An affinity cut cannot be defined. Rather, any concentration variation of a trace species advances at a rate given by the respective species velocity, which is independent of the properties and behavior of the other trace species [see Section III.B.1 and Eq. (3.18)]. Thus, the trace species are "uncoupled" in their dynamic as well as in their equilibrium behavior. Since the trace species differ in their velocities, a coherent boundary can involve a concentration variation of only one trace species (and of the bulk species).

The rules for coherent boundaries can be summarized as follows:

1. The concentrations and species velocities of *all* species vary at any point in a coherent boundary.
2. The variations of concentrations and species velocities across a coherent boundary are monotonic.
3. A coherent boundary has an affinity cut $j \mid k$ which divides the species into a high-affinity group $1, \ldots, j$ and a low-affinity group $k, \ldots, n$. The cut is between the species slower and faster than the composition (or step), and the concentrations of the species of the two groups vary in opposite directions.
4. No species other than those marking the affinity cut can be absent from one side of the boundary while being present on the other side.
5. In trace-component systems, all variations of trace concentrations advance independently at the respective trace-species velocities, and the rules above do not apply. An affinity cut cannot be defined. Only one trace concentration (and that of the bulk species) varies across a coherent boundary.

## C. Composition Paths

The compositions encountered in a coherent boundary can be mapped as a line in the composition space ($x$ space or $y$ space). Such lines will be called "composition paths."

Attention is called to the distinction made between *composition routes* and *composition paths*. The former represent sequences of compositions found in the column under given operating conditions (see Chapter 2, Section IV.A), and may or may not coincide with composition paths. The latter represent composition sequences which *may* occur in coherent boundaries and, like the equilibrium relations, are properties of the system and will prove to be independent of the operating conditions. Mathematically, the composition paths are the characteristics of the differential equations. They are an invaluable aid for deducing column responses (composition routes) in coherent as well as noncoherent systems.*

In the present section, the properties of composition paths are stated without proof. Mathematical derivations will be given in Section IV.D and in Appendix I. For convenience, we shall operate throughout with the composition "simplex" (see Chapter 2, Section II) rather than with orthogonal concentration coordinates, as this reduces the dimension of the composition space from $n$ to $n-1$; this is merely a matter of graphical representation and affects neither the conclusions nor the mathematical relations.

1. *Existence of Invariant Composition Paths*

A composition path as defined above corresponds to composition variations that can occur in a coherent boundary. A distinction between variations with time at fixed location, and with location at fixed time, need not be made; for a given boundary, both correspond to the same path because the compositions in a coherent boundary remain intact while traveling [see also condition (3.43)].

Arbitrary composition routes in general do not coincide with composition paths, as they are unlikely to meet the coherence requirements. Even if the concentration variations along such a route satisfy the affinity-cut rule, they do not necessarily conform to the differential coherence condition

$$(\partial \bar{C}_i/\partial C_i)_z = (\partial \bar{C}_j/\partial C_j)_z \qquad \text{for all } i \text{ and } j \tag{3.38}$$

This condition is obeyed only along distinct paths. In an $n$-component system, there are $n-1$ such paths through any given composition point, one for each possible affinity cut. A representative grid of composition paths for a three-component system is shown in Fig. 3.9.

---

* Composition paths (characteristics) were used by Glueckauf in one of his early papers [19] to illustrate column responses under conditions giving entirely coherent patterns. The possibility of using the paths for deducing noncoherent responses appears to have remained unnoticed.

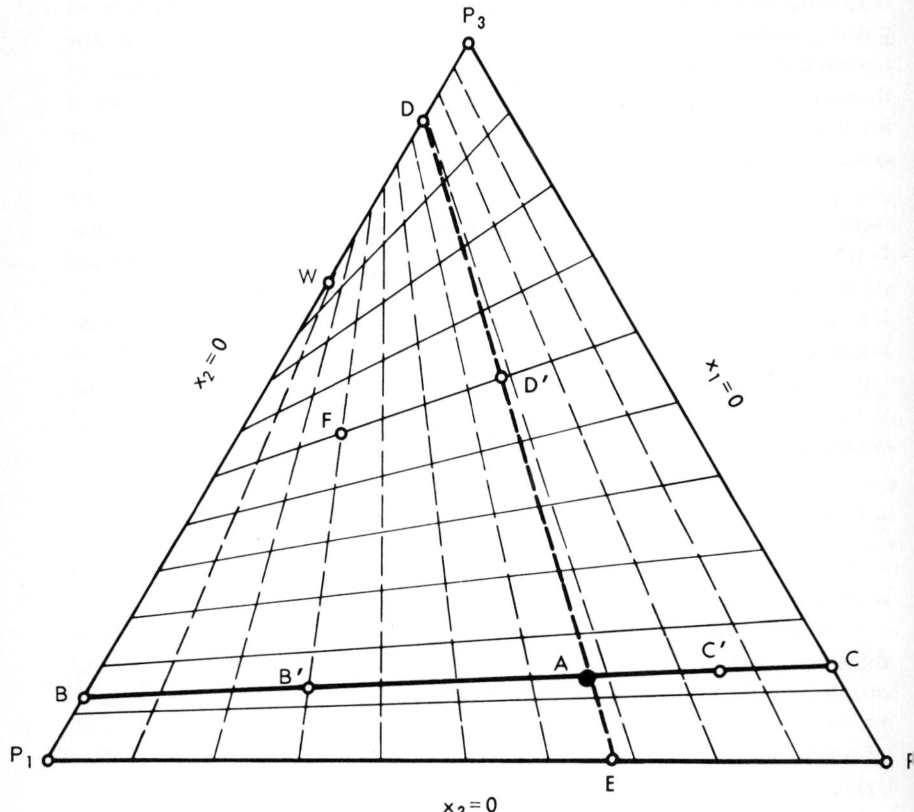


**Fig. 3.9.** Path grid in mobile-phase composition diagram of three-component system with separation factors $\alpha_{12}=2$, $\alpha_{13}=4$. Paths through point A are shown as heavy lines.

A boundary is coherent only if all its compositions, including those of the adjacent plateau zones, fall on the same composition path. However, the boundary need not extend over the full length of the path. In Fig. 3.9, for example, a coherent boundary with 1 | 2 cut may correspond to the path section B′C′ rather than to the entire path BC.

The composition paths are determined exclusively by the equilibrium properties of the system and the coherence requirement (3.38) and thus are independent of the operating conditions, time, and location in the column. Wherever a given composition exists in a coherent boundary, it is invariably preceded and followed by compositions on one of the $n-1$ paths leading through the given composition point. Because of the invariance of the composition paths, an examination of the properties of boundaries regardless of their origin is useful and can provide the basis of a more general description than previous theories for particular operating conditions have achieved.

The invariance of the composition paths still allows a given composition to exist in $n-1$ types of coherent boundaries with different affinity cuts. It is not possible to associate a particular affinity cut with particular types of operation. On the contrary, with the exception of singular cases, even the simplest operating conditions give rise to sets of boundaries representing all cuts (see Chapter 4, Section I). In any operation, a given composition may thus occur in one or several coherent boundaries of any cuts; also it may appear temporarily in noncoherent boundaries as they rearrange, it may constitute one or more plateau zones, or be entirely absent.

## 2. *Topology of Composition Paths*

A knowledge of the topology of composition paths will greatly facilitate later deductions of boundary interactions and column responses.

Composition paths can be mapped in mobile-phase composition diagrams ($x$ space) as in Fig. 3.9, or in stationary-phase ($y$ space) or overall composition diagrams. The general topology is the same in all cases. In fact, the $x$-space path grid in Fig. 3.9 is identical with the $y$-space path grid of a system with separation factors $\alpha_{12} = 1.25$, $\alpha_{13} = 2.5$. The paths in all grids are linear, provided the separation factors are independent of composition. (A path grid of a system with variable separation factors is shown in Fig. 5.10.) Of the infinite multitude of paths passing through all composition points, only a few at regular intervals will be shown in the illustrations.

*Three-component systems.* In three-component systems, for which Fig. 3.9 gives a typical example, there are two sets of paths which can be identified according to their concentration variations as having affinity cuts 1 | 2 (solid lines) and 2 | 3 (broken lines). Paths of the same set do not intersect one another. Each composition point is at an intersection of two paths, one from each set. The 1 | 2 paths connect the borders $x_1 = 0$ and $x_2 = 0$, the 2 | 3 paths connect $x_2 = 0$ and $x_3 = 0$, and paths connecting $x_1 = 0$ and $x_3 = 0$ do not exist. This reflects the rule that only the species marking the affinity cut can be absent from one side of a coherent boundary while being present on the other side (see Section IV.B). The intersections of paths with borders will be called "anchor points"; in Fig. 3.9, for example, the anchor points of the paths through A are B, C, D, and E.

An important property of the path grids is the existence of a "watershed point" W on the border $x_2 = 0$; all 1 | 2 paths originate from one side of the watershed, and all 2 | 3 paths from the other side. A point on any border, including $x_2 = 0$, thus is the anchor point of only one path. Also, because of the watershed, each path of one set intersects all paths of the other set. As a consequence, any two points not on the same path can be reached from one another by two different "one-turn routes," i.e., routes involving one switch of paths at an intersection. For example, point A in Fig. 3.9 can be reached from point F by the one-turn routes FB′A and FD′A. The position of the watershed point is determined by the separation factors and is given by

$$x_1 = \frac{\alpha_{12}-1}{\alpha_{13}-1}, \qquad x_2 = 0 \tag{3.47}$$

or, for $y$-space path grids,

$$y_1 = \frac{1-\alpha_{21}}{1-\alpha_{31}}, \qquad y_2 = 0 \tag{3.48}$$

The anchor points of any given path divide the lines $P_1W$ (or $WP_3$) and $P_2P_3$ (or $P_1P_2$) into segments which are in the same ratio to one another.* For example, in Fig. 3.9, the ratios $(P_1B):(BW)$ and $(P_2C):(CP_3)$ are equal. This "rule of equal intercept ratios" greatly facilitates the graphical construction of path grids.

The borders of the composition triangle are composition paths too. The borders $x_1 = 0$ and $x_3 = 0$ are paths with 2 | 3 and 1 | 2 cuts, respectively, and have all the properties of such paths. The border $x_2 = 0$ is exceptional in that it is the only path with 1 | 3 cut and intersects all other paths. The borders are also the only paths originating from the corner points $P_1$, $P_2$, and $P_3$. Paths from these points into the interior of the triangle do not exist because coherence does not allow two species to be absent from one side of a boundary if they are present on the other side (see Section IV.B).

*Four-component systems.* The topology of the composition paths in the tetrahedral composition simplex of a four-component system is indicated in Fig. 3.10. In the interior of the tetrahedron there are three sets of paths with affinity cuts 1 | 2, 2 | 3, and 3 | 4. No paths of the same set

---

* Here and in all later geometrical considerations, $P_i$ indicates the composition point at which $x_i = 1$.

intersect one another. Each composition point is at an intersection of three paths, one from each set.

In accordance with their affinity cuts, the paths of the three sets connect the borders (triangular faces of the tetrahedron) as follows: The 1 | 2 paths connect $x_1 = 0$ with $x_2 = 0$ ($P_2P_3P_4$ face with $P_1P_3P_4$ face), the 2 | 3 paths connect $x_2 = 0$ with $x_3 = 0$ ($P_1P_3P_4$ with $P_1P_2P_4$), and the 3 | 4 paths connect $x_3 = 0$ with $x_4 = 0$ ($P_1P_2P_4$ with $P_1P_2P_3$), as shown in Fig. 3.10.

The borders $x_1 = 0$ and $x_4 = 0$, both containing anchor points of only one set of paths, have no watersheds. On the other hand, the borders $x_2 = 0$ and $x_3 = 0$, both containing anchor points of two sets of paths, are divided by watershed lines into two regions of anchor points of one set only (see Fig. 3.10). Thus, as in three-component systems, any point on a border is the anchor point of only one path. The watersheds are straight lines connecting the watershed points of the four subordinate three-component systems, whose composition triangles constitute the faces of the tetrahedron. Regardless of the values of the separation factors, the watershed lines always connect the edge $P_1P_4$ with the edges $P_1P_3$ and $P_2P_4$ and are placed relative to one another in such a way that the quadrilaterals containing the anchor points of the 2 | 3 paths are not in contact (see Fig. 3.10b).

The borders (faces of the tetrahedron) are the composition triangles of the subordinate three-component systems and contain the respective path grids. For example, the border $x_3 = 0$ ($P_1P_2P_4$ face in Fig. 3.10) has paths with 1 | 2 and 2 | 4 affinity cuts, and their watershed point is at

$$x_1 = \frac{\alpha_{12}-1}{\alpha_{14}-1}, \qquad x_2 = 0, \qquad x_3 = 0 \tag{3.49}$$

All paths originating from the edges lead across the faces of the tetrahedron, none into its interior. The edges ("sub-borders," with two species absent) also are composition paths and are the only paths originating from the corner points.

A further important topological property, the existence of "common planes," will be discussed separately in Section IV.C.3.

*Generalized Rules for n-Component Systems.* The rules developed above can be stated in general form for systems with $n$ components. (1) In the interior of the $(n-1)$-dimensional composition simplex there are $n-1$ sets of paths with affinity cuts 1 |2, 2 | 3, ..., $n-1$ | $n$. (2) No paths of the same set intersect one another. (3) Each point in the interior is at an intersection of $n-1$ paths, one from each set. (4) Paths with $i \mid i+1$ cut connect the borders $x_i = 0$ and $x_{i+1} = 0$. (5) All borders except $x_1 = 0$ and $x_n = 0$ have

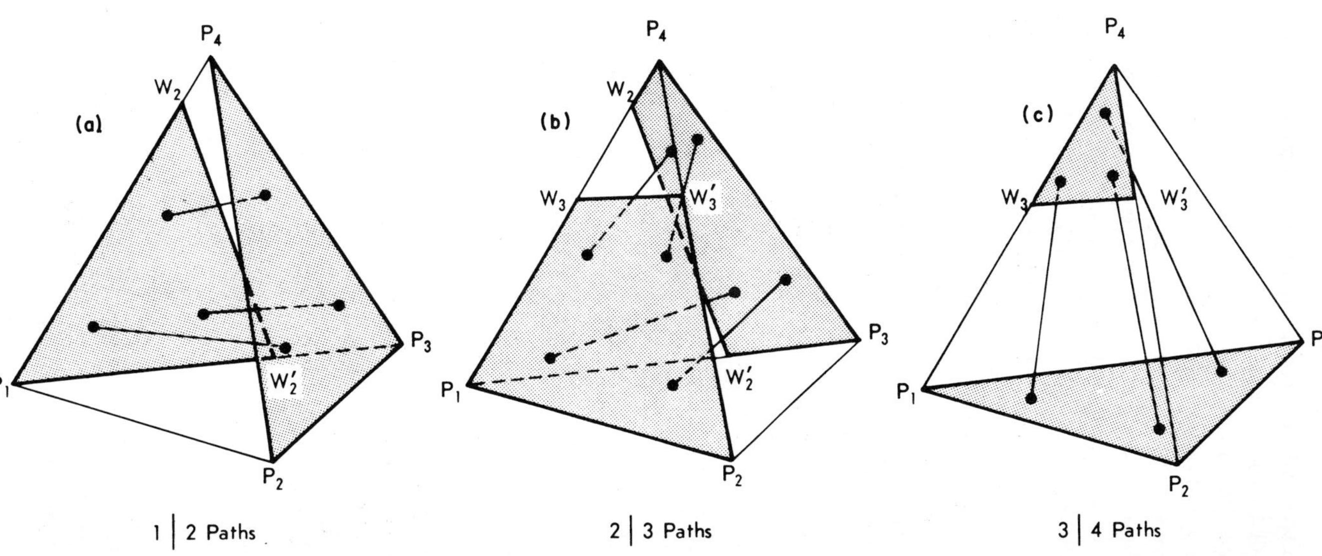


**Fig. 3.10.** Paths in four-component systems (calculated for separation factors $\alpha_{12} = 2$, $\alpha_{13} = 4$, $\alpha_{14} = 8$). Regions of anchor points of respective paths are shaded, paths are shown as heavy lines. Lines $W_2W_2'$ and $W_3W_3'$ are watersheds on borders $x_2 = 0$ and $x_3 = 0$, respectively.

watersheds, which are segments of $(n-3)$-dimensional hyperplanes and are bounded by the watersheds of the related subordinate systems with a second species absent. (6) The edge between the corner points $P_1$ and $P_n$ of the composition space intersects all watersheds of borders.* The intersections, counted from the point $P_n$, are in the sequence of decreasing affinity of the species absent from the respective border; that is, going along the edge from $P_n$ to $P_1$, one crosses first the watershed of the border $x_2=0$, then that of the border $x_3=0$, etc. (7) All paths on borders, sub-borders, etc., obey the rules for the respective subordinate systems with $n-1$, $n-2$, etc., components. (8) No paths lead from sub-borders into the interior of the composition space.

These rules are given mainly for the sake of completeness. It will be seen later that constructions in simple two-dimensional diagrams can account for the most important local phenomena, even if many species are present.

### 3. *Common Planes and Common Hyperplanes*

One of the most important topological properties of composition paths in systems with more than three components is the existence of what will be called "common planes" and "common hyperplanes".

The paths are oriented in space in such a way that all paths of one set intersecting a given path of another set are on one plane. Consequently, there are *common planes* of mutually intersecting paths of two sets. Such common planes are intersected by paths of all other sets, and only by paths of other sets.

In four-component systems, there are three classes of common planes, namely, of 1 | 2 and 2 | 3 paths, of 1 | 2 and 3 | 4 paths, and of 2 | 3 and 3 | 4 paths (see Figs. 3.11 and 3.12). Those of the first and third classes are triangular and have watershed points, and those of the second class are quadrilateral. The path grids on all common planes obey the rule of equal intercept ratios. Each composition point is at an intersection of three common planes, one from each class (see Figs. 3.13). Excepted are points on the edge $P_1P_4$, which are on only one common plane each; this edge is intersected by the 1 | 2-2 | 3 common planes between the point $P_1$ and the watershed of $x_3=0$, by the 1 | 2-3 | 4 common planes between the watersheds of $x_3=0$ and $x_2=0$, and by the 2 | 3-3 | 4 common planes between the watershed of $x_2=0$ and the point $P_4$ (see Fig. 3.12). A consequence of this topology is that each common plane of paths of two sets intersects

---

* See footnote to p. 64.

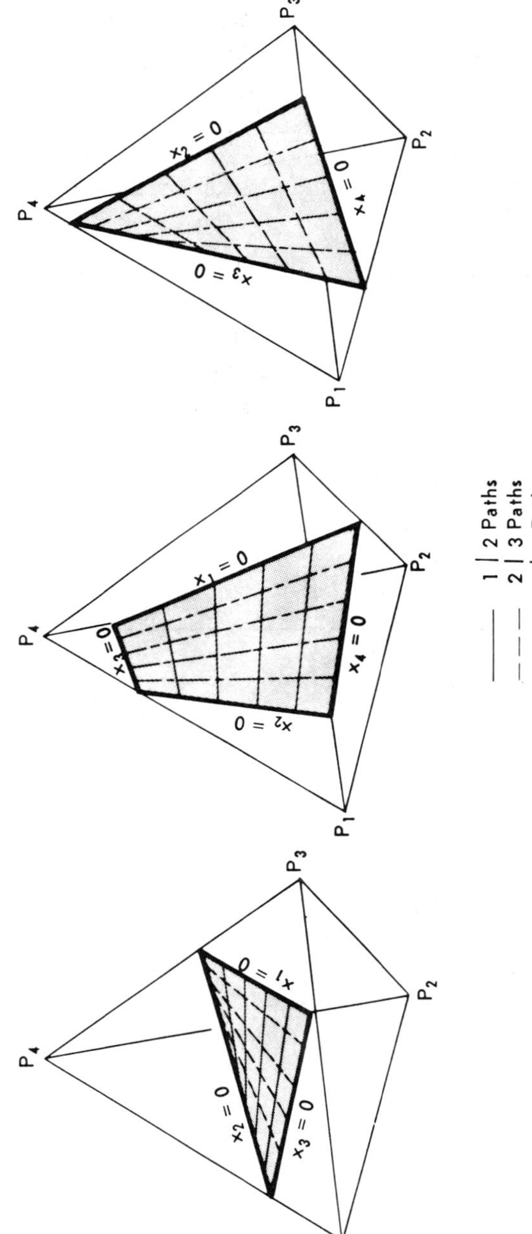


**Fig. 3.11.** Common planes (shaded) with path grids in four-component systems (calculated for separation factors $\alpha_{12} = 2, \alpha_{13} = 4, \alpha_{14} = 8$).

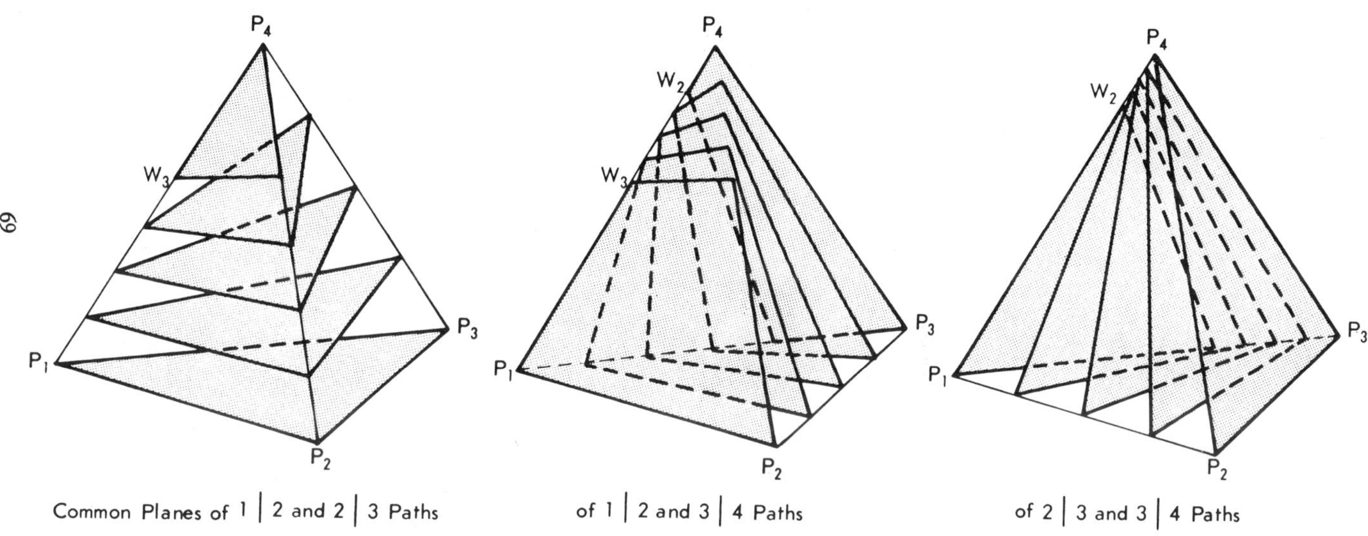


**Fig. 3.12.** Orientation of common planes (shaded) in four-component systems (calculated for separation factors $\alpha_{12} = 2$, $\alpha_{13} = 4$, $\alpha_{14} = 8$). Points $W_2$ and $W_3$ are intersections of watersheds of $x_2 = 0$ and $x_3 = 0$, respectively, with $P_1P_4$ edge.

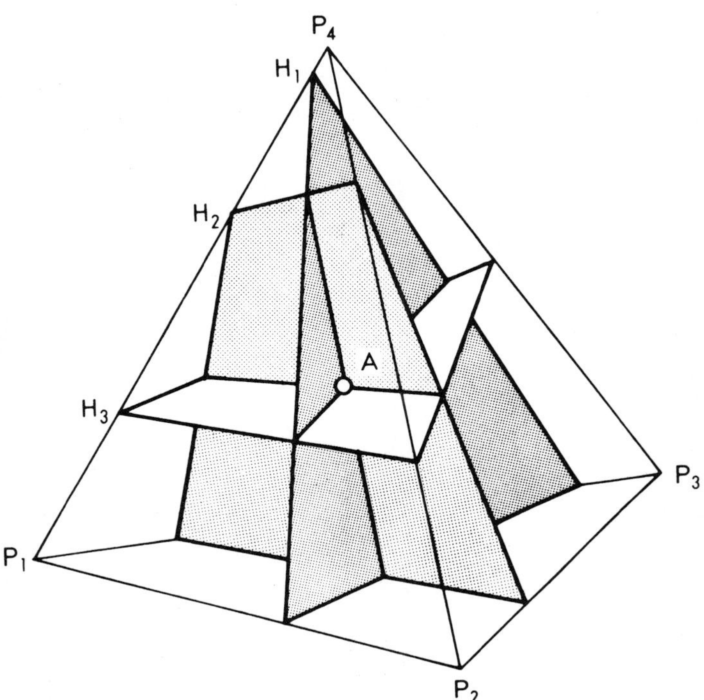


**Fig. 3.13.** Common planes of given composition point A in four-component system. Points $H_1$, $H_2$, and $H_3$ are intersections of common planes with $P_1P_4$ edge. (For $x_1 = 0.23$, $x_2 = 0.26$, $x_3 = 0.14$, $x_4 = 0.37$, and $\alpha_{12} = 2$, $\alpha_{13} = 4$, $\alpha_{14} = 8$.)

*all* paths of the third set. One can thus reach any point from any other point (not on the same common plane) by a "two-turn route" on paths in numerical sequence of cuts, e.g., taking first a 1 | 2 path to the 2 | 3-3 | 4 common plane of the target point and then a 2 | 3 and a 3 | 4 path on that plane. There is only one such route with numerical sequence of cuts between two given points.

Common planes also exist in systems with more than four components. Those of paths with cuts having a species in common are triangles and have watersheds; those with cuts not having a species in common are quadrilaterals (see Fig. 3.14). The rule of equal intercept ratios is obeyed on all.

Systems with more than four components also have *common hyperplanes*

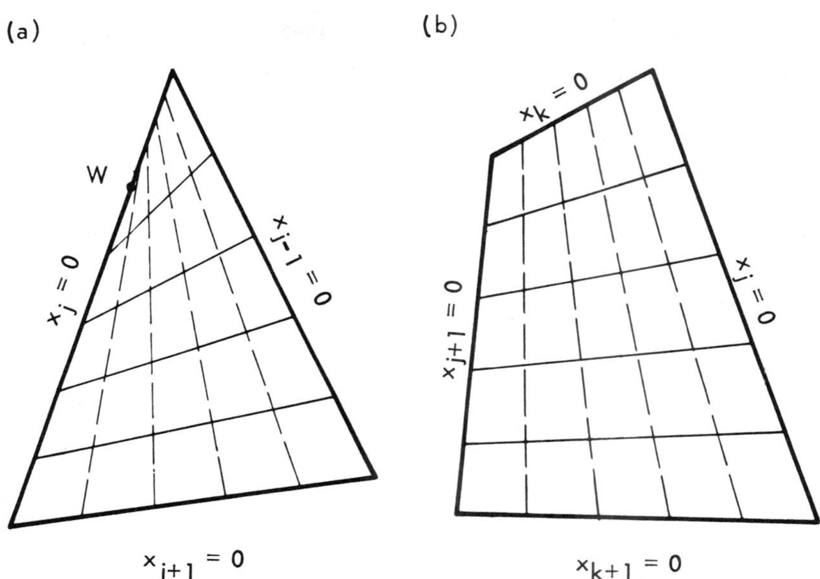


**Fig. 3.14.** Typical common planes in *n*-component system. (a) Affinity cuts with one common species, (b) affinity cuts without common species.

which contain mutually intersecting paths of three, four, ..., $n-2$ sets and are intersected by paths of the remaining sets. The $(n-2)$-dimensional common hyperplanes of paths of $n-2$ sets are important. Containing paths of all other sets, they intersect paths of one set only. We shall call the $(n-2)$-dimensional common hyperplanes intersecting 1 | 2 paths "1-hyperplanes" and, in general, those intersecting $i \mid i+1$ paths "*i*-hyperplanes."

The common planes in four-component systems as well as the composition paths in three-component systems can be considered as limiting cases of *i*-hyperplanes since they, too, intersect paths of one set only. For example, the 2 | 3-3 | 4 common planes in four-component systems and the 2 | 3 paths in three-component systems, both intersecting 1 | 2 paths, correspond to 1-hyperplanes. In fact, the following general hyperplane rules are also valid for common planes and composition paths in four- and three-component systems, respectively. (1) There are $n-1$ classes of *i*-hyperplanes, distinguished by the cut of the paths which they intersect.

(2) Each composition point (unless on the $P_1P_n$ edge*) is at an intersection of $n-1$ $i$-hyperplanes, one from each class. (3) Each $i$-hyperplane intersects *all* $i \mid i+1$ paths. (4) All $i$-hyperplanes intersect the edge $P_1P_n$; the 1-hyperplanes do so between $P_n$ and the watershed of $x_2 = 0$, the $j$-hyperplanes ($j = 2, \ldots, n-2$) between the watersheds of $x_j = 0$ and $x_{j+1} = 0$, and the $(n-1)$-hyperplanes between the watershed of $x_{n-1} = 0$ and $P_1$.

Two consequences of these rules warrant comment. First, rule (3) ensures that any point can be reached from any other point (not on the same hyperplane) by an $(n-2)$-turn route on $n-1$ paths in numerical sequence of affinity cuts; there is only one such route between two given points. Second, because of rule (4), any point on the edge $P_1P_n$ is on only one $i$-hyperplane; any point in the interior of the composition space can therefore be uniquely characterized by the $n-1$ intercepts of its $i$-hyperplanes with this edge (provided the separation factors are given). For example, point A in Fig. 3.13 is uniquely determined by the positions of points $H_1$, $H_2$, and $H_3$ on the $P_1P_4$ edge. It will be seen in the next sections that the method of characterizing composition points by such hyperplane intercepts rather than by concentration coordinates greatly facilitates the mathematical as well as conceptual treatment of multicomponent systems.

#### 4. *Trace-Component Systems*

In the immediate vicinity of a corner point $P_k$ of the composition simplex, where $k$ alone is present at a significant concentration and all other species are at trace levels, the various composition paths run virtually parallel to the composition-simplex edges. Thus, along any path in this region, only one trace-species concentration and that of $k$ vary while all others remain constant. This is in accordance with the behavior of coherent boundaries in such systems, discussed in Section IV.B.

### D. *H* Function and *h* Transformation

Mobile- and stationary-phase composition-path grids can be orthogonalized by a nonlinear transformation involving what will be called the "*H* function" (hyperplane function). This function and its roots will prove to be of central importance in the further discussion of coherent as well as noncoherent systems. Beyond providing simple mathematical proofs for the existence of universal composition paths and their topology as well as convenient equations for species velocities, composition-velocity eigen-

---

* See footnote to p. 64.

values, step velocities, etc., the $H$ function allows various rules and criteria to be stated in simple terms, leads to explicit solutions even for complex cases, and greatly facilitates practical calculations.

In essence, the transformation with the $H$ function amounts to replacing the set of $n$ mobile-phase or stationary-phase concentration $x_i$ or $y_i$ by a new set of $n-1$ composition variables $h_i$. (Only $n-1$ independent variables are needed to define a composition, since the $n$ concentrations are subject to the constraints $\Sigma_i x_i = 1$ and $\Sigma_i y_i = 1$.) The use of the $h_i$, which are the "natural coordinates" of the system, greatly reduces the mathematical complexity of derivations and solutions. After solutions in terms of the $h_i$ have been obtained, the latter are readily reconverted to concentrations $x_i$ or $y_i$ as desired.

In this section, the essential features of the $h$ transformation are described and equations needed for later applications are stated. Details and mathematical proofs are given in Appendix I.

Functions of the same general type as the $H$ function have been used in various areas of mathematics and physics. In differential geometry, such a function appears in the description of homofocal surfaces [30], and Binet is credited with having derived, in 1811, a theorem essentially equivalent to the transformation from $\mathbf{h}$ to $\mathbf{x}$ in the present approach [31]. A relation to one of Sylvester's formulas (Amundson's version [32]) is evident in Eqs. (3.53) and (3.57). Other equivalent or similar transformations have been used in the mathematical treatment of multicomponent electrodiffusion [33, 34], electrophoresis [35], and distillation [36]. Equations of the same general form also appear in a recent theory of multicomponent gas chromatography [37, 38] based on premises entirely different from those used here (see also Chapter 4, Section III.F). A variety of further multicomponent systems with interference, such as ultracentrifugation [39], seem to exhibit enough mathematical analogies to make the transformation applicable and potentially useful (see also Chapter 7).

### 1. *The h Space*

It is apparent that composition-path grids can be orthogonalized if they have the topological properties described in the previous section (watersheds, common planes, common hyperplanes, etc.). The orthogonalization will transform the triangular simplex of a three-component system into a rectangle with the watershed point as the fourth corner (see Fig. 3.15), the tetrahedral simplex of a four-component system into a rectangular parallelepiped with the two watershed lines and four watershed points as the additional edges and corners (see Fig. 3.16), etc., with all paths intersecting at right angles. Conversely, the essential topological properties of the path grids in the composition space can be established by proving

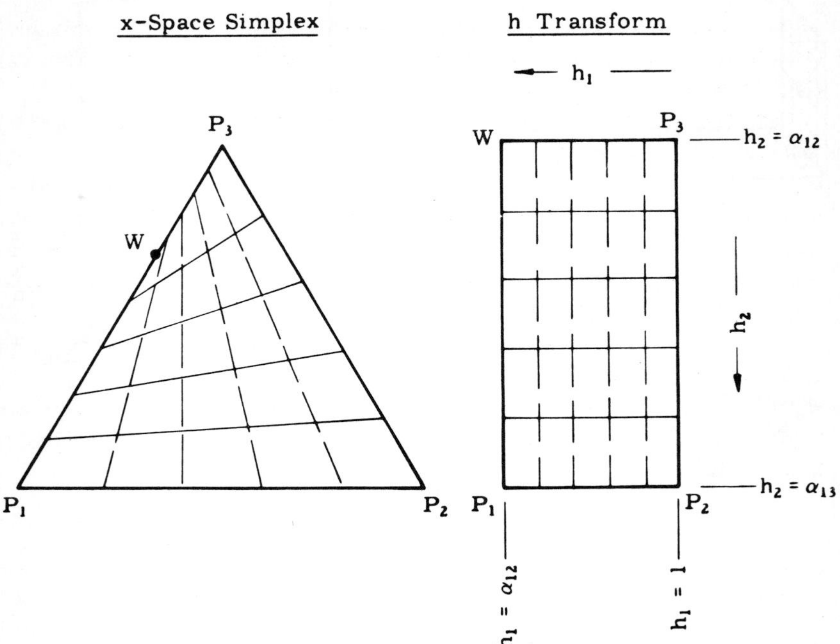


**Fig. 3.15.** Three-component composition simplex with path grid and its $h$ transform. Point W is watershed point. (For $\alpha_{12} = 2$, $\alpha_{13} = 4$.) (From F. Helfferich [41]. Reproduced from *Advances in Chemistry Series* No. 79 with permission of the American Chemical Society.)

that such an orthogonalization is possible. This proof is given in Appendix I.

The transformation orthogonalizing the path grid involves a change of coordinates. The $x$ space or $y$ space (i.e., mobile- or stationary-phase composition space) with species concentrations $x_i$ or $y_i$ as coordinates is transformed into the "$h$ space" with $n-1$ orthogonal coordinates $h_i$, each of which is parallel to the paths of one set. The $h$ coordinates could, of course, be normalized so that the transforms become squares, cubes, and hypercubes. However, all pertinent equations remain simpler without such normalization, namely, if the $h$ coordinates are taken to be the roots of the "$H$ function," a polynomial to be defined below. The coordinate values of physically meaningful compositions ($x_i \geqq 0$, $\Sigma_i x_i = 1$) then fall into the intervals

$$\alpha_{1i} \leqq h_i \leqq \alpha_{1,i+1} \qquad (i = 1, \ldots, n-1) \tag{3.50}$$

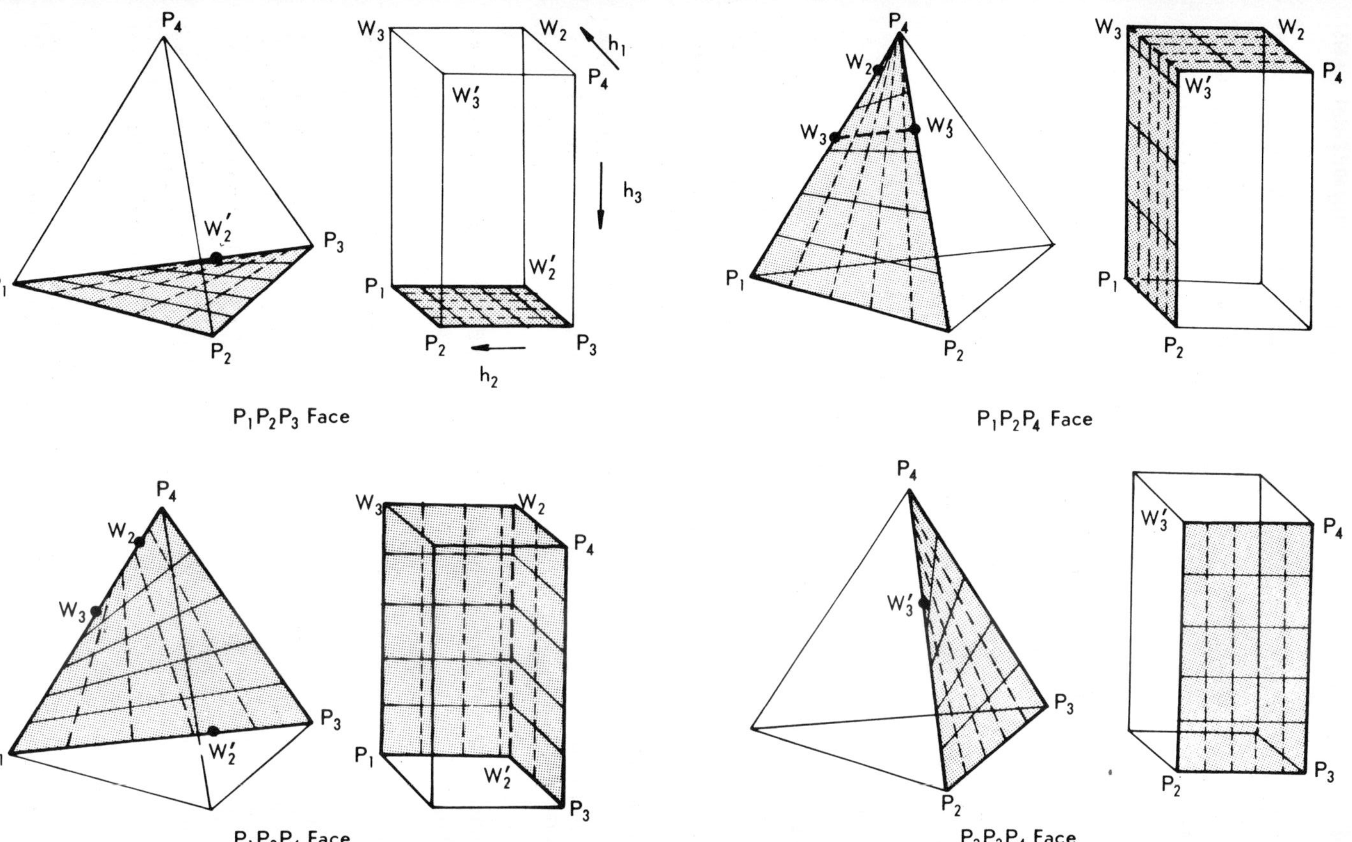


**Fig. 3.16.** Faces of four-component composition simplex and their *h* transforms (for $\alpha_{12} = 2$, $\alpha_{13} = 4$, $\alpha_{14} = 8$).

### 2. *H Function and h Transformation*

The new composition variables $h_1, \ldots, h_{n-1}$ are defined as the $n-1$ roots in $h$ of the equation

$$H(h, \mathbf{x}, \boldsymbol{\alpha}_{1i}) = 0 \tag{3.51}$$

or, in terms of $\mathbf{y}$, as the reciprocals of the $n-1$ roots in $1/h$ of

$$H(1/h, \mathbf{y}, \boldsymbol{\alpha}_{i1}) = 0 \tag{3.52}$$

where $h$ is a dummy variable appearing in the $H$ function, defined as the following polynomial of degree $(n-1)$*:

$$H(q, \mathbf{r}, \mathbf{s}) \equiv \sum_{i=1}^{n} \left[ \prod_{\substack{j=1 \\ j \neq i}}^{n} (q - s_j) r_i \right] \tag{3.53}$$

The $n-1$ roots are all real and fall into the intervals stated in condition (3.50). The sets of $h_i$ obtained with Eqs. (3.51) and (3.52) are identical for mobile- and stationary-phase compositions $\mathbf{x}$ and $\mathbf{y}$ in equilibrium with one another. A given set $\mathbf{h}$ thus characterizes a composition of both phases, and a distinction between the sets of roots calculated from $\mathbf{x}$ and $\mathbf{y}$ need not be made.

If one or several species $k$ are absent, each $h = \alpha_{1k}$ is an obvious solution of Eqs. (3.51) and (3.52): The term involving $x_k$ or $y_k$ then is zero, and all other terms involve $h - \alpha_{1k}$ or $1/h - 1/\alpha_{1k}$ as a factor. Such roots $h = \alpha_{1k}$ will be called *trivial roots.* The other, nontrivial roots—all roots if no species is absent—do not equal any separation factors $\alpha_{1i}$ $(i \neq k)$ and are more conveniently calculated from the equations

$$\sum_i \frac{x_i}{h - \alpha_{1i}} = 0 \quad \text{or} \quad \sum_i \frac{y_i}{1/h - \alpha_{i1}} = 0 \tag{3.54}$$

obtained when Eqs. (3.51) and (3.52), with the definition (3.53), are divided by $\Pi_j (h - \alpha_{1j})$ and $\Pi_j (1/h - \alpha_{j1})$, respectively. [These equations could be used to define the $h_i$, except that the sums are indeterminate at any $h = \alpha_{1k}$ if $x_k = 0$ or $y_k = 0$, so that the $h_i$ values corresponding to the trivial roots of Eq. (3.51) or (3.52) would remain undefined.] When the roots are indexed in the sequence $h_{i+1} > h_i$, any trivial root $h = \alpha_{1k}$, owing

---

* In earlier outlines of the present treatment [40, 41], a slightly different function having the same roots was defined as the $H$ function [see Eq. (A.33) in Appendix II].

its value to the absence of a species $k$, acquires the index number $k-1$ or $k$:

$$\begin{aligned} h_{k-1} &= \alpha_{1k} \quad \text{or} \quad h_k = \alpha_{1k} && \text{if} \quad x_k = 0 \qquad (k \neq 1, n) \\ h_1 &= \alpha_{11} \equiv 1 && \text{if} \quad x_1 = 0 \\ h_{n-1} &= \alpha_{1n} && \text{if} \quad x_n = 0 \end{aligned} \tag{3.55}$$

A nontrivial root may coincide with a trivial one; in this degenerate case one has $h_{k-1} = h_k = \alpha_{1k}$. The nontrivial roots are readily calculated from either of the two equations (3.54) by standard numerical procedures. Explicit equations for compositions with no more than three species present are given in Appendix I.

It is shown in Appendix I that the differential coherence condition (3.38) restricts any coherent composition variations to directions in which all but one $h_i$ remain constant. In terms of $H$ function roots, this condition is

$$dh_i = 0 \qquad \text{for all } i \neq k \tag{3.56}$$

where $k$ can be $1, \ldots, n-1$. This derivation proves the two principal postulates on which the presentation has been based, namely, that distinct composition paths exist and that they are orthogonalized by the $h$ transformation.

For the reverse transformation, from the $h$ space into the $x$ or $y$ space, explicit equations can be given. The mobile-phase and stationary-phase concentrations constituting a composition can be calculated from their $H$-function roots with the equations*

$$x_j = \prod_i (h_i - \alpha_{1j}) \Big/ \prod_{i \neq j} (\alpha_{1i} - \alpha_{1j}) \tag{3.57}$$

for all $j$

$$y_j = \prod_i \left( \frac{1}{h_i} - \alpha_{j1} \right) \Big/ \prod_{i \neq j} (\alpha_{i1} - \alpha_{j1}) \tag{3.58}$$

These relations, derived in Appendix I, are valid also for compositions with absent species.

From Eqs. (3.57) and (3.58) one obtains for the distribution ratio of the species $j$, in terms of $H$-function roots:

$$y_j / x_j = \alpha_{j1} \prod_i \alpha_{1i} \Big/ \prod_i h_i \tag{3.59}$$

---

* Sums $\Sigma_i$ and products $\Pi_i$ extend over $i = 1, \ldots, n$ if the index $i$ refers to species (as in $x_i$ or $\alpha_{1i}$), and over $1, \ldots, n-1$ if the index refers to roots $h_i$ (see Glossary of Symbols).

The $h$ transformation has certain symmetry properties, which provide convenient shortcuts in various derivations and calculations. As a comparison of Eqs. (3.51) and (3.52) shows, the functional relation between the $x_i$, $h_i$, $\alpha_{1i}$ is identical with that between the $y_i$, $1/h_i$, and $\alpha_{1i} = 1/\alpha_{i1}$. Thus, once a relation for the $x_i$ under any kind of operating conditions has been derived, that for the $y_i$ is readily obtained from it by replacement of all $H$-function roots and separation factors by their reciprocals. Examples reflecting this symmetry are Eq. (3.59) and the pairs of equations (3.57) and (3.58), (3.93) and (3.94), (4.47) and (4.48), and (4.60). Although this simple conversion formula is available, relations for both **x** and **y** will usually be given for ease of reference.

That the separation factors have the noted symmetry property—i.e., are replaced by their reciprocals if the $x_i$ and $y_i$ are interchanged—is a direct consequence of their definition [see Eqs. (2.8) and (2.9)]. That the $H$-function roots share this property is one of the reasons why their use so greatly reduces the mathematical complexity of derivations. The symmetry properties also extend to the independent variables $z$ and $\tau$ and to the adjusted velocities, as was already noted in Section III.C. An example illustrating the application of these symmetry considerations will be given in Chapter 4, Section IV.B.5 [see the substitutions (4.106)].

3. *Geometrical Significance of the H Function and its Roots*

The existence of common planes and common hyperplanes in the $h$ space is an obvious consequence of the orthogonality of the path grid. In the $h$ space, the $(n-2)$-dimensional "$j$-hyperplanes" (see Section IV.C.3) are parallel to all coordinates but one, namely $h_j$, where $j$ can be $1, \ldots, n-1$. Hence, $h_j$ is the only constant root on any $j$-hyperplane. The mobile- and stationary-phase concentrations on a given $j$-hyperplane thus obey the equations

$$\begin{aligned} H(h_j, \mathbf{x}, \boldsymbol{\alpha}_{1i}) &= 0 \\ H(1/h_j, \mathbf{y}, \boldsymbol{\alpha}_{i1}) &= 0 \end{aligned} \tag{3.60}$$

where $h_j$ is fixed and **x** and **y** are variable. In other words, the $H$ function with fixed argument $h$ describes an $(n-2)$-dimensional common hyperplane in the $x$ space or $y$ space.

The existence of, and equations for, $j$-hyperplanes in the $x$ space and $y$ space can be derived from the equilibrium relations and the differential coherence condition, without prior introduction of the $H$ function. Lengthy calculations not reproduced here lead to Eqs. (3.60) for the hyperplanes, with the $H$ function defined as in Eq. (3.53). In this manner, the $H$ function can be derived deductively.

A set of $n-1$ equations (3.60), with $j = 1, \ldots, n-1$, describes a composition **x** or **y** as the intersection of $n-1$ $j$-hyperplanes. This is shown in Fig. 3.17 for a four-component system, in which the $j$-hyperplanes are two-dimensional. Although each edge of the parallelpiped in the $h$ space intersects only one of the $j$-hyperplanes of a composition point, the $P_1P_n$ edge of the simplex in the $x$ or $y$ space, constituting the transform of $n-1$ $h$-space edges, intersects all $j$-hyperplanes.

The constant value of $h_j$ of any given $j$-hyperplane is linearly related to

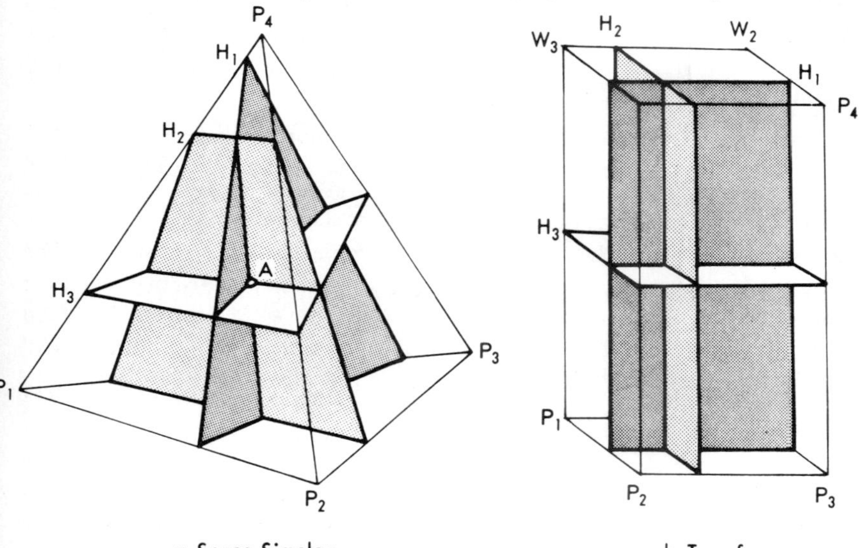


**Fig. 3.17.** Common planes of given composition point A in $x$ space and $h$ space (for same values as in Fig. 3.13).

the intercept $p_j$, measured from point $P_n$, of the hyperplane with the $P_1P_n$ edge of the $x$-space simplex:

$$h_j = (\alpha_{1n}-1)p_j+1 \tag{3.61}$$

as can be shown with Eqs. (3.51) and (3.53). To express a composition in terms of its $H$-function roots thus amounts to characterizing it by the intercepts of its $j$-hyperplanes with the $P_1P_n$ edge. (The $p_j$ instead of the $h_j$ could be used as coordinates for the orthogonalization of the path grid, but the resulting equations and functions are somewhat less convenient.)

### 4. *Borders and Watersheds*

The $h$ transform of the composition simplex is bounded by $n-1$ pairs of parallel borders $h_i = \alpha_{1i}$, $h_i = \alpha_{1,i+1}$ ($i = 1, \ldots, n-1$), since these values of the $h_i$ are the extremes permitted by condition (3.50). The total number of $h$-space borders thus is $2n-2$. In contrast, the $x$- or $y$-space simplex has only $n$ borders $x_i = 0$ or $y_i = 0$. This difference arises because all but two of the $x$- or $y$-space borders have watersheds, at which they are "folded" by the $h$ transformation, i.e., they are transformed each into two separate $h$-space borders at right angles (see Figs. 3.15 and 3.16). The two $h$-space borders are the transforms of the regions on the two sides of the watershed, and their intersection is the transform of the watershed. The two borders without watersheds are those with species 1 and $n$ absent.

Condition (3.55) identifies the two $h$-space borders on which a given species $k \neq 1, n$ is absent as those with $h_{k-1} = \alpha_{1k}$ and $h_k = \alpha_{1k}$. As their intersection, the watershed of $x_k = 0$ is given by

$$h_{k-1} = h_k = \alpha_{1k} \qquad (k \neq 1, n) \tag{3.62}$$

Equation (3.62) can be expressed in terms of $x_i$ or $y_i$:

$$\begin{aligned} &\sum_{i \neq k} \frac{x_i}{1-\alpha_{ki}} = 0, \qquad x_k = 0 \\ & \qquad\qquad\qquad\qquad\qquad\qquad (k \neq 1, n) \\ &\sum_{i \neq k} \frac{y_i}{1-\alpha_{ik}} = 0, \qquad y_k = 0 \end{aligned} \tag{3.63}$$

These equations describe the watershed in the $x$ space and $y$ space. The derivation is given in Appendix I, where it is also shown that

$$\begin{aligned} h_{k-1} = \alpha_{1k} \quad \text{if} \quad &\begin{cases} \displaystyle\sum_{i \neq k} \frac{x_i}{1-\alpha_{ki}} > 0, & x_k = 0 \\ \displaystyle\sum_{i \neq k} \frac{y_i}{1-\alpha_{ik}} < 0, & y_k = 0 \end{cases} \\ h_k = \alpha_{1k} \quad \text{if} \quad &\begin{cases} \displaystyle\sum_{i \neq k} \frac{x_i}{1-\alpha_{ki}} < 0, & x_k = 0 \\ \displaystyle\sum_{i \neq k} \frac{y_i}{1-\alpha_{ik}} > 0, & y_k = 0 \end{cases} \end{aligned} \tag{3.64}$$

This criterion can be used to establish, without calculation of $H$-function roots, whether a given composition point $\mathbf{x}$ or $\mathbf{y}$ with species $k$ absent is on the $h$-space border $h_{k-1}=\alpha_{1k}$ or $h_k=\alpha_{1k}$, and thus whether it is in the region from which paths with variable $h_k$ or $h_{k-1}$, respectively, originate.

### 5. *Composition Paths*

To orthogonalize the path grid, the $H$ function has been so defined that all but one $h_i$ remain constant along any path in the $h$ space [see condition (3.56)]. Along a path with $k \mid k+1$ affinity cut, the variable root is $h_k$. This is readily established with Eqs. (3.57) or (3.58) and condition (3.50), which show that the concentrations of species $1, \ldots, k$ and $k+1, \ldots, n$ vary in opposite directions if $h_k$ is varied. In terms of $H$-function roots, the general expression for paths with $k \mid k+1$ cuts thus is

$$h_i = \text{const} \qquad \text{for all } i \neq k, \qquad \alpha_{1k} \leqq h_k \leqq \alpha_{1,k+1} \tag{3.65}$$

where $k$ can be $1, \ldots, n-1$.

Paths with affinity cuts $k \mid k+2$, "bracketing" an absent species $k+1$, fall on the $x$- or $y$-space border $x_{k+1}=0$ (or $y_{k+1}=0$) and cross the watershed of this border (see Appendix I). As shown earlier, $x$- or $y$-space borders are "folded" at their watersheds by the $h$ transformation. The paths crossing the watershed thus are also folded, i.e., are transformed into two separate $h$-space paths at right angles and intersecting at the watershed (see Fig. 3.16). The variable root of a $k \mid k+2$ path is $h_k$ on one side of the watershed and $h_{k+1}$ on the other side, the trivial root equals $\alpha_{1,k+1}$ on both sides, and all other roots remain constant along the entire path. In terms of $H$-function roots, $k \mid k+2$ paths thus are given by

$$\begin{array}{ll} h_i = \text{const} & \left\{\begin{array}{lll} h_{k+1} = \alpha_{1,k+1}, & & \alpha_{1k} \leqq h_k \leqq \alpha_{1,k+1} \\ h_k = \alpha_{1,k+1}, & & \alpha_{1,k+1} \leqq h_{k+1} \leqq \alpha_{1,k+2} \end{array}\right. \\ \text{for all } i \neq k, k+1 & \end{array} \tag{3.66}$$

Paths with $k \mid l$ affinity cuts, bracketing several absent species $k+1, \ldots, l-1$, fall on subordinate borders in the $x$ or $y$ space and cross the various subordinate watersheds of these borders. Accordingly, each such path is transformed into several $h$-space paths which intersect at the watersheds. The extreme example is the $P_1P_n$ edge, a path with $1 \mid n$ cut, which is transformed into $n-1$ $h$-space paths (edges), as can be seen for a four-component system in Fig. 3.16. For the general case of a path with $k \mid l$ cut, the behavior of the roots $h_k, \ldots, h_{l-1}$ is shown in Table 3.1; the columns correspond to the various segments into which the path is divided by the watersheds. The variable root changes its index number from segment to segment but, along the entire path, varies continuously and monotonically from $\alpha_{1k}$ at one anchor point to $\alpha_{1l}$ at the other. In each segment, the trivial roots equal the separation factors $\alpha_{1,k+1}, \ldots, \alpha_{1,l-1}$. All roots not shown in the table remain constant along the entire path. The general behavior shown in Table 3.1 comprises expressions (3.65) and (3.66) as special cases with $l = k+1$ and $l = k+2$, respectively.

**Table 3.1**

VALUES OF $H$-FUNCTION ROOTS FOR THE VARIOUS SEGMENTS OF A COMPOSITION PATH WITH BRACKETING AFFINITY CUT $k \mid l$.

| Root | Root values | | | | | |
|---|---|---|---|---|---|---|
| $h_k$ | Varies from $\alpha_{1k}$ to $\alpha_{1,k+1}$ | $\alpha_{1,k+1}$ | $\alpha_{1,k+1}$ | ... | $\alpha_{1,k+1}$ | $\alpha_{1,k+1}$ |
| $h_{k+1}$ | $\alpha_{1,k+1}$ | Varies from $\alpha_{1,k+1}$ to $\alpha_{1,k+2}$ | $\alpha_{1,k+2}$ | ... | $\alpha_{1,k+2}$ | $\alpha_{1,k+2}$ |
| $h_{k+2}$ | $\alpha_{1,k+2}$ | $\alpha_{1,k+2}$ | Varies from $\alpha_{1,k+2}$ to $\alpha_{1,k+3}$ | ... | $\alpha_{1,k+3}$ | $\alpha_{1,k+3}$ |
| ... | ... | ... | ... | ... | ... | ... |
| $h_{l-2}$ | $\alpha_{1,l-2}$ | $\alpha_{1,l-2}$ | $\alpha_{1,l-2}$ | ... | Varies from $\alpha_{1,l-2}$ to $\alpha_{1,l-1}$ | $\alpha_{1,l-1}$ |
| $h_{l-1}$ | $\alpha_{1,l-1}$ | $\alpha_{1,l-1}$ | $\alpha_{1,l-1}$ | ... | $\alpha_{1,l-1}$ | Varies from $\alpha_{1,l-1}$ to $\alpha_{1l}$ |

6. *Species Velocities and Composition-Velocity Eigenvalues*

In later applications it will be convenient to express species velocities and composition-velocity eigenvalues in terms of $H$-function roots. The respective relations for the adjusted velocities are given below (for definition of adjusted velocities, see Chapter 2, Section IV.B).

For the adjusted *species velocities* of the species $j$ one obtains with Eqs. (3.35) and (3.59)

$$u_j = x_j/y_j = \alpha_{1j}\prod_i h_i \prod_i \alpha_{i1} \qquad \text{for all } j \tag{3.67}$$

The *composition velocity* is defined only for coherent boundaries. Any composition point is at an intersection of $n-1$ composition paths with different variable roots, and has $n-1$ composition-velocity eigenvalues, one for each path. As shown in Appendix I, the eigenvalues are

$$u = h_k \prod_i h_i \prod_i \alpha_{i1} \qquad (k = 1, \ldots, n-1) \tag{3.68}$$

where $h_k$ is the root which varies along the respective path.

The composition-velocity eigenvalues as expressed by Eqs. (3.68) are "point properties," i.e., are completely determined by local, momentary values of variables and parameters and are independent of gradients. Thus, as far as coherent systems are concerned, Eqs. (3.68) provide a solution of the problem, pointed out in Section III.B.1, of finding practicable functional relations with which the velocities of composition variables can be connected with equilibrium properties and operating conditions.

In *trace-component systems* (i.e., with all but one species at trace levels), the $H$-function roots are virtually equal to the separation factors $\alpha_{1i}$ of the trace species [compare condition (3.55)]. With $k$ as the bulk species, Eq. (3.67) then reduces to

$$\lim_{x_k \to 1} u_j = \alpha_{kj} \qquad \text{for all } j \tag{3.69}$$

Also, in the composition region $x_k \to 1$, only one trace concentration (and the concentration of $k$) varies along any composition path (see Section IV.C.4). For paths with variable $x_j$, the variable root is $h_j$ if $j < k$, and $h_{j-1}$ if $j > k$. In either case, Eq. (3.68) reduces to

$$\lim_{x_k \to 1} u = \alpha_{kj} \qquad (\mathrm{d}x_j \neq 0) \tag{3.70}$$

As is readily verified with Eq. (2.20) or (2.21) for the interconversion of true and adjusted velocities, Eqs. (3.69) and (3.70) are equivalent to Eq. (3.18). The equality of the right-hand sides of Eqs. (3.69) and (3.70) reflects the fact, derived in Section III.B.1, that the concentration variations of trace species advance at the respective trace-species velocities.

### 7. *Integral Coherence Condition, Composition Steps, and Step Velocities*

Sharp composition steps are coherent if, and only if, the concentration differences between their upstream and downstream sides obey the integral coherence condition (3.40) (see Section IV.A). As shown in Appendix I, this condition is equivalent to the requirement that the compositions upstream and downstream fall on the same composition path. Diffuse coherent boundaries also meet this requirement since, by definition, all their compositions are on one path. The integral coherence condition thus is a necessary condition for coherence of any boundaries, whether entirely or partially sharp or diffuse, but is a sufficient condition for coherence only in the case of entirely sharp boundaries.

Equations for adjusted step velocities in terms of $H$-function roots are derived in Appendix I. For a step with $k \mid k+1$ affinity cut:

$$u_\Delta = h_k' h_k'' \prod_{i \neq k} h_i \prod_i \alpha_{i1} \tag{3.71}$$

where $h_k'$ and $h_k''$ are the values of the variable root, $h_k$, at the upstream and downstream sides of the step. For a step with $k \mid l$ affinity cut, bracketing species $k+1, \ldots, l-1$:

$$u_\Delta = h'h'' \prod_{i=1}^{k-1} h_i \prod_{i=l}^{n-1} h_i \prod_{i=1}^{k} \alpha_{i1} \prod_{i=l}^{n} \alpha_{i1} \tag{3.72}$$

where $h'$ and $h''$ are the values of the variable root on the two sides of the step and may have any index numbers $k, \ldots, l-1$, since the upstream and downstream compositions may fall on any of the path segments listed in Table 3.1. Equation (3.72) comprises Eq. (3.71) as the special case $l = k+1$.

A simple relation, valid for bracketing and nonbracketing affinity cuts, exists between step velocities and composition-velocity eigenvalues:

$$u_\Delta = (u'u'')^{1/2} \tag{3.73}$$

where $u'$ and $u''$ are the velocity eigenvalues, for the composition path of the step, of the compositions on the upstream and downstream sides. This relation is obtained from Eqs. (3.68) and (3.71) or (3.72) after the separation factors $\alpha_{1,k+1}, \ldots, \alpha_{1,l-1}$ of the bracketed species, if any, have been substituted for the trivial roots according to Table 3.1.

The step velocity is a physically meaningful quantity only if the step remains sharp when traveling. To do so, the step must be "self-sharpening," as will be discussed in the next section.

## E. Sharpening Characteristics

According to their behavior upon development, coherent boundaries can be classified as "self-sharpening" and "nonsharpening." Self-sharpening boundaries, when diffuse, sharpen while traveling; if artificially disturbed, they regain their sharpness when development is continued. In contrast, nonsharpening boundaries, whether sharp or diffuse, become more diffuse while traveling.* This behavior will now be examined, and criteria for predicting it will be derived.

Sharpening effects are caused by concentration-dependent partition coefficients (nonlinear isotherms) and are by no means confined to multicomponent systems. They were first described quantitatively by DeVault [2] and are well established [7–9] as the principal feature distinguishing "nonlinear" from "linear" chromatography. The terms "self-sharpening" and "nonsharpening," borrowed from electrophoresis, have become standard usage. As far as coherent boundaries are concerned, the present approach merely shows that no additional complications arise from the presence of a multiplicity of interfering species.

It is interesting to note that the formal analogy between chromatographic boundaries and waves in mechanics of compressible fluids, pointed out in Section III.B.3, extends to the sharpening characteristics. Sharpening of a self-sharpening boundary corresponds, in fluid mechanics, to formation of a shock front [42].

### 1. *Self-sharpening and Nonsharpening Boundaries*

Granted the premises of local equilibrium, absence of axial diffusion and dispersion, and uniform flow rate, any sharpening or spreading of a coherent boundary is caused exclusively by the composition dependence of the composition velocity. If this velocity increases in the direction of flow, compositions farther downstream travel faster than their upstream neighbors. The compositions within the boundary then draw farther apart as they advance, and the boundary spreads. On the other hand, if the composition velocity decreases in the direction of flow, the compositions behave in the opposite manner, and the boundary sharpens. If such a boundary travels a sufficient distance, it becomes ideally sharp, namely, when its upstream end has caught up with its downstream end. (In practice, disturbances neglected here prevent the boundary from becoming ideally sharp. Rather, a so-called "constant pattern" is attained; see Chapter 5, Section V.A.)

As examples, distance-time diagrams and concentration profiles of a self-sharpening and a nonsharpening boundary are given in Fig. 3.18. The lines in the distance-time diagrams are trajectories of fixed compositions. The profiles are shown for only one species; since compositions are

* The terms "self-sharpening" and "nonsharpening" are used to describe a trait rather than a behavior. Thus, a boundary which has become ideally sharp (or has attained its constant pattern, see Chapter 5, Section V.A) is still called self-sharpening, although it does not sharpen any further.

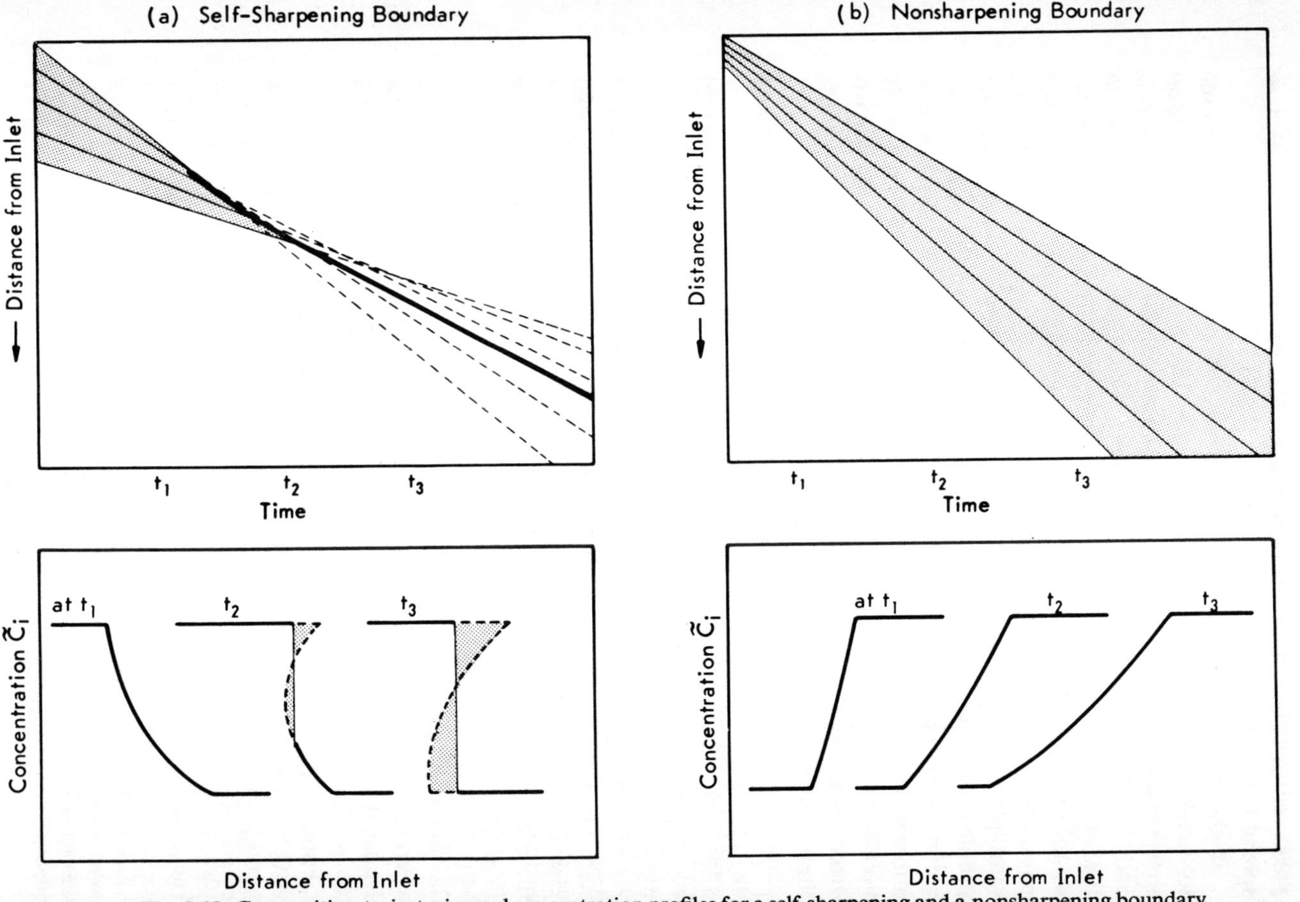


**Fig. 3.18.** Composition trajectories and concentration profiles for a self-sharpening and a nonsharpening boundary.

preserved in coherent boundaries, the profiles of all species sharpen or spread together.

If the compositions in a self-sharpening boundary would continue to advance at their composition velocities beyond attainment of ideal sharpness, their trajectories in the distance-time diagram would cross, and "overhanging" concentration profiles would result (broken lines in Fig. 3.18a). These trajectories and profiles constitute a mathematical solution of the differential material-balance equation (3.16), but are not physically meaningful because different compositions cannot coexist at the same place and time in the column. Of the mathematical concentration profile at any given time, only the parts shown as heavy lines in Fig. 3.18a are realized, and a discontinuity corresponding to a composition step appears. At the step, the differential material balance (3.16) does not apply because the derivatives do not exist. Instead, the location of the step is determined by an integral material balance, which requires the area under the physically realized step profile to be equal to that under the mathematical "overhanging" profile.* The shaded areas in the profile diagrams in Fig. 3.18a must thus be equal. In the distance-time diagram in Fig. 3.18a, the trajectory of the composition step is shown as a heavy line. In accordance with Eq. (3.19), the step velocity does not become constant until the entire boundary has become completely sharp.

Coherence is not impaired when a diffuse boundary sharpens into a step. Since all compositions in the boundary and those of the adjacent plateau zones fall on the same composition path, a step between any two of these compositions obeys the integral coherence condition (see Section IV.D.7).

### 2. *Sharpening Criteria*

Whether a coherent boundary is self-sharpening or nonsharpening is readily established. Boundaries with nonbracketing affinity cuts $k \mid k+1$

---

* Strictly speaking, the requirement of equal areas under the two curves applies to overall concentrations $\tilde{C}_i \equiv \bar{C}_i + C_i$ in concentration profiles (plots of concentration versus distance at fixed time), to mobile-phase concentrations $x_i$ or $C_i$ in concentration histories (plots of concentration versus time at fixed distance), and to stationary-phase concentrations $y_i$ or $\bar{C}_i$ in plots of concentration versus distance at fixed adjusted time. The proof is implicit in the conventional theories based on DeVault's work [2] and is not reproduced here because it is not needed in the present approach, in which "overhanging" profiles are used merely for qualitative illustration while the positions of composition steps are calculated from their step velocities rather than from the equal-area requirement. This procedure avoids complications which would otherwise arise in noncoherent situations (e.g., see Section V.D.1).

will be considered first. As shown in Section IV.D.5, all compositions in the boundary fall on the same composition path, along which one $H$-function root, $h_k$, varies while all others remain constant. As discussed above, the boundary is self-sharpening if the composition velocity decreases in the direction of flow, and is nonsharpening in the opposite case. Equations (3.68) and (3.57), with condition (3.50), show that the composition velocity, the variable root $h_k$, and the concentrations of the high-affinity species $1, \ldots, k$ all increase or decrease together,* while the concentrations of the low-affinity species $k+1, \ldots, n$ vary in the opposite manner. Therefore, *a coherent boundary is self-sharpening if the concentrations of the high-affinity species decrease in the direction of flow, and is nonsharpening in the opposite case.*

To establish the sharpening characteristics with this rule, only the compositions on the upstream and downstream sides of the boundary need be known, since the concentrations vary monotonically across coherent boundaries (see Section IV.B; "composition pulses" with concentration maxima and minima will be discussed in Section IV.G).

The rule is also valid for boundaries with bracketing affinity cuts. It is true that the variable root then changes its index number from segment to segment of the composition path. However, the variable root, the composition velocity, and the concentrations of the high-affinity species nevertheless vary continuously and monotonically with one another because the sets of constant root values, despite index-number changes, are the same for all path segments (see Section IV.D.5, Table 3.1).

For later applications, a criterion in terms of $H$-function roots rather than concentrations will prove convenient. It is evident from the discussion above that a coherent boundary is

$$\begin{array}{ll} \text{self-sharpening} & \text{if } h' > h'' \\ \text{nonsharpening} & \text{if } h' < h'' \end{array} \tag{3.74}$$

where $h'$ and $h''$ are the values of its variable $H$-function root at the upstream and downstream side, respectively.

The criteria above have been derived with the premises listed in Section I. In particular, it has been assumed that the separation factors are independent of composition and that the flow rate is uniform. Complications

---

* Equation (3.68) only shows the *adjusted* composition velocity to increase or decrease together with $h_k$. That the true composition velocity does the same is readily verified with Eq. (2.20) and conditions (2.27), the adjusted composition velocity being positive under any conditions.

arising from relaxation of these premises will be discussed briefly in Chapter 5.

### 3. *Boundaries Involving Infinitesimal Composition Changes. Trace-Component Systems*

It has been shown that, granted the premises in Section I, sharpening and spreading of coherent boundaries is caused by the variation of the composition velocity with composition. If the composition change is infinitesimal, then the changes of the variable root and of the composition velocity are also infinitesimal. This is the case, for example, in systems with all species but one at trace levels. Since the composition velocity then is virtually constant across a boundary, the latter has neither self-sharpening nor nonsharpening character, but retains constant width as it travels. (In practice, disturbances neglected here, and not combatted by self-sharpening, cause the boundary to spread, as is well known from theories of linear chromatography; see also Chapter 5, Section V).

### 4. *Summary of Rules*

For ease of reference, the rules relating to sharpening characteristics are summarized below.

1. A coherent boundary is self-sharpening if the concentrations of the high-affinity species decrease in the direction of flow, and is nonsharpening in the opposite case.
2. In terms of $H$-function roots, a coherent boundary is self-sharpening if the value of its variable root decreases in the direction of flow, and is nonsharpening in the opposite case.
3. A coherent boundary involving only infinitesimal composition variations travels without sharpening or spreading (granted that disturbances are absent).

## F. Plateau Zones between Coherent Boundaries

*Plateau zones* have been defined as finite regions of uniform composition (see Chapter 2, Section IV.A). The purpose of the present section is to establish under which conditions such zones, between coherent boundaries, will lengthen, shorten, or retain constant length upon development. The conclusions are summarized at the end of this section. The derivations below, which do not contribute general insight, may be skipped without loss of continuity.

1. *Velocities of Zone End Points*

Whether a plateau zone lengthens or shortens upon development obviously depends on the relative velocities of its "end points," at which it is bounded by the adjacent boundaries. If, and as long as, the boundaries are diffuse at least in the immediate vicinity of the end points, the velocities of the latter are the composition-velocity eigenvalues of the plateau-zone composition and for the affinity cuts of the boundaries. The same is true if the boundaries are sharp but nonsharpening, so that they become diffuse immediately. On the other hand, if the zone is bounded by sharp and self-sharpening composition steps, the end-point velocities are the respective step velocities. (The steps may constitute only portions of self-sharpening boundaries, which may still be diffuse in other portions not directly adjacent to the plateau zone.)

The following notation will be adopted for the derivations in this section. Affinity cuts are stated as subscripts, and asterisks refer to the plateau zone including its end points. Thus, end-point velocities may appear as

$(u^*)_{j|k}$ velocity eigenvalue of plateau-zone composition and for $j \mid k$ cut,

$(u_\Delta^*)_{j|k}$ velocity of step at end point, and with $j \mid k$ cut.

Furthermore, $j$, $k$, $l$, and $m$ will signify arbitrary species which, in the affinity sequence, appear in this order but are not necessarily adjacent (i.e., $j < k < l < m$).

2. *Affinity-Cut Combinations*

The possible combinations of the affinity cuts of the two boundaries bounding a plateau zone can be established and classified as follows. In the composition space, a plateau zone corresponds to a point, and its two boundaries to segments of two composition paths which both pass through the zone point. Thus, in principle, a plateau zone can exist between boundaries of any affinity cuts, provided their paths either coincide or intersect. This restriction rules out the existence of plateau zones between boundaries with $j \mid l$ and $k \mid m$ cuts, and with $j \mid m$ and $k \mid l$ cuts, since paths with such cuts can be shown not to intersect one another. The possible combinations of cuts can be divided into four groups:

| | | | |
|---|---|---|---|
| 1. Cuts without common species: | | $j \mid k$ and $l \mid m$ |
| 2. "Adjacent" cuts: | | $j \mid k$ and $k \mid l$ |
| 3. Cuts with common high- or low-affinity species: | | $j \mid k$ and $j \mid l$ |
| | or | $j \mid l$ and $k \mid l$ |
| 4. Identical cuts: | | $j \mid k$ and $j \mid k$ |

In all cases, either boundary can be upstream or downstream of the plateau zone and, with exceptions in group 3, can be self-sharpening or nonsharpening.

3. *Behavior of Plateau Zones*

The development behavior of plateau zones between boundaries with the affinity-cut combinations stated above will now be examined. For groups 1, 2, and 4, the behavior can be derived from material-balance considerations which are not restricted to constant separation factors. Here, the weaker assumptions of competitive sorption and invariant selectivity sequence are sufficient. In cases of group 3, the $h$ transformation will be invoked, which implies that the separation factors are independent of composition.

*Group* 1*: Affinity Cuts without Common Species*. The affinity cuts are $j \mid k$ and $l \mid m$, and either boundary can be self-sharpening or nonsharpening. Condition (3.45), applied to sharp self-sharpening steps at the zone end points, shows that

$$(u_\Delta^*)_{j|k} \leqq u_k^* \qquad \text{and} \qquad (u_\Delta^*)_{l|m} \geqq u_l^* \tag{3.75}$$

where $u_k^*$ and $u_l^*$ are the species velocities of $k$ and $l$ in the plateau zone. Similarly, condition (3.46), applied to boundaries which are nonsharpening or are diffuse at the zone end points, shows that

$$(u^*)_{j|k} \leqq u_k^* \qquad \text{and} \qquad (u^*)_{l|m} \geqq u_l^* \tag{3.76}$$

According to condition (3.11), $u_k^* < u_l^*$. Therefore, irrespective of the sharpness and sharpening characteristics of the boundaries, conditions (3.75) and (3.76) establish that the velocity of the zone end point is lower at the boundary with $j \mid k$ cut than at that with $l \mid m$ cut. The plateau zone thus lengthens if the boundary with $j \mid k$ cut is on the upstream side, and shortens in the opposite case.

*Group* 2*: "Adjacent" Affinity Cuts*. The paths of both boundaries and the composition point of the plateau zone are on a $j \mid k-k \mid l$ common plane, as shown in Fig. 3.19. The paths of the two sets intersect the border $x_k = 0$ on different sides of the watershed point W. Therefore, the composition point of the plateau zone cannot be on this border, and species $k$ must be present in the plateau zone.

For sharp self-sharpening steps at the zone end points, condition (3.45) gives

$$(u_\Delta^*)_{j|k} \leqq u_k^* \qquad \text{and} \qquad (u_\Delta^*)_{k|l} \geqq u_k^* \tag{3.77}$$

For diffuse or nonsharpening boundaries, however, the velocity limits are

$$(u^*)_{j|k} < u_k^* \qquad \text{and} \qquad (u^*)_{k|l} > u_k^* \tag{3.78}$$

because the presence of $k$ in the plateau zone does not permit the composition velocities to equal $u_k^*$. [Condition (3.46) allows $u = u_k$ only for $x_k \to 0$, see Section IV.B.]

Conditions (3.77) and (3.78) show that, as in cases of group 1, the plateau zone lengthens if the boundary with $j \mid k$ cut is on the upstream side, and shortens in the opposite case (see Fig. 3.19a and b), with only the following exception. As shown in Section IV.B, the velocity of a step equals $u_k^*$ if, and only if, $k$ is absent from one of its sides. Accordingly, both step velocities in condition (3.77) equal $u_k^*$, and the plateau zone thus retains constant length, if both boundaries are self-sharpening and entirely sharp and if species $k$ is absent from both adjacent plateau zone upstream and downstream (see Fig. 3.19c).

*Group* 3*: Affinity Cuts with Common High- or Low-Affinity Species*. The combinations involve one boundary with $j \mid l$ cut and one with $j \mid k$ or $k \mid l$ cut. Again, the paths of both boundaries and the composition point of the plateau zone are on a $j \mid k-k \mid l$ common plane (see Fig. 3.20). Species $k$ must be absent from the plateau zone, because $j \mid l$ paths exist only on the border $x_k = 0$.

To simplify bookkeeping of root numbers, the derivations below are given for cases in which $j$, $k$, and $l$ are adjacent in the affinity sequence. The results are the same if species of intermediate affinities exist, as can be verified with the equations in Sections IV.D.5–7. Such species, bracketed by the $j \mid l$ cut, would have to be absent from the plateau zone and both boundaries and can therefore not be expected to affect the results.

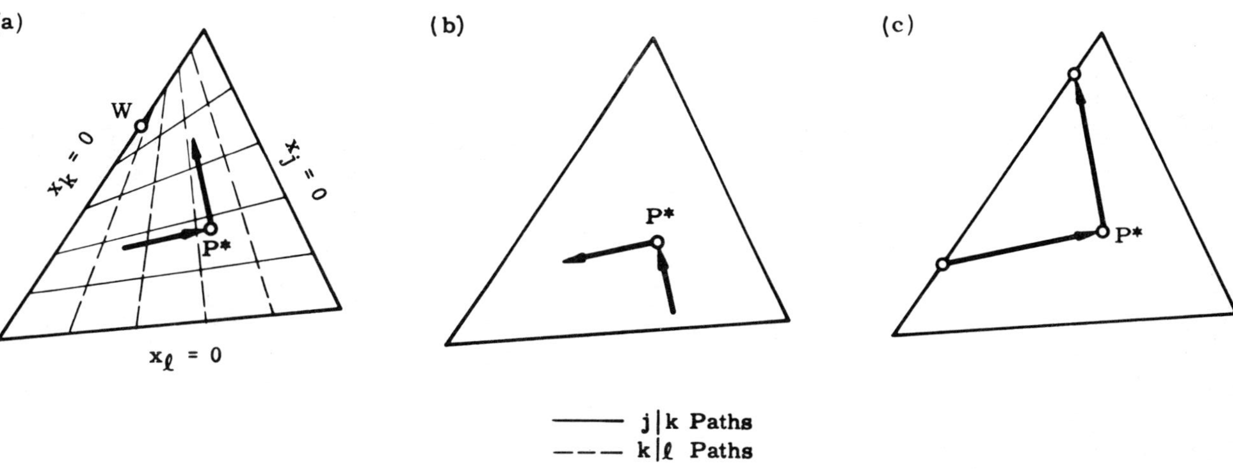


**Fig. 3.19.** Plateau zone P* between boundaries with $j \mid k$ and $k \mid l$ cuts. Composition routes are shown on $j \mid k - k \mid l$ common plane, arrows point in direction of flow. (a) Zone lengthens, (b) zone shortens, (c) zone retains constant length.

If the plateau zone is between boundaries with $j \mid l$ and $j \mid k$ cuts, its composition point is on the $h$-space border $h_j = \alpha_{1k}$, and the variable $H$-function roots are $h_j$ for the $j \mid k$ path and $h_k$ for the $j \mid l$ path at the zone point (see Fig. 3.20). On the other hand, if the zone is between boundaries with $j \mid l$ and $k \mid l$ cuts, its composition point is on $h_k = \alpha_{1k}$, and the variable roots are $h_k$ for the $k \mid l$ path and $h_j$ for the $j \mid l$ path at the zone

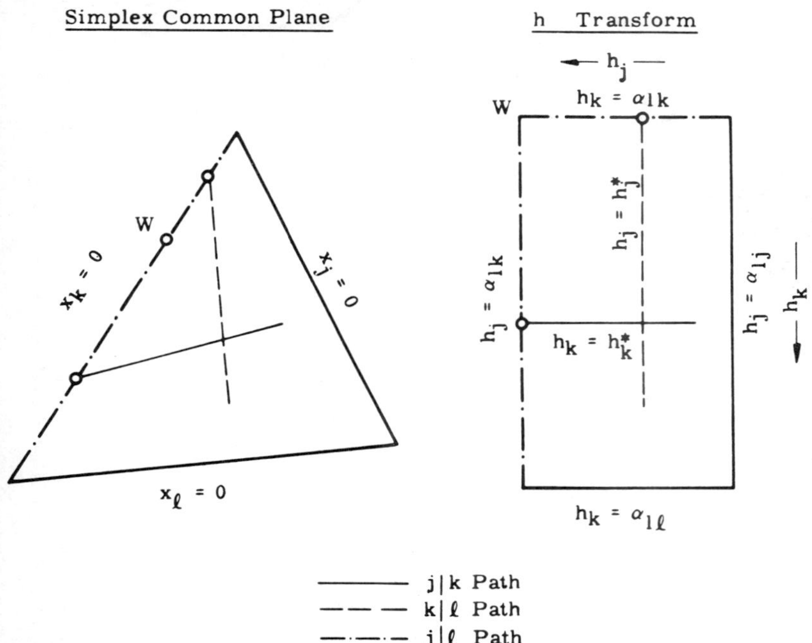


**Fig. 3.20.** Paths with $j \mid l$, $j \mid k$, and $k \mid l$ cuts on common plane in $x$ space and $h$ space.

point (see Fig. 3.20). For boundaries which are nonsharpening or are diffuse at the zone end points, Eq. (3.68) thus gives for the combination of $j \mid l$ and $j \mid k$ cuts

$$(u^*)_{j|l} = \alpha_{1k} h_k^{*2} P \tag{3.79}$$

$$(u^*)_{j|k} = \alpha_{1k}^2 h_k^* P \tag{3.80}$$

and for the combination of $j \mid l$ and $k \mid l$ cuts

$$(u^*)_{j|l} = \alpha_{1k} h_j^{*2} P \tag{3.81}$$

$$(u^*)_{k|l} = \alpha_{1k}^2 h_j^* P \tag{3.82}$$

where asterisks refer to the plateau-zone composition, and where

$$P \equiv \prod_{i \neq j,k} h_i \prod_i \alpha_{i1} \tag{3.83}$$

is a constant which depends only on the position of the common plane. Since $h_j^* < \alpha_{1k} < h_k^*$, conditions (3.79) to (3.82) show that the plateau zone, between diffuse

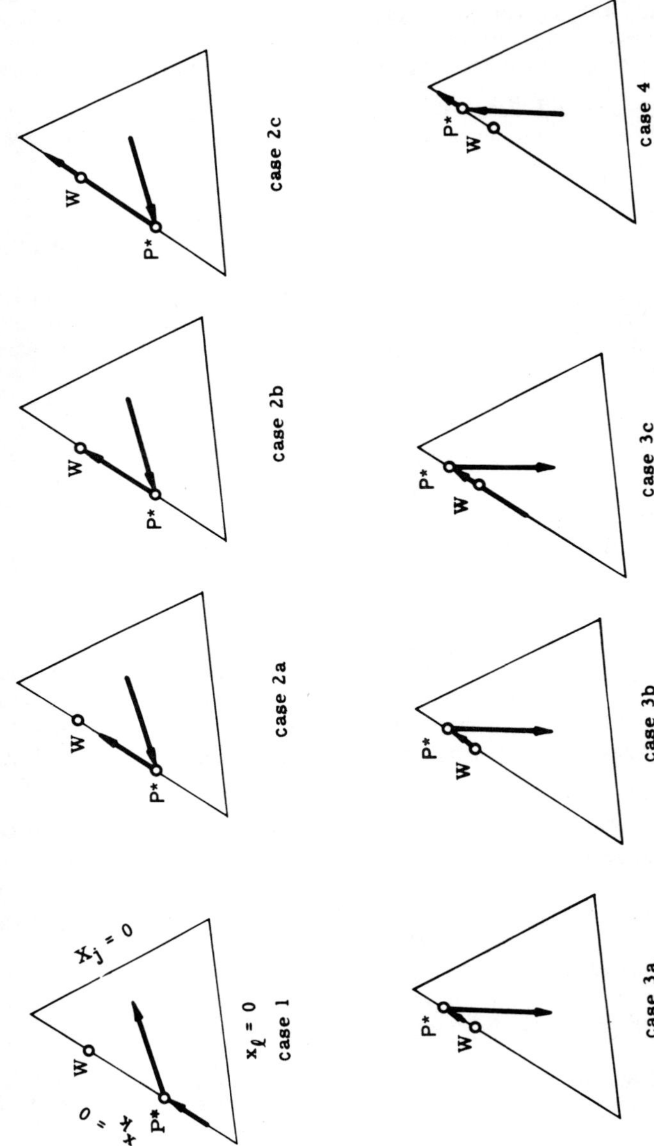


**Fig. 3.21.** Possible route configurations for plateau zone P* between self-sharpening boundary with bracketing $j \mid l$ cut and boundary with $j \mid k$ or $k \mid l$ cut. Routes are shown on $j \mid k - k \mid l$ common plane.

or nonsharpening boundaries, lengthens if the boundary with $j \mid l$ cut is downstream of that with $j \mid k$ cut or upstream of that with $k \mid l$ cut, and shortens in the opposite cases.†

This rule also holds if the boundary with $j \mid k$ or $k \mid l$ cut is sharp and self-sharpening, since

$$(u_\Delta^*)_j|_k < (u^*)_j|_k \quad \text{and} \quad (u_\Delta^*)_k|_l > (u^*)_k|_l \tag{3.84}$$

This is readily shown as follows. Along the path of the boundary with $j \mid k$ cut, the variable root $h_j$ and thus the composition-velocity eigenvalue decrease with increasing distance from the composition point of the plateau zone, where $h_j = \alpha_{1k}$. The step velocity therefore is lower than the eigenvalue at the zone point [see Eq. (3.73)]. Similarly, along the path of the boundary with $k \mid l$ cut, the variable root $h_k$ increases with distance from the zone point, so that the step velocity is higher than the eigenvalue at the zone point.

Still to be considered are the combinations involving a self-sharpening step with $j \mid l$ cut. Composition routes for the four possible combinations of this type are shown in Fig. 3.21. In cases 2 and 3, the route section corresponding to the $j \mid l$-cut step may (a) be confined to one side of the watershed, (b) extend to the watershed, or (c) extend across the watershed. The path topology requires the boundary with $j \mid k$ or $k \mid l$ cut to be self-sharpening in cases 1 and 4, and nonsharpening in cases 2 and 3 (see Section IV.E for sharpening criteria). According to Eq. (3.72), the velocity of the step with $j|l$ cut is

$$(u_\Delta^*)_j|_l = \alpha_{1k} h' h'' P \tag{3.85}$$

where $h'$ and $h''$ are the values of the variable root at the upstream and downstream sides of the step, and $P$ is given by Eq. (3.83). With Eqs. and conditions (3.85), (3.80), (3.82), and (3.84), the behavior of the plateau zone can be established for each case, after the appropriate values or limits for $h'$ and $h''$ have been inserted. This is shown in Table 3.2, where the affinity cuts, sharpening characteristics, values or limits of $h'$ and $h''$, and resulting conditions for the end-point velocities are listed. The rule that the plateau zone lengthens if the $j \mid k$-cut boundary is upstream or the $k \mid l$-cut boundary is downstream, and shortens if the opposite is true, is seen to be obeyed, except if the step with $j \mid l$ cut extends to or across the watershed (cases 2b, 2c, 3b, and 3c).

The behavior of the plateau zone may change during development if the composition route corresponds to case 2c or 3c in Fig. 3.21 and if the boundary with $j \mid l$ cut is entirely or partly diffuse initially. As long as this boundary has not sharpened, at the zone end point, into a step extending across the watershed, the plateau zone lengthens; thereafter, the zone shortens.

*Group 4: Identical Affinity Cuts.* The behavior of plateau zones between boundaries having the same affinity cut is readily established.

If both boundaries are diffuse at the zone end points or are nonsharpening, both end-point velocities are velocity eigenvalues of the same composition and for the same cut, and thus are identical. Accordingly, the plateau zone retains constant length upon development.

If bounded on either or both sides by a self-sharpening composition step, the plateau zone shortens upon development. This can be shown as follows. Since sharpening results

---

† Since all adjusted composition and step velocities are positive, inequalities are not altered by the transition from adjusted to true velocities, as is readily verified with Eq. (2.20) and conditions (2.27).

**Table 3.2**

BEHAVIOR OF PLATEAU ZONES BETWEEN SELF-SHARPENING COMPOSITION STEP WITH $j|l$ AFFINITY CUT AND BOUNDARY WITH $j|k$ OR $k|l$ AFFINITY CUT

| Case | Cuts and sharpening characteristics of boundaries[a] | | Values of variable root of step with $j\|l$ cut | | Conditions for end-point velocities | Plateau-zone behavior |
|---|---|---|---|---|---|---|
| | Upstream | Downstream | | | | |
| 1 | $j\|l$ (S) | $j\|k$ (S) | $h' > h_k^*$ | $h'' = h_k^*$ | $(u_\Delta^*)_j\|_l > (u^*)_j\|_k > (u_\Delta^*)_j\|_k$ | Shortens |
| 2a | $j\|k$ (N) | $j\|l$ (S) | $h' = h_k^*$ | $h_k^* > h'' > \alpha_{1k}$ | $(u^*)_j\|_k < (u_\Delta^*)_j\|_l$ | Lengthens |
| b | | | | $h'' = \alpha_{1k}$ | $(u^*)_j\|_k = (u_\Delta^*)_j\|_l$ | Retains constant length |
| c | | | | $h'' < \alpha_{1k}$ | $(u^*)_j\|_k > (u_\Delta^*)_j\|_l$ | Shortens |
| 3a | $j\|l$ (S) | $k\|l$ (N) | $h_j^* < h' < \alpha_{1k}$ | $h'' = h_j^*$ | $(u_\Delta^*)_j\|_l < (u^*)_k\|_l$ | Lengthens |
| b | | | $h' = \alpha_{1k}$ | | $(u_\Delta^*)_j\|_l = (u^*)_k\|_l$ | Retains constant length |
| c | | | $h' > \alpha_{1k}$ | | $(u_\Delta^*)_j\|_l > (u^*)_k\|_l$ | Shortens |
| 4 | $k\|l$ (S) | $j\|l$ (S) | $h' = h_j^*$ | $h'' < h_j^*$ | $(u_\Delta^*)_k\|_l > (u^*)_k\|_l > (u_\Delta^*)_j\|_l$ | Shortens |

[a] S = self-sharpening, N = nonsharpening

from decrease of the composition velocity in the direction of flow (see Section IV.E), the velocity of a self-sharpening step is lower than the velocity eigenvalue of the composition on its downstream side, and higher than that of the composition on its upstream side. Thus, the velocity of a step on the upstream side of the plateau zone is higher, and that of a step on the downstream side is lower, than the composition velocity eigenvalue of the plateau-zone composition, and the zone necessarily shortens.

The plateau zone will change its behavior upon development if either or both boundaries are self-sharpening but are initially diffuse at the zone end points. The zone then retains constant length at first, but shortens after a step has formed at either end point.

### 4. *Conclusions*

The results of the derivations can be consolidated into the following rules:

1. In general, a plateau zone lengthens if the affinity cut of its upstream boundary is between species of higher affinity than that of its downstream boundary, and shortens in the opposite case. In this context, a $j \mid l$ cut counts as between species of lower affinity than a $j \mid k$ cut, and as between species of higher affinity than a $k \mid l$ cut. (Here and below, $j$, $k$, and $l$ are arbitrary species which, in the affinity sequence, appear in that order but need not be adjacent.)
2. The rule above is not obeyed in the following cases:

   (*a*) A plateau zone between entirely sharp and self-sharpening boundaries with $j \mid k$ cut (upstream) and $k \mid l$ cut (downstream) retains constant length if species $k$, present in the zone, is absent from both adjacent zones upstream and downstream (see Fig. 3.19c).

   (*b*) A plateau zone between a self-sharpening step with $j \mid l$ cut and a nonsharpening boundary with $j \mid k$ cut (upstream) or $k \mid l$ cut (downstream) shortens if the route section corresponding to the step extends across the watershed of $x_k = 0$, and retains constant length if this route section extends to the watershed (see Fig. 3.21 and Table 3.2, cases 2b, 2c, 3b, and 3c).
3. A plateau zone between boundaries having the same affinity cut retains constant length if both boundaries are nonsharpening or are diffuse at the zone end points, and shortens if either or both boundaries are self-sharpening and sharp at the zone end points.

## G. Coherent Composition Pulses

By definition, a necessary and sufficient condition for coherence in an individual region of nonuniform composition is that all compositions

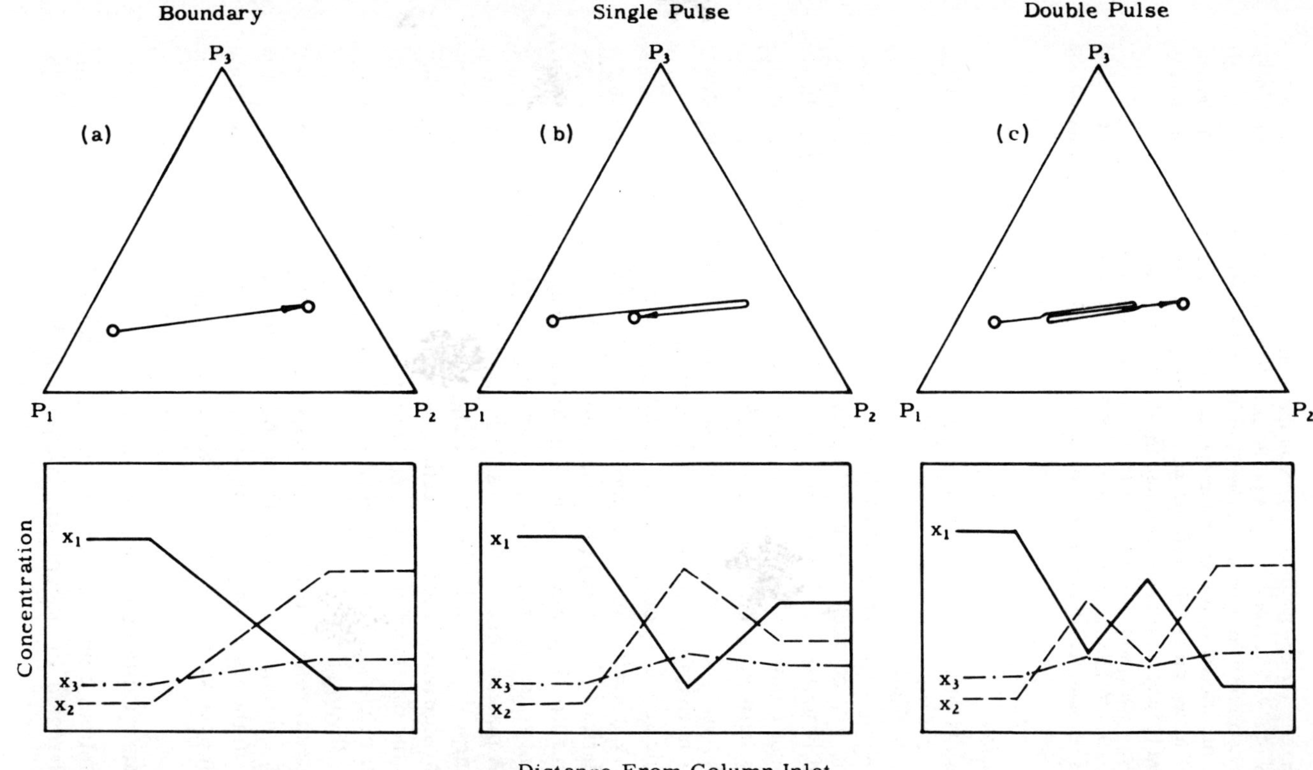


**Fig. 3.22.** Composition routes and concentration profiles for coherent boundary, coherent single pulse, and coherent double pulse with 1 | 2 cut in three-component system.

fall on the same composition path. The composition route of such a region may either follow the respective path in one direction, or may reverse its direction at one or several points along the path (see Fig. 3.22). In the first case, the species concentrations vary monotonically across the region, which then constitutes a coherent boundary. In the second case, all species concentrations have maxima and minima where the route reverses its direction. Such composition profiles or histories, which will be called coherent *composition pulses*, are discussed in the present section.

### 1. *Single and Multiple Pulses*

A coherent pulse may have one or more loci of concentration maxima and minima. For example, the pulse shown in Fig. 3.22b has one such locus ("single" pulse) and that in Fig. 3.22c has two ("double" pulse). In multiple pulses, maxima and minima in the concentration profiles or histories alternate.

Coherent pulses are composites of ordinary coherent boundaries with monotonic concentration changes. A single pulse consists of two such boundaries which constitute its two flanks, a double pulse consists of three, etc. All boundaries constituting a pulse have the same affinity cut, since their routes run along the same composition path.

The rules derived earlier for coherent boundaries are also valid if the boundaries are parts of a pulse. Application of the sharpening criteria (see Section IV.E) shows that a single pulse with concentration maxima of the high-affinity species and concentration minima of the low-affinity species has a self-sharpening front and nonsharpening rear flank, and that a pulse with opposite concentration behavior has the opposite sharpening characteristics. In multiple pulses, self-sharpening and nonsharpening boundaries alternate.

Upon development, the amplitude of a pulse remains constant as long as the self-sharpening flank is diffuse, but decreases after the latter has become sharp. Figure 3.23 illustrates this for a single pulse with self-sharpening front flank. It was shown in Section IV.E.1 that "overhanging" profiles not realized physically (broken lines in Figs. 3.18a and 3.23) would result if the compositions would continue to advance at their composition velocities after a boundary has become sharp, and that, instead, a step appears whose location is determined by the requirement that the areas under the step profile and the "overhanging" profile must be equal. Figure 3.23 shows that, as a result, the concentration maximum of the physically realized step profile decreases upon development. The

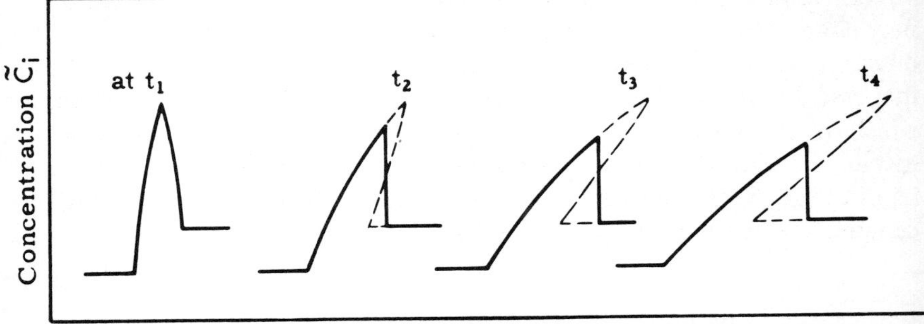


**Fig. 3.23.** Successive overall concentration profiles of high-affinity species in coherent single pulse with self-sharpening front flank. Physically not realized portions of profile are shown as broken lines. (Schematic.)

situation is analogous in pulses with self-sharpening rear flanks and in multiple pulses.

Coherence is not impaired by decreasing pulse amplitudes. Since all compositions in a pulse are on the same composition path, the self-sharpening flanks continue to obey the integral coherence condition even though the composition on one or both sides varies.

2. *Transient and Permanent Pulses*

A distinction may be made between "transient" and "permanent" pulses. A transient pulse, having traveled a sufficient distance, degenerates into a single coherent boundary with monotonic concentration variations. A permanent pulse does not.

The development of various types of permanent and transient pulses is illustrated in Fig. 3.24 by concentration profiles at successive times $t_1$ and $t_2$. The profiles are those of a high-affinity species, so that the composition-velocity eigenvalue is higher at higher concentrations. The profile portions not realized physically are shown as broken lines.

Single pulses between plateau zones of identical compositions (Fig. 3.24a) are readily seen to be permanent. Points of equal concentrations on the two flanks have equal velocity eigenvalues, so that the area ABC of the pulse remains constant upon development. The pulse thus does not disappear, although it continues indefinitely to broaden and flatten.

Single pulses are also permanent if the composition variation is greater at their nonsharpening than at their self-sharpening flanks (Fig. 3.24b).

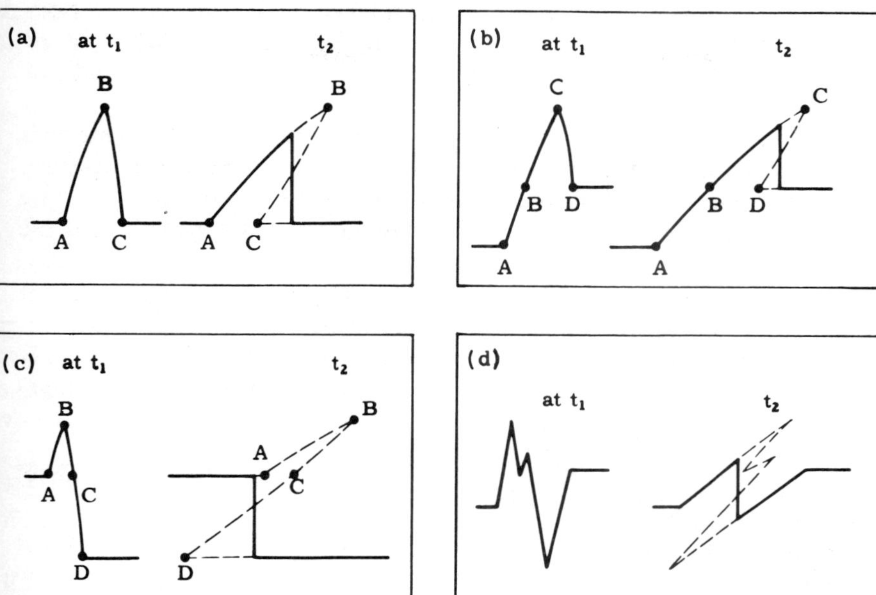


**Fig. 3.24.** Successive overall concentration profiles of high-affinity species in various types of coherent pulses. (a) Permanent single pulse between plateau zones of identical compositions, (b) permanent single pulse between plateau zones of different compositions, (c) transient single pulse, (d) permanent multiple pulse undergoing reduction in multiplicity.

Such a pulse may be viewed as a composite of a pulse between identical compositions (profile portion BCD) and a residual nonsharpening boundary (portion AB). The latter does not interfere with the partial pulse BCD because, in the entire nonsharpening flank, compositions farther upstream advance more slowly than their downstream neighbors. Thus, the partial pulse is permanent, and the concentration maxima and minima do not disappear. However, upon prolonged development the composition variation at the self-sharpening flank becomes very small compared to that at the nonsharpening flank, so that the profile becomes more and more similar to that of a single nonsharpening boundary.

On the other hand, single pulses are transient if the composition variation is greater at their self-sharpening than at their nonsharpening flanks. This is illustrated in Fig. 3.24c for a pulse with self-sharpening front flank. Although the area ABC again remains constant upon development, point D eventually falls so far behind that the physically realized portion of the profile no longer includes the pulse. A single sharp and self-sharpening boundary remains.

The extension to multiple pulses will not be discussed in detail. One finds that such pulses, too, are transient if a single boundary between the plateau zones upstream and downstream would be self-sharpening, and are permanent if such a single boundary would be nonsharpening or if the zones have identical compositions. Permanent multiple pulses may, however, undergo a reduction of multiplicity upon development. For example, the quadruple pulse in Fig. 3.24d reduces to a double pulse.

With condition (3.74), a criterion in terms of $H$-function roots is readily formulated from the considerations above. A coherent (single or multiple) pulse is

$$\begin{aligned} &\text{permanent if} \quad h' \leqq h'' \\ &\text{transient if} \quad h' > h'' \end{aligned} \tag{3.86}$$

where $h'$ and $h''$ are the values of the variable root at the upstream and downstream side, respectively, of the pulse.

### 3. *Infinitesimal Pulses*

Granted the simplifying premises stated in Section I, coherent boundaries involving infinitesimal composition variations neither sharpen nor spread, but retain constant width upon development (see Section IV.E.3). Accordingly, coherent infinitesimal pulses, composed of such boundaries, travel with constant widths and amplitudes. (In reality, the pulses broaden and flatten because of disturbances neglected here; see Chapter 5, Section V.A.3).

The adjusted velocity of an infinitesimal coherent pulse equals its virtually uniform composition velocity:

$$u = h_k \prod_i h_i \prod_i \alpha_{i1} \tag{3.68}$$

where $h_k$ is the variable $H$-function root (whose variation is infinitesimal).

Infinitesimal coherent pulses are not exempt from the requirement that all their compositions fall on the same composition path. The concentration variations of all species in such a pulse thus are of comparable magnitude if the species are present in finite concentrations. However, if all but one species are at trace levels, the path directions in the respective region of the composition space are such that only one trace concentration (and that of the bulk species) varies across a coherent boundary or pulse (see Section IV.C.4). The adjusted velocity of a pulse with varying concentration of trace species $j$ in such a system is given by

$$\lim_{x_k \to 1} u = \alpha_{kj} \qquad (\mathrm{d}x_j \neq 0) \tag{3.70}$$

where $k$ is the bulk species (see Section IV.D.7).

### 4. *Summary of Rules*

The results of the derivations can be summarized as follows:

1. Coherent pulses contain one or more loci at which the concentrations of all species present have maxima or minima.
2. Coherent pulses are composites of two or more coherent boundaries with monotonic concentration changes. The composition routes of all boundaries constituting a pulse are on the same composition path.
3. Adjacent boundaries in a pulse have opposite sharpening characteristics. A single pulse with concentration maxima of the high-affinity species has a self-sharpening front flank, and one with maxima of the low-affinity species has a nonsharpening front flank.
4. Pulse amplitudes decrease upon development after self-sharpening flanks have become sharp.
5. A coherent pulse degenerates into a single coherent boundary if this boundary is self-sharpening, and does not do so if such a single boundary, did it develop, were nonsharpening.
6. Exceptions for infinitesimal pulses are that (a) coherent infinitesimal pulses travel with unchanged widths and amplitudes, and (b) in coherent pulses in trace-component systems the concentrations of all but one trace species are uniform.

## H. Review and Perspective

In this section, the initially more general scope of the discussion has been narrowed to a particular type of behavior, called *coherence*. The state of coherence—defined by the condition that the coexisting concentrations of all species at any given location and time advance at the same rate—bears some resemblance to a steady state: Both coherence and steady state are "stable" in that they are attained from arbitrary initial conditions and then continue until disturbed. Here, however, the analogy ends; while the variables become time-independent in a steady state, they remain time-dependent in a state of coherence. (Even the use of a moving frame of reference cannot in general eliminate the dependence on time, since coherent boundaries may sharpen or spread as they travel.)

While the state of coherence has been examined in detail, the proof of its attainment has been relegated to the next section for reasons of expedience. As with a steady state, the proof of attainment is more complex than the description of the final state itself, and is easier to give after the latter has been fully characterized.

For better perspective, a brief look at the scopes of the concepts introduced in conjunction with coherence is indicated. Coherence itself is a quite general concept, i.e., is not contingent on any premises or particular properties of the system. The same is largely true for the concept of composition paths, defined as representing composition sequences which conform to the coherence criteria; here, a generalized treatment will, however, have to make allowances for the difference between the differential and integral coherence conditions and for deviations from local equilibrium. The affinity cut, defined as dividing the species into high- and low-affinity groups with concentrations varying in opposite directions, is a more limited concept in that it presupposes an invariant affinity sequence and competitive sorption; for systems with other equilibrium properties, the derivation of the affinity-cut rules can only serve as a recipe showing how the implications of coherence can be deduced. Even more limited is the concept of the $H$-function roots; local equilibrium, absence of disturbances, and constancy of the separation factors are prerequisites if the $H$ function, as defined here, is to give the desired orthogonalization of the composition-path grid. The problem of qualifying and extending the concepts to overcome their limitations will be taken up in Chapter 5.

Aside from the introduction of coherence and related concepts, the two principal developments in this section are the analysis, focused on physical causes and effects, of the concerted behavior of interfering species in coherent systems, and the introduction of the $h$ transformation and $H$-function roots. The particular effects which interference produces in coherent systems, most notably the existence of affinity cuts, are seen to arise as direct consequences of the continuity restrictions and equilibrium properties and can be understood in terms of the interplay of species velocities and concentration velocities, discussed for single species in Section III. This conceptual approach, mainly based on the material balance (3.22) and its implications (3.26), (3.28), and (3.30), is qualitative. It is supplemented by the quantitative treatment based on the $h$ transformation. The two main results achieved with this transformation are the orthogonalization of the composition-path grid and the establishment of a relation for the composition velocity as a point property (i.e., a quantity independent of gradients). Inherent in the orthogonalization is that only one $H$-function root varies along any composition path [condition (3.56)]. The composition velocities are found as solutions of an eigenvalue problem: For any composition there are $n-1$ velocity eigenvalues, one for each affinity cut. Giving the eigenvalues as point properties in terms of $H$-function roots, Eq. (3.68) establishes practicable functional relations between composition variables and their velocities in coherent systems.

The rules deduced and summarized in the various subsections and the equations and criteria derived with the $h$ transformation are all that is needed for a qualitative and quantitative treatment of column responses under operating conditions giving entirely coherent behavior from the outset, provided the as yet unproven attainment of coherence is accepted as a postulate. Although some of the cases of greatest practical interest are of this type, their discussion will be postponed until noncoherent behavior and its development toward coherence have been examined.

## V. Noncoherent Boundaries

The importance and usefulness of the concept of coherence hinges on the fact that coherence is a "stable" condition which, given sufficient time and distance for undisturbed development, is eventually attained from any arbitrary initial conditions. How and why coherence develops from initially noncoherent composition profiles will be examined in this section.

### A. Dynamic Properties of $H$-Function Roots

In noncoherent boundaries the various species concentrations at any fixed location and time advance at different rates, so that given compositions exist only momentarily. Composition velocities, extensively used in the previous section, thus are meaningless here, and the individual behavior of all species must be considered. The pertinent simultaneous differential equations in terms of concentrations $x_i$ were derived from the material balances (3.16) as early as 1943 by DeVault [2]. Unfortunately, their interpretation is difficult, and neither DeVault nor later authors were able to deduce general rules (except when postulating entirely coherent behavior).

A more convenient set of equations is obtained when compositions are expressed in terms of their $H$-function roots instead of concentrations:

$$\left(\frac{\partial h_k}{\partial \tau}\right)_z = -h_k \prod_i h_i \prod_i \alpha_{i1} \left(\frac{\partial h_k}{\partial z}\right)_\tau \qquad (k = 1, \ldots, n-1) \tag{3.87}$$

where $\tau$ is the adjusted time (for derivation, see Appendix II). In contrast to DeVault's, the differential equations of this set are partially uncoupled: the time derivative of each root is independent of gradients of other roots. Also, all coefficients of the set (3.87) are negative and finite since all roots and separation factors are positive and finite.

An immediately apparent property of Eqs. (3.87) is

$$(\partial h_k/\partial \tau)_z = 0 \qquad \text{if} \quad (\partial h_k/\partial z)_\tau = 0 \tag{3.88}$$

so that no variation of $h_k$ is induced by variations of other roots where no gradient of $h_k$ exists. Furthermore, solutions of differential equations of the form of Eqs. (3.87), with negative and finite coefficients, have the important property that the values of the dependent variables are "conserved": No $h_i$ value appears or disappears within the system, that is, the complete solution $\mathbf{h}(z, \tau)$ is composed exclusively of $h_i$ values which occur in the initial and boundary conditions, and none of these values cease to exist as $z$ or $\tau$ is increased. In a distance-time diagram, each $h_i$ value existing at some point along the borders $z = 0$ or $\tau = 0$ traverses the diagram on a trajectory pointing between "south" and "east," none of these trajectories end in the interior of the diagram, and no trajectories originate from any points other than on $z = 0$ or $\tau = 0$ (e.g., see Fig. 3.25). [These rules must be qualified on two counts. First, any step in the variation of a root along $z = 0$, $\tau = 0$, or both is the origin of trajectories of the intermediate root values, so that the latter, too, occur in the solution $\mathbf{h}(z, \tau)$. Second, the statement that no root value ceases to exist as $z$ or $\tau$ is increased, so that no trajectory ends in the interior of the distance-time diagram, refers to mathematical solutions. Portions of these may not be realized physically (see Sections IV.E.1 and V.D.1). Root values and their trajectories may thus disappear from the physically realized part of the solutions $\mathbf{h}(z, \tau)$ at points within the distance-time diagram. This happens whenever a root variation has sharpened into a step.]

A *root velocity* can be defined as the rate of advance of a given root value in the direction of flow. For the adjusted velocity of a given $h_k$ value one obtains from Eq. (3.87) with the general rule (3.17):

$$u_{h_k} \equiv (\partial z/\partial \tau)_{h_k} = h_k \prod_i h_i \prod_i \alpha_{i1} \qquad \text{for all } k \tag{3.89}$$

The adjusted root velocities are positive, depend only on composition and separation factors, and coincide with the composition-velocity eigenvalues [see Eqs. (3.68)]. Like the latter, they are point properties (i.e., independent of gradients). Since the roots are confined to the intervals

$$\alpha_{1i} \leqq h_i \leqq \alpha_{1,i+1} \qquad (i = 1, \ldots, n-1) \tag{3.50}$$

the velocities of the root values coexisting at any given time and location obey the condition

$$u_{h_j} \leqq u_{h_k} \qquad \text{if} \quad j < k \tag{3.90}$$

Thus, as a rule, *roots with lower index numbers have lower velocities*. (The equality applies only in exceptional cases, namely, two coexisting roots $h_{k-1}$ and $h_k$ have the same velocity if $h_{k-1} = h_k = \alpha_{1k}$, so that the respective composition point falls on the watershed of $x_k = 0$; see Section IV.D.4.)

Discontinuities (steps) of roots may exist, or arise from self-sharpening, as was shown for coherent boundaries in Section IV.E.1. A step of $h_k$, provided it does not coincide with a step of another root, obeys the integral coherence conditions since it involves variation of one root only. This is true even if the profiles on either side are noncoherent. Therefore, even in otherwise noncoherent systems, a step of $h_k$ constitutes a coherent composition step with variable $h_k$, and Eq. (3.71) for the adjusted step velocity, originally derived for coherent systems, remains applicable:

$$u_{\Delta h_k} \equiv h_k' h_k'' \prod_{i \neq k} h_i \prod_i \alpha_{i1} \tag{3.91}$$

where $h_k'$ and $h_k''$ are the values of $h_k$ at the two sides of the step.

The following dynamic properties of the $H$-function roots are now apparent:

1. In any operation, only the root values existing initially and introduced with the influent occur in the column. Any compositions in the column thus arise exclusively from combinations of these root values (including intermediate values of any steps in the initial profile and influent history of a root).
2. The root variations existing initially or induced by influent changes are propagated in the direction of flow at finite adjusted velocities. Variations of each root are propagated without inducing variations of other roots. The only interaction between variations of different roots is that they affect one another's velocities of propagation.
3. Any given composition has a unique set of root velocities. These coincide with the composition-velocity eigenvalues, but are not restricted to coherence. Roots with lower index numbers have lower velocities (except on watersheds, where two root velocities are equal).

Coherent boundaries and plateau zones constitute limiting cases of this general behavior. In coherent boundaries, only one root varies. Since all other roots are constant, their advance does not manifest itself, so that any composition variations are propagated at the velocity of the variable root. Also, since all other roots are constant, any advancing given value of the variable root is accompanied by an invariant composition. Within plateau zones, no root variations exist, none are induced, and the composition thus remains constant.

None of the properties listed above is shared by concentrations $x_i$ or $y_i$. Concentrations are not "conserved," their velocities may be negative or infinite and depend on gradients as well as on composition, and a variation of one concentration induces variations of the others. Thus, it is the substitution of $H$-function roots for concentrations that makes the following simple derivations possible.

The significant simplification achieved with the $h$ transformation is that a practicable functional relation, not involving gradients nor restricted to coherence, has been established between composition variables ($H$-function roots) and their velocities, and that the "conservation" of the roots permits the entire response to be computed by tracing the trajectories of the initial and influent root values. On the other hand, the nonlinearity of the differential equations has not been removed, so that numerical procedures are in general still needed to obtain solutions. However, the essential features of the development behavior of noncoherent boundaries can be deduced without such calculations, as will now be shown.

### B. Resolution of Noncoherent Boundaries. Attainment of Coherence

Development of a noncoherent boundary between plateau zones, undisturbed by other boundaries farther upstream or downstream, will now be examined. For the time being, exceptional cases involving compositions on watersheds are excluded. (Mutual interference of boundaries will be discussed in Section VI, and compositions on watersheds in Section V.F.)

The principal qualitative feature of development becomes readily apparent when trajectories of fixed root values are traced in a schematic distance-time diagram, as shown in Fig. 3.25. The trajectories, shown only where gradients of the respective roots exist, represent contour lines of the roots. At any point in the diagram, the roots with higher index numbers have steeper trajectories [see condition (3.90); compositions on watersheds are excluded]. As a result, the roots "sort themselves out" according to

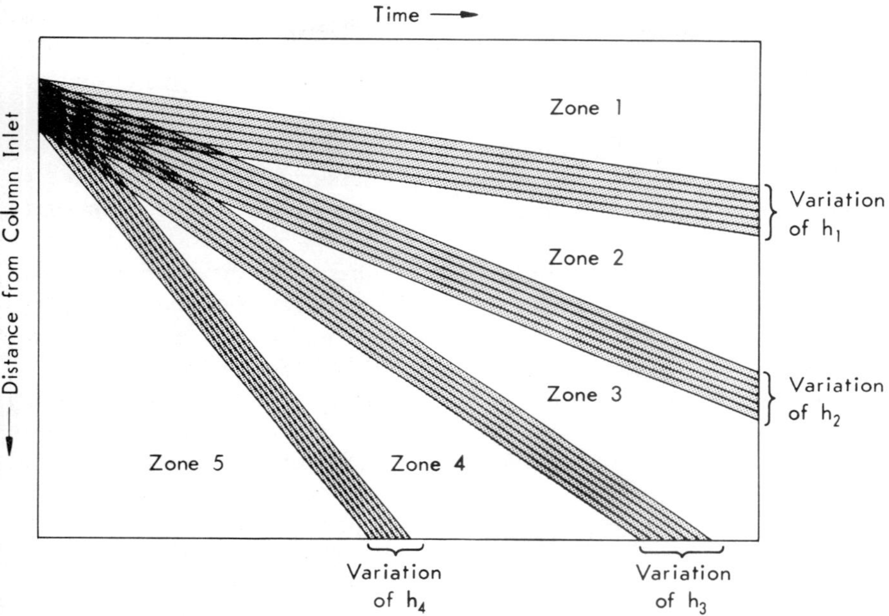


**Fig. 3.25.** Root trajectories for development of noncoherent boundary in five-component system, illustrating resolution into coherent boundaries in numerical sequence of index numbers of variable roots. Regions of gradual root variations are shaded, region of noncoherence darkly shaded. (Schematic.) (From F. Helfferich [41]. Reproduced from *Advances in Chemistry Series* No. 79 with permission of the American Chemical Society.)

their index numbers. The trajectories are resolved into separate bundles, each consisting of trajectories of values of one root only. After resolution, each bundle corresponds to a coherent boundary or pulse since it involves variation of only one root. The coherent boundaries or pulses are in the sequence of increasing index number of their variable roots (counted in the direction of flow), and become separated by lengthening plateau zones (zones 2, 3, and 4 in Fig. 3.25). Thus, given sufficient distance and time for undisturbed development, *a noncoherent boundary is resolved into coherent boundaries or pulses separated by plateau zones.*

For simplicity, the trajectories of each root have been drawn schematically as parallel straight lines in Fig. 3.25. This is correct only if the composition variation across the initial boundary is infinitesimal, so that all composition-velocity eigenvalues are virtually constant. In all other cases the trajectories, while being linear in regions of coherence, are curved in the region of noncoherence and are not parallel (e.g., see Fig. 3.30). However, the proof of attainment of coherence, based on the inequality (3.90), is not invalidated.

It is interesting to note that in this general description, applicable to any arbitrary noncoherent boundaries, variations of $H$-function roots are resolved in the same manner as species are resolved in conventional chromatography.

The schematic distance-time diagram also shows how the transition from noncoherence to coherence takes place. The region of noncoherence is that in which more than one root varies, i.e., in which trajectories of different roots cross (darkly shaded in Fig. 3.25). Upon development, coherent boundary portions with variable $h_1$ and $h_{n-1}$ start growing immediately at the upstream and downstream ends of the overall boundary. Later, portions of coherent boundaries with variable roots of intermediate index numbers start growing from within the still noncoherent interior of the overall boundary. Intermediate plateau zones appear where shrinking noncoherent regions disappear. Thus, from the time of its appearance, each intermediate plateau zone is flanked by coherent boundaries. The sequence of appearance of intermediate coherent boundaries and plateau zones depends on the initial conditions and cannot be predicted by qualitative considerations.

## C. Physical Causes

The physical causes for attainment of coherence become apparent when the continuity considerations developed in Section III.B.3 are applied to noncoherent systems. For simplicity, the argument that follows will be restricted to three-component systems, although extension to more components is possible.

As an example for the type of effects to be explained, Fig. 3.26 shows the composition profile routes (compositions in sequence of increasing distance from column inlet) and concentration profiles before and after resolution of a noncoherent boundary. In the case shown, species 2 has a concentration minimum in the initial boundary. Development not only eliminates this minimum but produces, in the new intermediate plateau zone, a much higher concentration of species 2 than had existed anywhere in the initial profile.

Why and in which direction the composition route is deformed upon

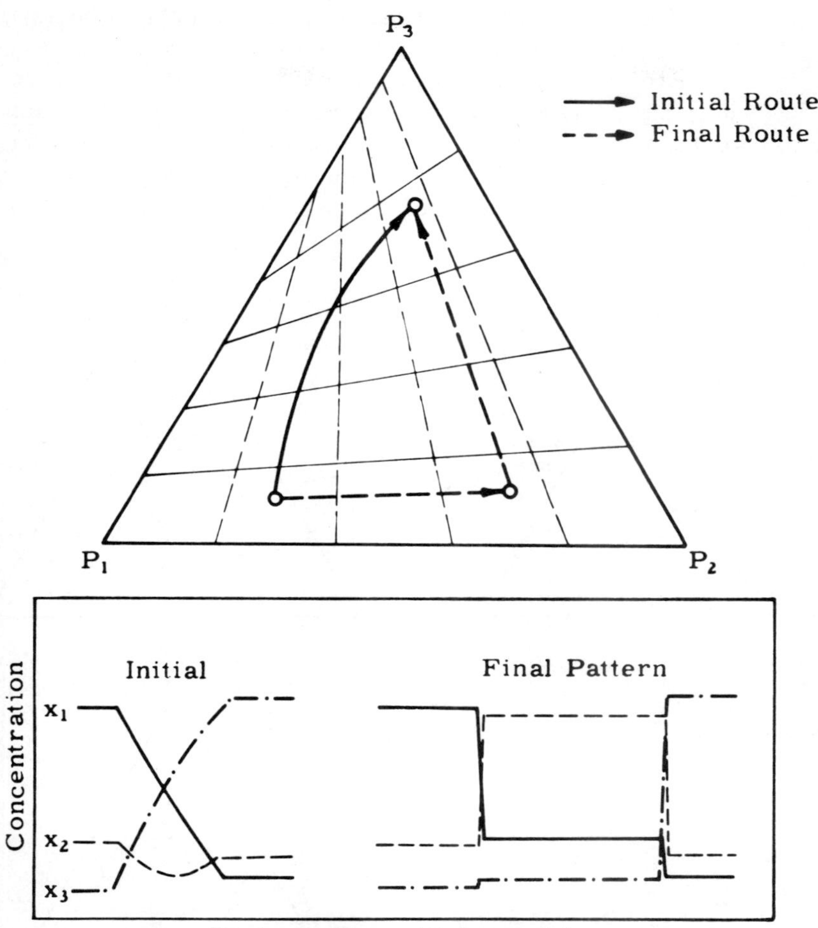


**Fig. 3.26.** Composition profile routes and concentration profiles before and after resolution of noncoherent boundary in three-component system.

development will first be shown for special cases, namely, for initial routes along which the overall concentration of one species is constant. [The overall concentration is defined as $\tilde{C}_i \equiv \bar{C}_i + C_i$; see Eq. (2.2).]

Suppose that, along the initial route, $\tilde{C}_2$ is constant while $\tilde{C}_1$ decreases and $\tilde{C}_3$ increases in the direction of flow (see Fig. 3.27a; note that the simplex is drawn for overall rather than mobile-phase concentrations).

In this case, all species velocities decrease in the direction of flow because, in this direction, the concentration of a species with lower affinity increases at the expense of one with higher affinity (see Section III.A). Since the velocity of species 2 thus is higher in the upstream than in the downstream portions of the boundary, continuity does not allow $\tilde{C}_2$ to remain uniform. Rather, this species accumulates within the boundary so that its concentration profile develops a maximum, as was shown in Fig. 3.6a. This

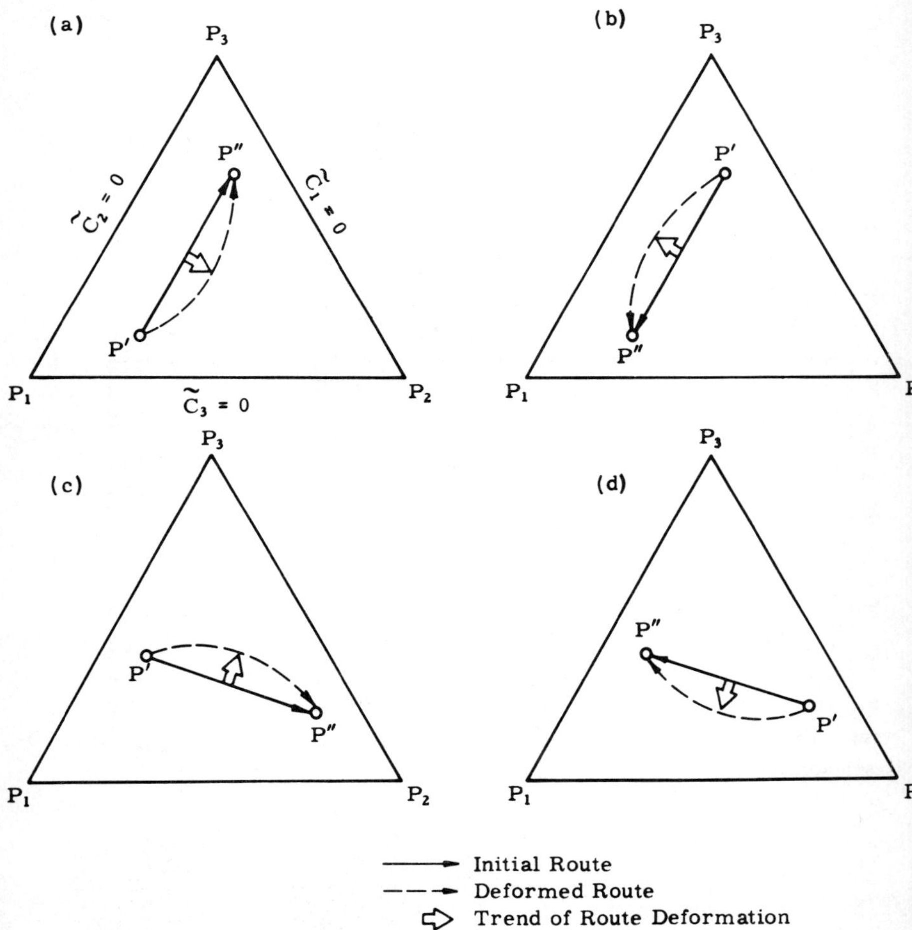


**Fig. 3.27.** Trends of deformation of composition profile route of noncoherent boundary between plateau zones P′ and P″ in $\tilde{C}$ simplex of three-component system. Initial route is along line of constant $\tilde{C}_2$ (a and b) or of constant species velocities (c and d). (Schematic.)

corresponds to a deformation of the composition route in the direction indicated schematically in Fig. 3.27a. The qualitative behavior in the similar case shown in Fig. 3.26 can now be understood as resulting mainly from such an accumulation of species 2 within the noncoherent boundary.

If the initial route with constant $\tilde{C}_2$ runs in the opposite direction (see Fig. 3.27b), the species velocities increase in the direction of flow. The resulting depletion of species 2 within the boundary, as shown in Fig. 3.6b, then produces the route deformation indicated in Fig. 3.27b.

Analogous considerations apply to routes along which $\tilde{C}_1$ or $\tilde{C}_3$ are constant. The results are shown in Fig. 3.28.

Route deformation has a different cause if the initial route follows a line of constant species velocities. Along such lines, $\tilde{C}_2$ varies in the opposite direction as $\tilde{C}_1$ and $\tilde{C}_3$ (see Fig. 3.2). The continuity considerations in Section III.B.3 have shown that the overall concentration velocity of a species equals the species velocity if the latter does not vary with composition [see conditions (3.26)]. Thus, if the route is along a line of constant species velocities, concentrations $\tilde{C}_3$ advance at the highest rate and tend to progress from the upstream portions of the boundary in the direction of flow, while concentrations $\tilde{C}_1$ advance at the lowest rate and tend to fall behind. Therefore, if $\tilde{C}_2$ increases and $\tilde{C}_1$ and $\tilde{C}_3$ decrease in the direction of flow along the initial route, development causes the concentrations to shift relative to one another in such a manner that $\tilde{C}_3$ rises at the expense of $\tilde{C}_1$ in the middle portion of the boundary (see Fig. 3.27c). If the initial route has the opposite direction, the composition shifts in the opposite manner (see Fig. 3.27d).

The trends of route deformation for the various distinct route directions discussed so far are shown collectively in Fig. 3.28. If the route direction deviates slightly from one of those shown, the effect causing the indicated deformation is weaker but is not entirely eliminated. Moreover, it can be shown that, for intermediate route directions, the effects of concentration-velocity inequalities reinforce rather than counteract those of accumulation or depletion of a species in the boundary. The overall result thus is that, regardless of route direction, the upstream portion of the route is deflected toward the 1 | 2 path through the upstream end point while the downstream portion is deflected toward the 2 | 3 path through the downstream end point. That this deflection continues until coherence is attained is not surprising, since the species concentrations are known to shift relative to one another as long as the route is still noncoherent.

These considerations show that attainment of coherence can be attributed to two effects. First, continuity does not allow a concentration to

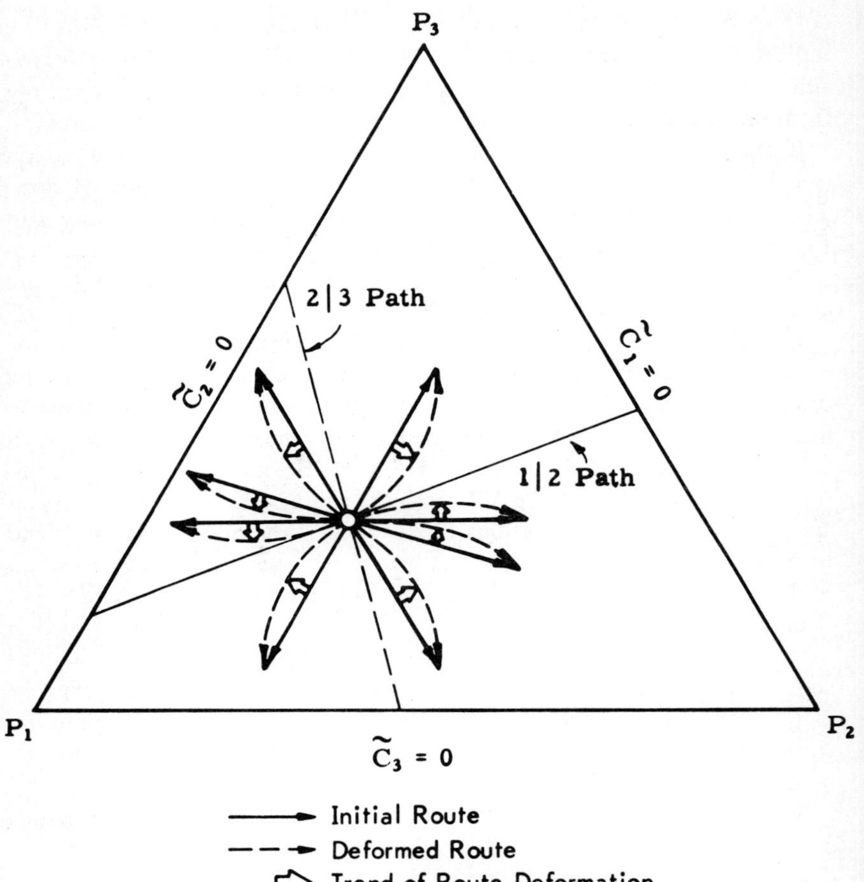


**Fig. 3.28.** Trends of deformation of noncoherent composition profile routes of various directions in $\tilde{C}$ simplex of three-component system (schematic).

remain uniform (or approximately uniform) within a region in which the species velocities vary. Second, unequal concentration velocities cause the concentration profiles to shift relative to one another. Both effects are seen to supplement one another in deforming the composition route until coherence is attained.

The premise of composition-independent separation factors has not been invoked. The continuity considerations outlined here can thus serve to prove attainment of coherence in systems with variable separation factors, to which the simpler proof based on $H$-function roots is not applicable.

## D. Final Patterns

The essential properties of the final coherent patterns resulting from undisturbed development of given noncoherent boundaries can be assessed without solving the differential equations. It will now be shown how the rules for $H$-function roots, derived in Section V.A, can be applied to establish the number and sequence of boundaries and pulses, their sharpening characteristics, and the compositions of the plateau zones in such final patterns.

### 1. *Boundaries and Pulses*

The qualitative properties—sharpening characteristics, multiplicity, transient or permanent character—of the boundary or pulse with (arbitrarily selected) variable root $h_k$ in the coherent pattern depend only on the type of variation of $h_k$, since variations of other roots are not involved. Even before coherence is attained, variations of other roots do not induce or alter a variation of $h_k$, but merely affect its propagation rate, as shown in Section V.A. The qualitative properties of the resulting coherent boundary or pulse with variable $h_k$ thus are determined exclusively by the initial profile of this root and are independent of the behavior of the other roots. Accordingly, a given initial profile of $h_k$ in a noncoherent boundary, regardless of variations of other roots, produces the same type of final boundary or pulse as it would in an entirely coherent system with $h_k$ as the only variable root. The properties of the final boundary or pulse can therefore be established with the rules for development of coherent boundaries and pulses.

Various types of initial profiles of $h_k$ are shown schematically in Fig. 3.29. The properties of the resulting boundaries or pulses with variable $h_k$, listed in this figure, are readily deduced with the rules and criteria in Sections IV.E–G. Transient responses may degenerate into the final ones before or after resolution from other boundaries. Responses arising from other types of initial root profiles are predicted with equal ease.

As an example, the root trajectories and the initial and final root and concentration profiles of a fairly complex six-component system are shown in Fig. 3.30. As can be predicted from the initial root profiles, the coherent pattern consists of two nonsharpening boundaries (with variable $h_1$)

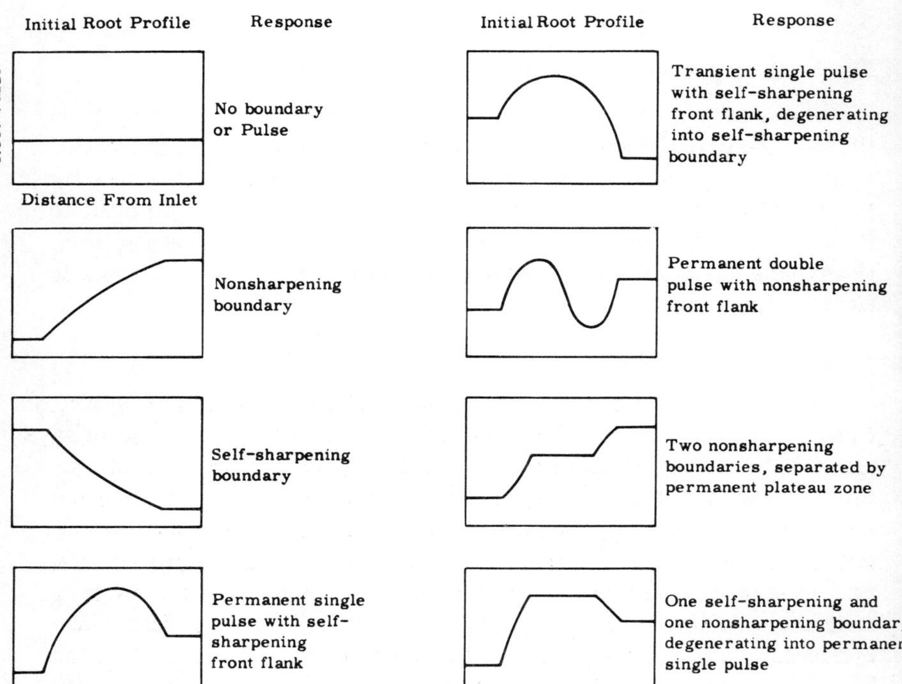


**Fig. 3.29.** Various types of initial profiles of a root in noncoherent boundary, and resulting responses (schematic).

separated by a plateau zone of constant length, one permanent single pulse with nonsharpening front flank (variable $h_2$), one self-sharpening boundary (variable $h_4$), and one transient single pulse with self-sharpening front flank (variable $h_5$); no boundaries or pulses with variable $h_3$ appear.

Various applications of practical interest will be discussed in Chapter 4. Two general conclusions, however, are stated here. First, a noncoherent boundary whose initial route is linear in the $x$ or $y$ space, and thus involves monotonic root variations only, is resolved exclusively into boundaries each with a different variable root, i.e., the final pattern contains neither pulses nor separate boundaries having the same variable root. Second, pulses in final patterns arise exclusively from maxima or minima in the initial root profiles and cannot be attributed to maxima

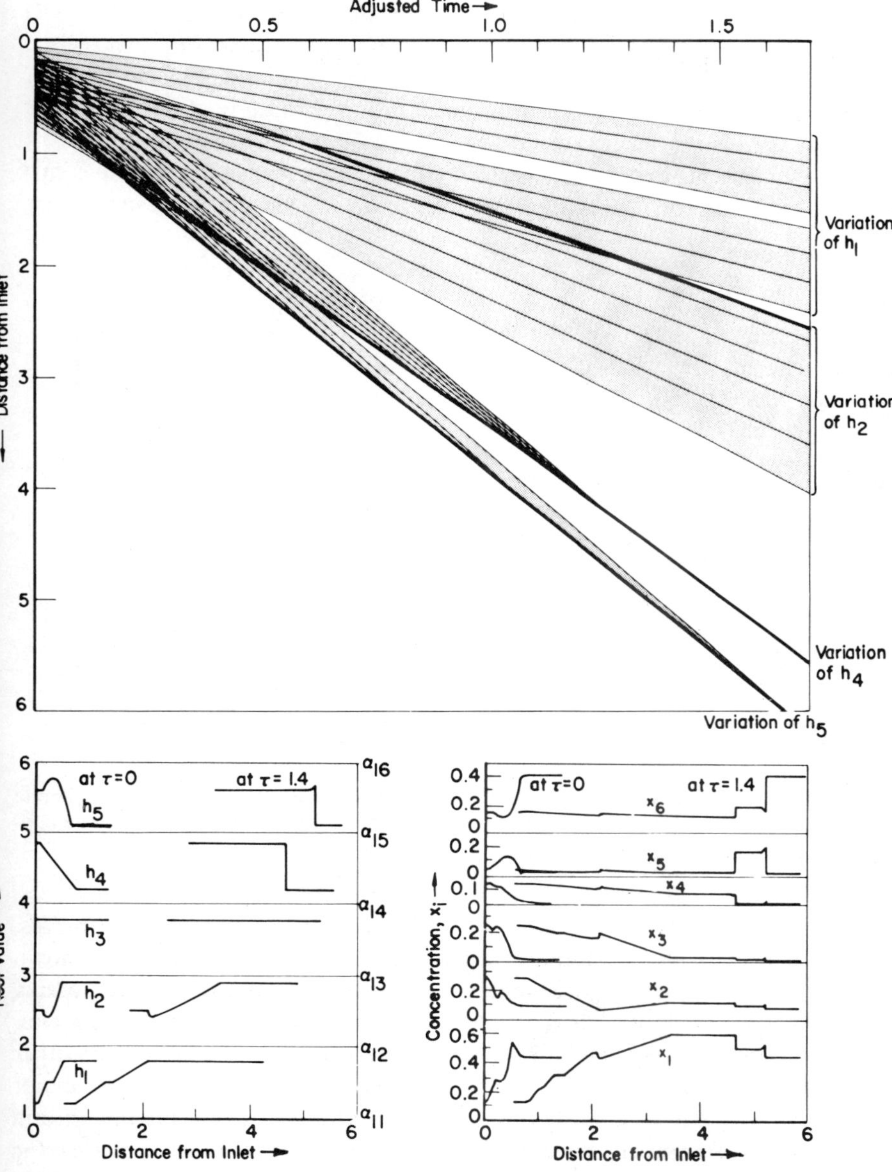


**Fig. 3.30.** Root trajectories, root profiles, and concentration profiles of developing noncoherent boundary in six-component system (computer calculation with $\alpha_{12} = 2$, $\alpha_{13} = 3$, $\alpha_{14} = 4$, $\alpha_{15} = 5$, $\alpha_{16} = 6$).

or minima in the initial concentration profiles; to illustrate this, Fig. 3.31 shows two initial routes in a three-component system, one involving a concentration maximum of species 2 but not producing a pulse (solid lines), the other involving monotonic concentration variations of all species but nevertheless producing a pulse (broken lines).

A comment regarding crossovers of trajectories of different values of a root is called for. The differential equations (3.87) lead to such crossovers,

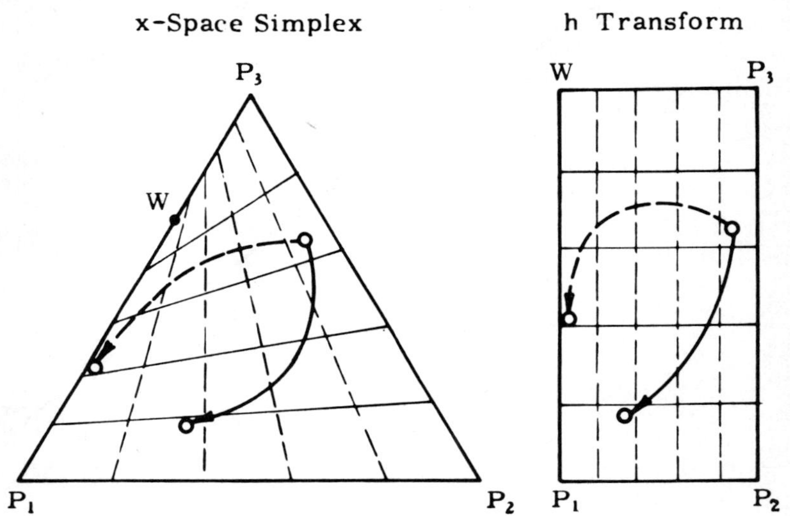


**Fig. 3.31.** Three-component composition routes involving maximum of $x_2$ despite monotonic root variations (solid lines), and minimum of $h_1$ despite monotonic concentration variations (broken lines).

and thus to "overhanging" root profiles and concentration profiles, if the respective root initially decreases in the direction of flow. As shown in Fig. 3.18a, the overhanging profile is not physically realized, and a step appears in its stead. In development of noncoherent boundaries, $h_k$ trajectories may cross in the noncoherent region, in which roots other than $h_k$ also vary. In such cases, a unique solution of Eqs. (3.87) is not obtained; $h_k(z, \tau)$ is no longer single-valued after such a crossover, and since $h_k$ appears in the coefficients of the differential equations of the other roots also, all roots then have multiple velocities. The physically realized solution is that in which the value of $h_k$ in the coefficients is taken as that of the realized step profile of $h_k$. Incidentally, the behavior of the differential equations given by DeVault [2] is similar: they have multiple solutions, even in coherent systems, if trajectories of different concentrations of the same species cross.

In the event of multiple solutions of the differential equations, the physically realized solution is that obeying the integral material balances of all species. To find this solution through simultaneous application of all material balances is difficult, if at all possible. In the present approach, this difficulty is circumvented as follows. Trajectories of root values are calculated only to where they cross others of the same root. From there on, the trajectory of the physically realized step, formed through the crossovers, is calculated with Eq. (3.91), which gives the step velocity. In this way, all functions $h_i(z, \tau)$ are kept single-valued.

### 2. *Compositions of Intermediate Plateau Zones*

The compositions of the intermediate plateau zones which separate the various coherent boundaries or pulses in the final pattern are readily calculated from the sets of $H$-function roots $\mathbf{h}'$ and $\mathbf{h}''$ of the two plateau zones upstream and downstream, respectively, of the initial boundary.

As the schematic distance-time diagram in Fig. 3.25 shows, the entire variation of any root $h_k$ in the final coherent pattern takes place exclusively across the boundary (or pulse) with $h_k$ as variable root. Thus, $h_k$ has the value $h_k'$ in all plateau zones upstream of this boundary and the value $h_k''$ in all plateau zones downstream of it. For convenience, the zones will be numbered consecutively in the direction of flow in such a way that zone $k$ is between the boundaries with variable $h_{k-1}$ and $h_k$. The sets of root values of the various zones then are

$$
\begin{array}{lll}
\text{Zone } 1 & : & h_1', h_2', \ldots, h_{k-1}', h_k', \ldots, h_{n-2}', h_{n-1}' \\
\text{Zone } 2 & : & h_1'', h_2', \ldots, h_{k-1}', h_k', \ldots, h_{n-2}', h_{n-1}' \\
\vdots & & \qquad\qquad\qquad \vdots \\
\text{Zone } k & : & h_1'', h_2'', \ldots, h_{k-1}'', h_k', \ldots, h_{n-2}', h_{n-1}' \\
\vdots & & \qquad\qquad\qquad \vdots \\
\text{Zone } n-1 & : & h_1'', h_2'', \ldots, h_{k-1}'', h_k'', \ldots, h_{n-2}'', h_{n-1}' \\
\text{Zone } n & : & h_1'', h_2'', \ldots, h_{k-1}'', h_k'', \ldots, h_{n-2}'', h_{n-1}''
\end{array}
\tag{3.92}
$$

(Zones 1 and $n$ are those upstream and downstream of the initial boundary. If the pattern does not contain a boundary with variable $h_k$, zones $k$ and $k+1$ become identical.) The mobile- and stationary-phase concentrations

$x_{jk}$ and $y_{jk}$ of any arbitrary species $j$ in the arbitrary zone $k$ can now be obtained with Eqs. (3.57) and(3.58):

$$x_{jk} = \prod_{i=1}^{k-1} (h''_i - \alpha_{1j}) \prod_{i=k}^{n-1} (h'_i - \alpha_{1j}) \Big/ \prod_{i \neq j} (\alpha_{1i} - \alpha_{1j}) \tag{3.93}$$

$$y_{jk} = \prod_{i=1}^{k-1} \left(\frac{1}{h''_i} - \alpha_{j1}\right) \prod_{i=k}^{n-1} \left(\frac{1}{h'_i} - \alpha_{j1}\right) \Big/ \prod_{i \neq j} (\alpha_{i1} - \alpha_{j1}) \tag{3.94}$$

As shown earlier, the coherent pattern may also contain additional plateau zones between boundaries having the same variable root if the initial profiles of the respective roots have plateaus (see Figs. 3.29 and 3.30). In such an additional zone between boundaries with variable $h_k$, the value of $h_k$ is the same as that of the plateau in the initial profile of this root. Hence, if this initial value is called $h_k^*$, the set of root values of the plateau zone in the final pattern is $h''_1, \ldots, h''_{k-1}, h_k^*, h'_{k+1}, \ldots, h'_{n-1}$. The composition of the zone can thus be calculated with Eqs. (3.93) and (3.94) after $h'_k$ has been replaced by $h_k^*$.

According to Eqs. (3.93) and (3.94), the compositions of the intermediate plateau zones (except those between boundaries having the same variable root) are determined exclusively by the compositions upstream and downstream of the initial boundary, and thus are independent of the concentration profiles within this boundary.

## E. Transient Behavior

A quantitative description of the transient development behavior of a noncoherent boundary is in general best achieved by numerical integration of the differential equations.† The quantitative aspects of transient behavior, however, can be deduced in a simpler manner, at least for three-component systems, as will now be shown.

The graphical constructions in this section can be carried out quantitatively and then present an alternative to numerical computation [43]. Extension of the method to more than three components is possible in principle, but appears to be too cumbersome for practical use.

### 1. *Transient Composition Profile Routes. Development Times*

The development of a noncoherent boundary can be described by transient composition profile routes, which represent sequences of com-

† A computer program based on root trajectories is operative at the Sea Water Conversion Laboratory, University of California.

positions in the column at fixed times before attainment of the coherent pattern. As can be seen in distance-time diagrams (e.g., Fig. 3.25), the compositions along transient routes result from shifts of the root profiles relative to one another. In a three-component system, any transient composition has root values $h_1$ and $h_2$ which existed in previous profiles, but of which the "faster" $h_2$ value was previously farther upstream than the "slower" $h_1$ value. Accordingly, in the composition simplex, a composition point on a transient route is at the intersection of two composition

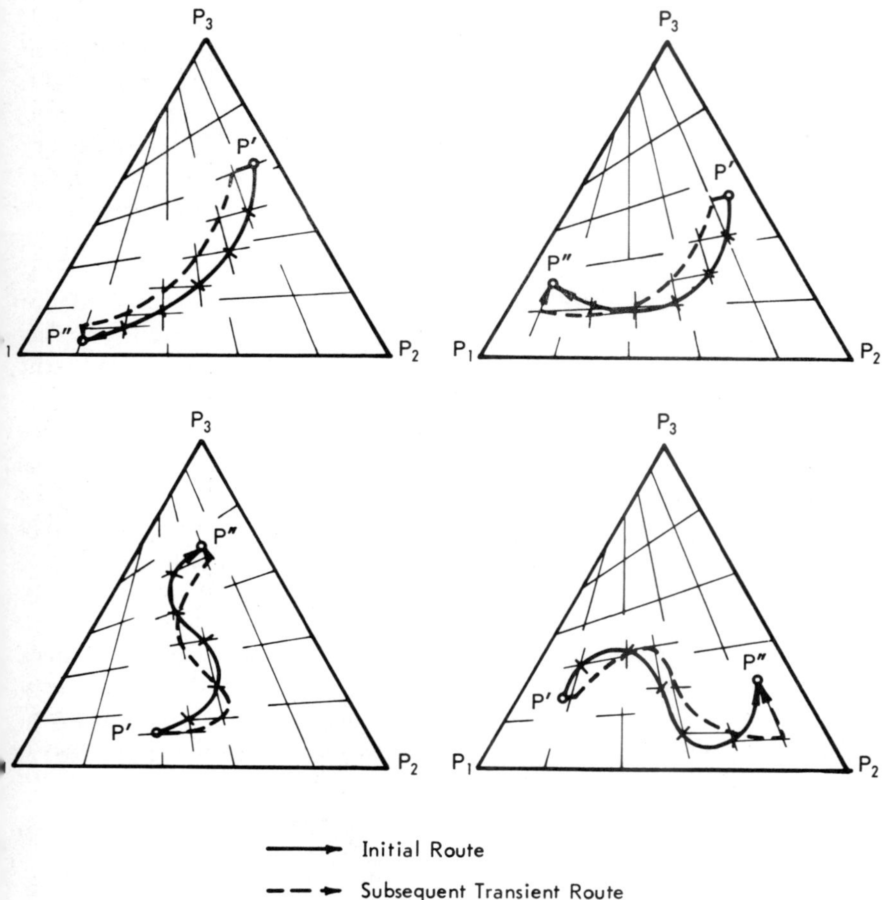


**Fig. 3.32.** Initial and subsequent transient composition profile routes of various noncoherent boundaries in three-component system (schematic).

paths originating from two points on an earlier transient route; the paths are a 2 | 3 path (line of constant $h_1$) and a 1 | 2 path (line of constant $h_2$), the former originating from a point farther downstream than the latter.

Graphical constructions based on this rule provide a qualitative idea of how successive transient routes develop from the initial route in any given case. Points of origin of paths can be selected on the initial route, and points on an approximate subsequent route are obtained as intersections of paths from these points, as described above. The procedure can then be repeated to find later transient routes from prior ones. Four examples are shown in Fig. 3.32.

These constructions reflect a most important general principle, that of *local trend toward coherence under arbitrary, variable boundary conditions.* That coherence develops if the boundary values of the variables (concentrations) are fixed has been shown earlier. But a local, momentary event has no means of knowing whether or not the values at its boundaries are constant, and, hence, its direction is independent of their constancy. Accordingly, the local, momentary deformation of the composition route at any point and time under arbitrary, variable boundary conditions is always toward coherence, that is, in the direction that would produce coherence locally if the neighboring points of the route were fixed. This principle provides a key to understanding and predicting noncoherent response behavior under any arbitrary conditions.

Constructions with arbitrarily selected points on prior transient routes can serve only for qualitative orientation. Attainment of any particular transient composition requires a certain "development time," since the $h_2$ value of that composition must catch up with the $h_1$ value. An accurate transient route, representing the sequence of compositions at a fixed time, thus is obtained only if the construction is "synchronized" in such a way that the development time is the same for all its composition points.

A relation for this development time is readily derived. Root values $h_1$ and $h_2$, traveling at their root velocities $u_{h_1}$ and $u_{h_2}$, advance distances $u_{h_1}\mathrm{d}\tau$ and $u_{h_2}\mathrm{d}\tau$ in the (adjusted) time interval $\mathrm{d}\tau$. The development time $\mathrm{d}\tau$ required for an $h_2$ value to catch up with an $h_1$ value, previously a distance $\mathrm{d}z$ ahead, thus is

$$\mathrm{d}\tau = \mathrm{d}z/(u_{h_2} - u_{h_1}) \tag{3.95}$$

(see Fig. 3.33a; in constructions with finite time and distance intervals, appropriate mean values of the composition-dependent root velocities must be used). For the construction of a synchronous transient route, all pairs of composition points on the previous route must be selected in such

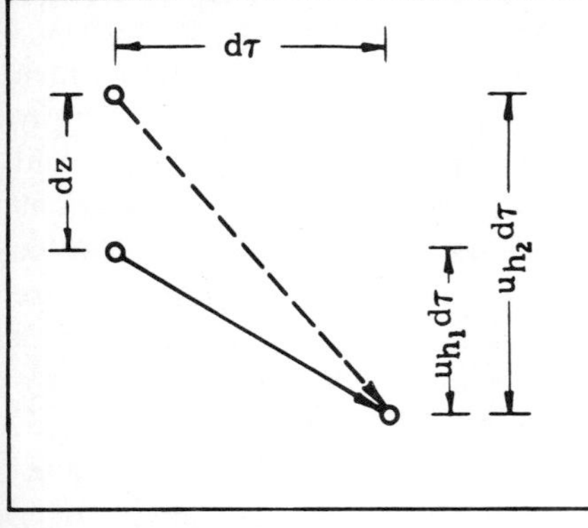


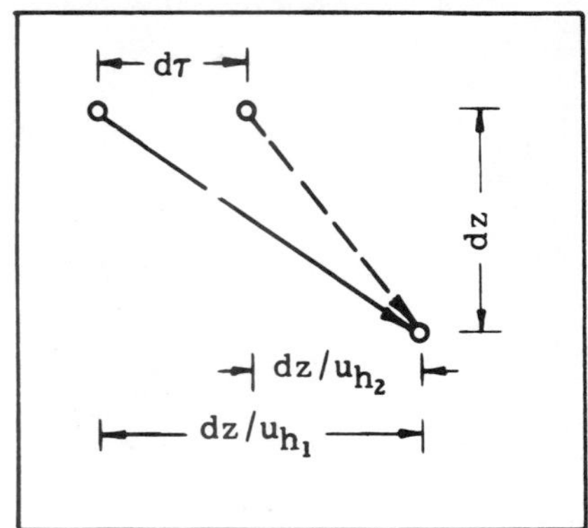


Trajectory of $h_1$
Trajectory of $h_2$

**Fig. 3.33.** Development time and development distance for construction of transient profile and history routes.

a manner that their distances $dz$ in the column give, with Eq. (3.95), the same development time $d\tau$.

The development time is of interest as a measure of the relative speed of resolution. For example, Eq. (3.95) shows that a noncoherent boundary with given initial route is resolved more quickly if it is sharper initially: Any two compositions in the initial boundary are, in the column, a lesser distance apart if the boundary is sharper, and Eq. (3.95) then gives a shorter development time. In the extreme case of an ideally sharp noncoherent composition step, the distance between the compositions on the two sides is zero, and resolution thus is instantaneous.

The speed of resolution also depends on composition, since Eq. (3.95) involves the composition-dependent root velocities. The greater the difference between the root velocities, the shorter is the development time. A contour-line diagram illustrating the dependence of the root-velocity difference on composition is shown in Fig. 3.34 (solid lines). Development is particularly slow for routes near the watershed point. At this point itself, the two root velocities are equal, and no resolution is achieved. The ensuing complications will be discussed in Section V.F.

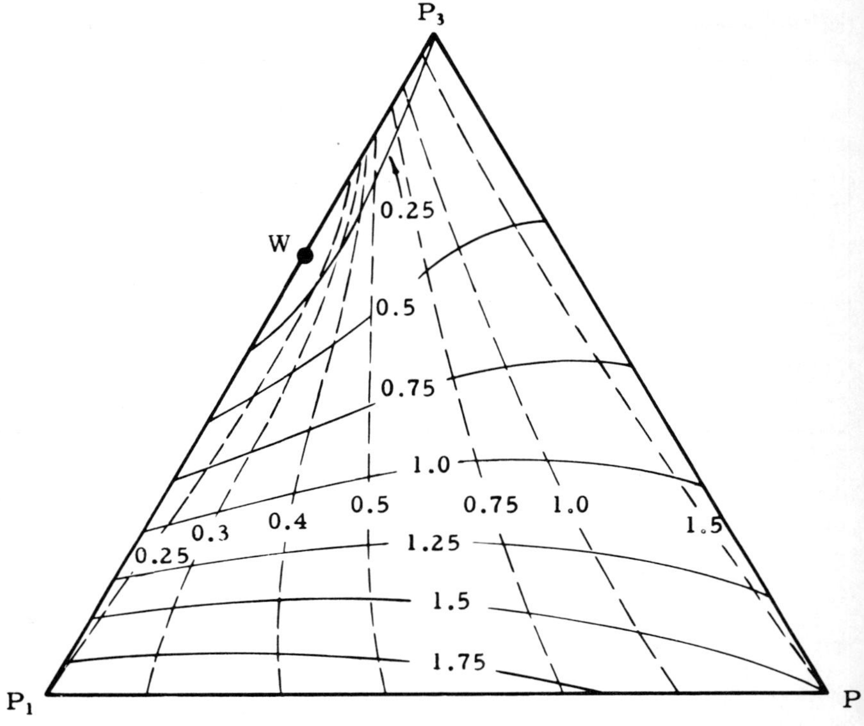


**Fig. 3.34.** Contour lines of constant $u_{h_2} - u_{h_1}$ (solid lines) and $1/u_{h_1} - 1/u_{h_2}$ (broken lines) in $x$ simplex of three-component systems (for $\alpha_{12} = 2, \alpha_{13} = 4$).

2. *Transient Composition History Routes. Development Distances*

The composition routes discussed so far are *profile routes*, defined as representing sequences of compositions in the column in the direction of flow at fixed times. For certain later applications it will be more convenient to operate with *history routes*, defined as representing chronological sequences of compositions at fixed column levels. In distance-time diagrams, compositions along profile routes fall on vertical lines (constant $t$), and those along history routes, on horizontal lines (constant $z$).

In coherent systems, the distinction between profile and history routes is of no consequence because they coincide with one another. It must merely be noted that they have opposite directions, since a composition farther upstream at a given time reaches a given column level later than do its downstream neighbors. In noncoherent boundaries, however, compositions exist only momentarily, so that those constituting a profile at a given time are not the same as those passing through a given column level.

As distance-time diagrams show, the root values of $h_1$ and $h_2$ of a composition along a transient history route at a given column level have both passed through a column level farther upstream, the $h_1$ value earlier than the $h_2$ value. Successive transient history routes can thus be obtained in a similar manner as transient profiles routes, and the qualitative behavior of both is the same. The routes shown in Fig. 3.32 could therefore be history routes instead of profile routes, except that the arrows would have to be reversed to point in the direction of increasing time.

An accurate transient history route, representing the successive compositions at a fixed column level, is obtained from a given history route at or nearer the column inlet if the "development distance" is made the same for all composition points. A relation for the development distance can be derived as follows. Root values $h_1$ and $h_2$ require times $dz/u_{h_1}$ and $dz/u_{h_2}$ to advance a given distance $dz$. Accordingly, the development distance $dz$ needed for an $h_2$ value to catch up with an $h_1$ value which, at a given column level, leads by $d\tau$ is

$$dz = d\tau/(1/u_{h_1} - 1/u_{h_2}) \tag{3.96}$$

(see Fig. 3.33b).

Contour lines of constant values of the difference $1/u_{h_1} - 1/u_{h_2}$ are shown in Fig. 3.34 as broken lines. The differences of both the root velocities and their reciprocals are very small near the watershed point, so that resolution in this composition region requires both very long development times and distances. Otherwise, the two sets of contour lines in Fig. 3.34 differ in a characteristic manner. In a fixed period of time, the roots advance farther if their velocities are higher. Therefore, along any line of constant development time (constant velocity difference, solid line), the development distance increases with increasing root velocities. For example, near point $P_1$, the root velocities are high and their difference is large, so that the development distance is long although the (adjusted) development time is short. The opposite is true at point $P_3$.

### F. Compositions on or near Watersheds. Permanent Noncoherence

Exceptional cases involving compositions on or near watersheds, excluded in the proof for attainment of coherence in Section V.B, will now be considered.

The complication arising at watersheds is that two roots with adjacent index numbers have equal velocities. Resolution of variations of these two roots from one another must therefore be reexamined. Only the behavior of these two roots need be considered, because resolution from variations of other roots, which have different velocities, proceeds in a normal manner.

In this section, the roots whose values are equal at the watershed are taken to be $h_{k-1}$ and $h_k$, where $k$ can be 2, ..., $n-1$. The watershed thus is on the border $x_k = 0$ and is given by

$$h_{k-1} = h_k = \alpha_{1k} \tag{3.62}$$

(see Section IV.D.4). As before, primes and double primes will refer to the plateau-

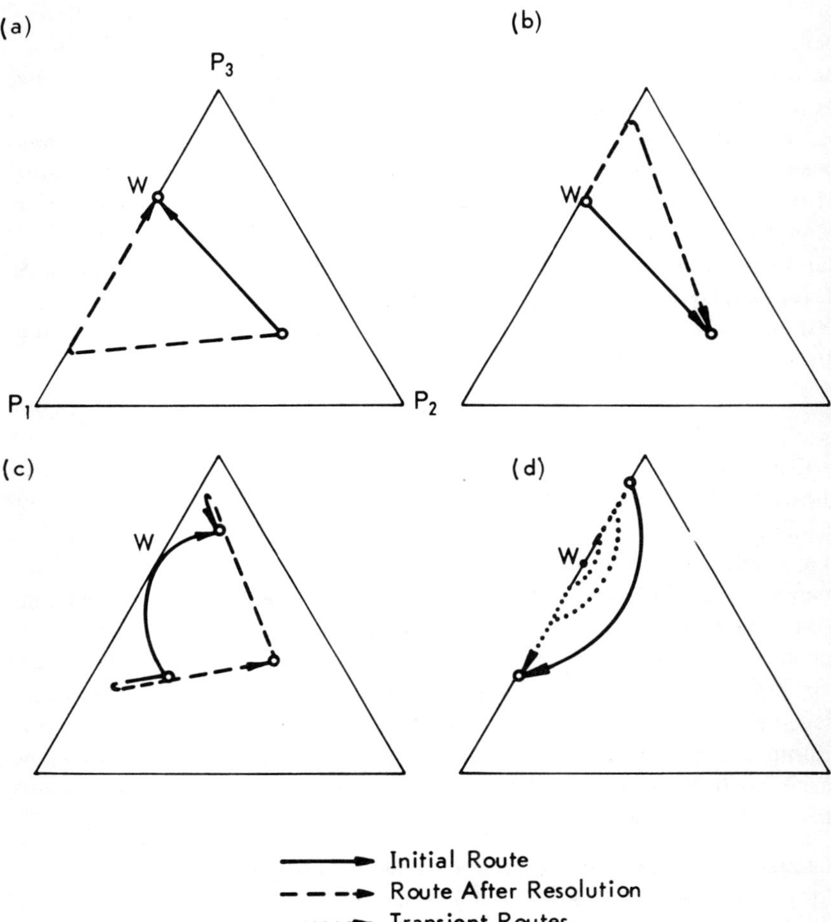


**Fig. 3.35.** Development of profile routes of noncoherent boundaries involving compositions on or near watershed in three-component system. (a) Watershed point at downstream end, (b) watershed point at upstream end, (c) watershed point within initial boundary, (d) permanent noncoherence.

zone compositions upstream and downstream, respectively, of the initial noncoherent boundary.

1. *Initial Routes Involving Compositions on Watersheds*

In principle, a watershed composition can occur at either end or within the initial noncoherent boundary, as shown for a three-component system in Fig. 3.35a–c.

If the composition point of the plateau zone downstream of the initial boundary falls on the watershed (see Fig. 3.35a), the profiles of $h_{k-1}$ and $h_k$ develop as shown in Fig. 3.36a. The root values at the downstream end point W of the root variations are

$$h''_{k-1} = h_k'' = \alpha_{1k} \tag{3.97}$$

According to Eq. (3.89), both values advance at the same rate. The variation of $h_{k-1}$ is nonsharpening while that of $h_k$ is self-sharpening and eventually becomes ideally sharp (see Section V.D.1), so that an $h_k$ step with values $h_k{}^*$ and $h_k''$ on its upstream and downstream sides forms at W. ($h_k{}^*$ will equal $h_k'$ eventually, but need not do so in the earlier stages, because a step at point W may form while the variation farther upstream is still gradual; see profile at time $t_2$ in Fig. 3.36a.) With Eqs. (3.91), (3.89), and (3.97) one obtains for the step velocity of $h_k$ and the root velocity of $h_{k-1}$ at point W, where $h_k \rightarrow h_k{}^*$ and $h_{k-1} \rightarrow h''_{k-1}$,

$$u_{\Delta h_k} = h_k{}^* h_k'' h''_{k-1} P = h_k{}^* \alpha_{1k}^2 P \tag{3.98}$$

$$u_{h_{k-1}} = h''^{2}_{k-1} h_k{}^* P = h_k{}^* \alpha_{1k}^2 P \tag{3.99}$$

where

$$P \equiv \prod_{i \neq k,k-1} h_i \prod_i \alpha_{i1} \tag{3.100}$$

Both velocities are equal and remain so after the entire variation of $h_k$ has become discontinuous ($h_k{}^* = h_k'$). The overall result is that the two boundaries with variable $h_{k-1}$ and $h_k$ are completely resolved but, in the final pattern, are not separated by a plateau zone.

The situation is analogous if the composition point of the plateau zone upstream of the initial boundary falls on the watershed, as shown for a three-component system in Fig. 3.35b. Here, the variation of $h_{k-1}$ becomes discontinuous and is resolved, but not separated by a plateau zone, from the spreading variation of $h_k$ (see Fig. 3.36b).

A watershed composition may also occur within the initial boundary, as shown for a three-component system in Fig. 3.35c. In such cases, $h_{k-1}$ has a maximum and $h_k$ has a minimum at the watershed composition. Both root variations thus constitute pulses, whose amplitudes diminish once the self-sharpening flanks have become sharp. The watershed composition then is no longer a part of the physically realized profiles (see Fig. 3.36c). As a result, the variations of $h_{k-1}$ and $h_k$ are completely resolved into transient or permanent coherent pulses, separated by a plateau zone.

The considerations above show that a completely coherent final pattern is attained even if the initial route involves a composition point which falls on a watershed.

2. *Permanent Noncoherence*

Under exceptional circumstances, a completely coherent final pattern is not attained although the route of the initial boundary does not involve a watershed composition. This situation, shown for a three-component system in Fig. 3.35d, arises if progressing

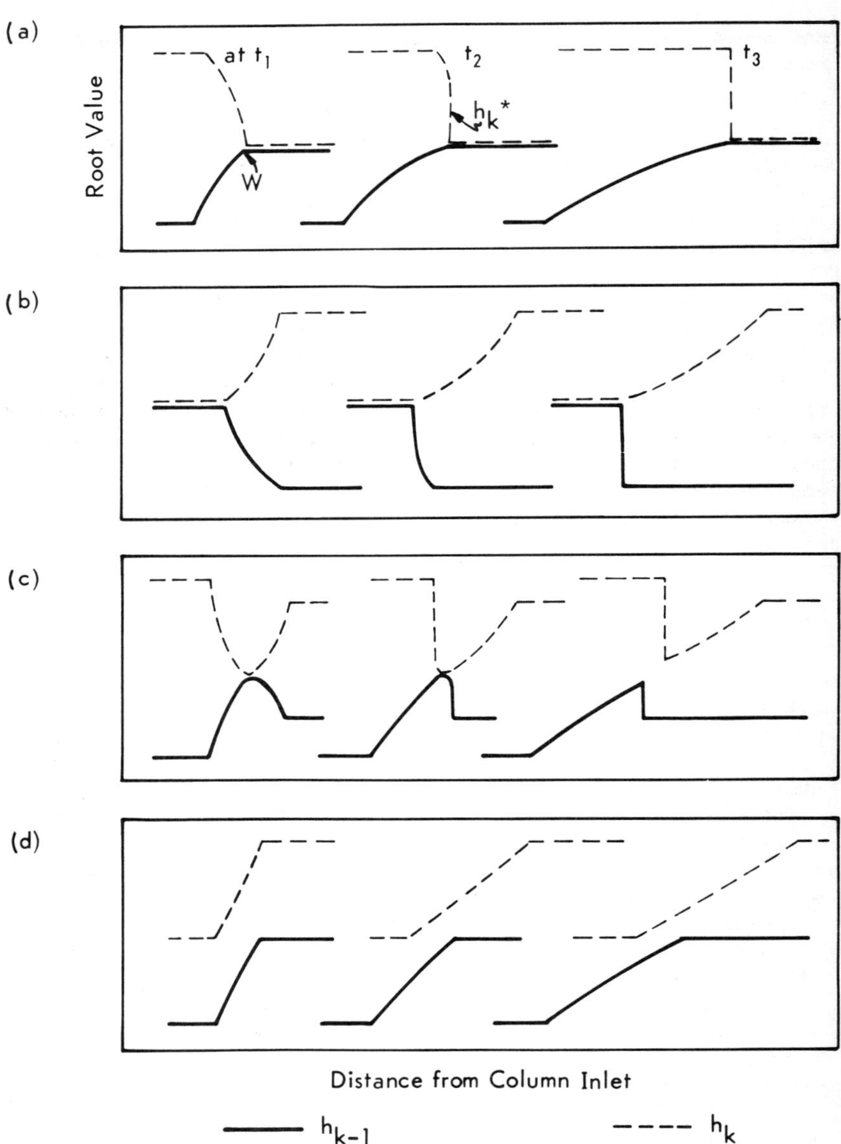


**Fig. 3.36.** Root profiles of $h_{k-1}$ and $h_k$ at three successive times during development of noncoherent boundary involving composition on or near watershed of $x_k = 0$. (a) Watershed composition at downstream end, (b) watershed composition at upstream end, (c) watershed composition within initial boundary, (d) permanent noncoherence. (Schematic.)

resolution of nonsharpening variations of $h_{k-1}$ and $h_k$ brings the composition route into the immediate vicinity of the watershed, where the development time is infinite. In terms of $H$-function roots, the condition for such "permanent noncoherence" is

$$h''_{k-1} = h_k' = \alpha_{1k}, \qquad h_{k-1}, h_k \neq \text{const} \tag{3.101}$$

This condition requires species $k$ to be present in the initial boundary, but absent from the plateau zones on both sides.

The development of the root profiles in such a boundary is shown in Fig. 3.36d. In the early stages, development proceeds in a normal manner. Since the initial route is not near the watershed, the velocity of $h_k$ is higher than that of $h_{k-1}$ at any given time and location, so that partial resolution is achieved. However, $h_k'$ cannot catch up with $h''_{k-1}$, as would be required for complete resolution, because

$$u_{h'_k} = h_k'^2 h_{k-1} P = h_{k-1}\alpha_{1k}^2 P \tag{3.102}$$

$$u_{h''_{k-1}} = h_{k-1}''^2 h_k P = h_k \alpha_{1k}^2 P \tag{3.103}$$

and $h_{k-1} \leqq h_k$. [Here, $P$ is defined as in Eq. (3.100).] As a result, the two developing nonsharpening boundaries with variable $h_{k-1}$ and $h_k$ continue indefinitely to overlap, so that a small portion of the overall pattern remains permanently noncoherent.

That resolution must remain incomplete can also be shown with a material-balance argument. If an entirely coherent pattern with boundaries or pulses in the sequence of increasing index numbers of their variable roots were attained, the final route would run exclusively along paths on the $h$-space borders $h_{k-1} = \alpha_{1k}$ and $h_k = \alpha_{1k}$, which are the $h$ transforms of the border $x_k = 0$ (see Section IV.D.4). Species $k$ would thus be completely absent from the final pattern. This species, however, is present in the initial boundary and cannot vanish from the system.

## G. Summary of Rules

For ease of reference, the rules derived for undisturbed development of noncoherent boundaries are summarized below.

1. A noncoherent boundary is resolved into coherent boundaries or pulses, separated by plateau zones. The boundaries and pulses in the final coherent pattern are in the sequence of increasing index numbers of their variable roots (counted in the direction of flow). Exceptions are noted below.
2. The qualitative properties—sharpening characteristics, multiplicity, permanent or transient character—of each boundary or pulse in the resulting coherent pattern are determined exclusively by the initial profile of its variable root, and are the same as would result from this initial profile in an entirely coherent system.
3. The compositions of the plateau zones separating boundaries with different variable roots in the final pattern depend only on the compositions of the two plateau zones upstream and downstream of the initial boundary.

4. The rate of resolution depends on sharpness and compositions. The sharper an initial boundary with given composition route, the faster it is resolved. Ideally sharp boundaries are resolved instantaneously. Resolution is relatively slow if compositions near watersheds are involved.

Exceptions to rule 1 above are (a) no plateau zone appears between the boundaries with variable $h_{k-1}$ and $h_k$ if the composition point of the plateau zone upstream or downstream of the initial boundary falls on the watershed of a border $x_k = 0$; (b) resolution of the boundaries with variable $h_{k-1}$ and $h_k$ remains incomplete if both are nonsharpening and if species $k$ is present in the initial boundary but is absent from both its sides.

### H. Review and Perspective

The principal developments in this section are the proof of attainment of coherence from arbitrary initial conditions, the establishment of a qualitative and quantitative theory of noncoherent behavior, and the recasting of the fundamental differential equations in terms of $H$-function roots.

Attainment of coherence has been proved mathematically by means of the $h$ transformation: Upon development, variations of different $H$-function roots "sort themselves out" because roots with lower index numbers have lower velocities. This proof is limited by the premises on which the transformation is based (see Section IV.H). The explanation of the physical causes, however, is on a more general level and can be broadened into a proof for systems with arbitrary equilibrium properties. Under exceptional conditions, a final coherent pattern is not attained, but even then the composition route keeps developing in the direction of coherence.

The universal trend toward coherence from arbitrary initial conditions provides the key to understanding the behavior of noncoherent systems. On this basis, the qualitative behavior of noncoherent boundaries upon development and the chief features of the eventual coherent patterns are readily predicted. Moreover, the principle of development toward coherence remains locally valid even under variable boundary conditions.

For rigor and quantitative solutions, the treatment of noncoherent boundaries relies mainly on the $h$ transformation, which converts the differential equations into a simpler form [Eqs. (3.87)]. In particular, the $H$-function roots are "conserved," that is, compositions at any time in the column are composed exclusively of root values which can be traced back to the initial profile or to the influent. Moreover, the velocities of given root values can be expressed as point properties, and a practicable functional relation, not restricted to coherence, between composition variables and their velocities is thus established [Eqs. (3.89)]. This relation and the conservation properties of the roots permit responses under any conditions to be derived with relative ease.

The treatment of noncoherent behavior in this section completes the theoretical basis of the present approach. All elements essential for applications to column behavior under specified conditions have now been developed. Such applications will be considered in the next chapter, after a few rules for interference of coherent boundaries, a frequent occurrence in various cases of practical interest, have been derived for later reference.

## VI. Interference of Boundaries

Boundaries travel at different rates and may thus encounter one another on their way through the column. The interference arising in such encounters will now be examined.

The basic rules for the dynamic behavior of $H$-function roots have been deduced from the differential equations (3.87). Since no particular initial or boundary conditions were invoked, these rules remain valid if interference occurs. In particular, the "sorting out" of root values according to their index numbers, as illustrated for a case without interference in Fig. 3.25, takes place in the same manner. As an example, schematic root trajectories of two interfering noncoherent boundaries, initially separated by a plateau zone, are shown in Fig. 3.37. (With regard to the linear and parallel course of the trajectories the same remarks as for Fig. 3.25 apply.) Transient and final routes and patterns in cases with interference can thus be established with the same methods as in the previous section. The fact that all initial root profiles have intermediate plateaus if the initial route involves more than one noncoherent boundary makes for a greater variety of possible patterns, but does not introduce complications in principle.

While no new insight is to be gained from further examination of the general case, rules useful for later applications can be derived for the much simpler special case of mutual interference of coherent boundaries. The remainder of the present section deals exclusively with such interference.

### A. Crossover of Coherent Boundaries or Pulses

Interference of coherent boundaries or pulses having different variable $H$-function roots usually results in a particularly simple behavior, to be

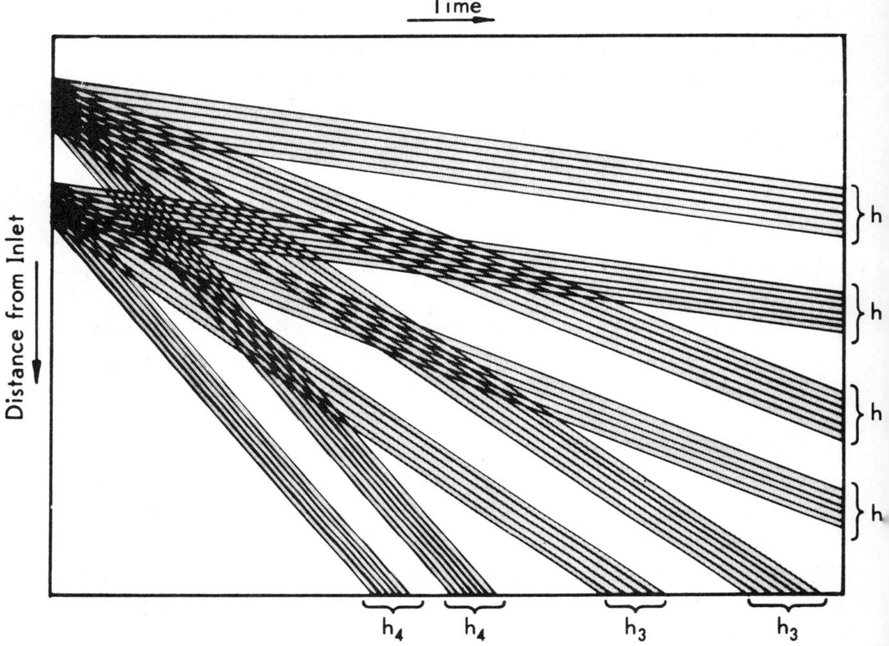


**Fig. 3.37.** Root trajectories for development of two interfering noncoherent boundaries in five-component system (schematic). Regions of gradual root variations are shaded.

called a "crossover" for reasons which will immediately become apparent. Such crossovers will be examined in this section.

The following notation will be adopted. The variable roots of the two interfering boundaries or pulses are $h_j$ and $h_k$ $(j<k)$ and vary from $h_j'$ to $h_j''$ and from $h_k'$ to $h_k''$ in the direction of flow. Initially, the boundary or pulse with variable $h_k$ is upstream of that with variable $h_j$. The three plateau zones upstream of, between, and downstream of the two initial boundaries or pulses are numbered 1, 2, and 3, as indicated in Fig. 3.38.

### 1. *Conservation of Roots and General Properties*

Root trajectories for a crossover are shown schematically in Fig. 3.38. Since the values of the root with the higher index number, $k$, have the higher velocities, the trajectory bundles of the $h_j$ and $h_k$ values cross one another [see condition (3.90) and Section V.B; exceptional cases involving watershed compositions with $u_{h_j} = u_{h_k}$ will be discussed in Section VI.B].

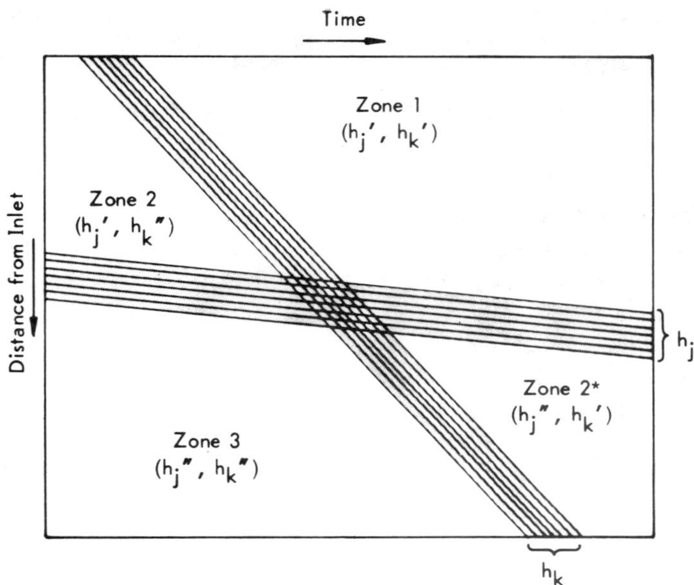


**Fig. 3.38.** Root trajectories for crossover of two coherent boundaries (shaded) with variable $h_j$ and $h_k$ (schematic).

All other roots, having uniform values initially, remain constant regardless of the behavior of $h_j$ and $h_k$ [see condition (3.88)]. The crossover thus does not generate any additional boundaries.

As shown in Section V for the general case of noncoherent boundaries, the behavior of a trajectory bundle of values of a given root is not affected by variations of other roots, except for root-velocity changes. Accordingly, in the special case of a crossover, too, the crossing trajectory bundles only affect one another's root velocities while retaining their other properties. Thus, the sharpening characteristics of boundaries and the transient or permanent character and multiplicity of pulses, being determined exclusively by the initial profile of the respective variable root, remain unaffected by crossovers, and so do the values of the variable root on the upstream and downstream sides. Of course, events bound to occur regardless of interference, such as formation of a step from an initially diffuse self-sharpening boundary, degeneration of a transient pulse into a single boundary, or reduction of transient pulse multiplicity, may happen to take place during a crossover.

These considerations can be summarized as follows: A coherent boundary or pulse, when crossing another with a different variable root, retains the index number and the upstream and downstream values of its variable root as well as its general properties such as sharpening characteristics, transient or permanent character, and multiplicity.

## 2. *Plateau-Zone Compositions*

In the course of a crossover, the initial plateau zone 2 between the two boundaries or pulses disappears and a new plateau zone, to be called 2*, appears in its place (see Fig. 3.38). The composition of the zone 2* is readily predicted from those of the initial zones, as will now be shown.

The sets of $H$-function roots of the initial three plateau zones are, by virtue of the convention as to notation,

$$\begin{aligned} &\text{Zone 1:} \quad h_1, \ldots, h'_j, \ldots, h'_k, \ldots, h_{n-1} \\ &\text{Zone 2:} \quad h_1, \ldots, h'_j, \ldots, h''_k, \ldots, h_{n-1} \\ &\text{Zone 3:} \quad h_1, \ldots, h''_j, \ldots, h''_k, \ldots, h_{n-1} \end{aligned} \tag{3.104}$$

All roots other than $h_j$ and $h_k$ have the same values in all zones including 2*. As is apparent from Fig. 3.38, zone 2* is downstream of the boundary or pulse with variable $h_j$ and upstream of that with variable $h_k$. Its set of $H$-function roots thus is

$$\text{Zone 2*:} \quad h_1, \ldots, h''_j, \ldots, h'_k, \ldots, h_{n-1} \tag{3.105}$$

The values $h''_j$ and $h'_k$ are the same as in the sets (3.104), since crossover does not alter the upstream and downstream values of the variable root of a boundary or pulse (see Section VI.A.1). All roots can thus be calculated from the compositions of zones 1 and 3, and the composition of zone 2* can then be calculated from the roots by means of Eqs. (3.57) and (3.58).

The following qualitative predictions can be made without calculation of $H$-function roots:

1. Species having lower concentrations in zone 2 than in zones 1 and 3 will have higher concentrations in zone 2* than in zones 1 and 3. Conversely, species having higher concentrations in zone 2 than in 1 and 3 will have lower concentrations in zone 2* than in 1 and 3.
2. Species having intermediate concentrations in zone 2 relative to zones 1 and 3 will also have intermediate concentrations in zone 2*.
3. Species present in only one initial zone, either 1 or 3, but absent from the two others, will be present in zone 2*. Conversely, species absent

from one initial zone, either 1 or 3, but present in the two others, will be absent from zone 2*.

4. If the compositions of zones 1 and 2 are equal, those of zones 2* and 3 will be equal. Similarly, if the compositions of zones 2 and 3 are equal, those of 2* and 1 will be equal. If the compositions of zones 1, 2, and 3 are equal, zone 2* will also have the same composition.

These rules apply for mobile- as well as stationary-phase compositions and are readily derived from the root sets (3.104) and (3.105) and Eqs. (3.57) and (3.58).

### 3. *Root- and Step-Velocity Variations*

In the schematic diagrams in Figs. 3.37 and 3.38, any root-velocity variations have been ignored, so that the root trajectories appear as parallel straight lines. Such behavior is approached in the case of boundaries or pulses merely involving infinitesimal composition changes. Here, all root variations remain infinitesimal, and, according to Eqs. (3.89) and (3.91), the root- or step-velocity variations then do the same. On the other hand, crossovers of boundaries or pulses involving finite composition changes entail finite root- or step-velocity variations, which will now be examined.

As a coherent boundary crosses another, the values of its variable root pass through a region in which another root varies. According to Eqs. (3.89) and (3.91), such variation of another root causes the velocities of given root values (or steps) to change in the same direction. As Fig. 3.38 shows, the topology of the trajectories is such that, along the trajectories of the "faster" $h_k$ values or steps, $h_j$ varies from $h'_j$ before crossover to $h''_j$ after. Hence, values or steps of $h_k$ accelerate if $h'_j < h''_j$, and decelerate if $h'_j > h''_j$. Similarly, along the trajectories of the "slower" $h_j$ values or steps, $h_k$ varies from $h''_k$ before crossover to $h'_k$ after. Values or steps of $h_j$ thus accelerate if $h''_k < h'_k$, and decelerate if $h''_k > h'_k$. Invoking the sharpening criteria (3.74) one can state the following rule: Root values or steps accelerate when crossing a slower nonsharpening or a faster self-sharpening boundary, and decelerate when crossing a faster nonsharpening or a slower self-sharpening boundary.

To illustrate the variations of root and step velocities, root trajectories for four cases involving crossover of sharp and self-sharpening, and diffuse and nonsharpening, boundaries are shown in Fig. 3.39. The velocity variations, reflected by the changes of the trajectory slopes, are gradual when the trajectories cross a diffuse boundary, and abrupt when they cross

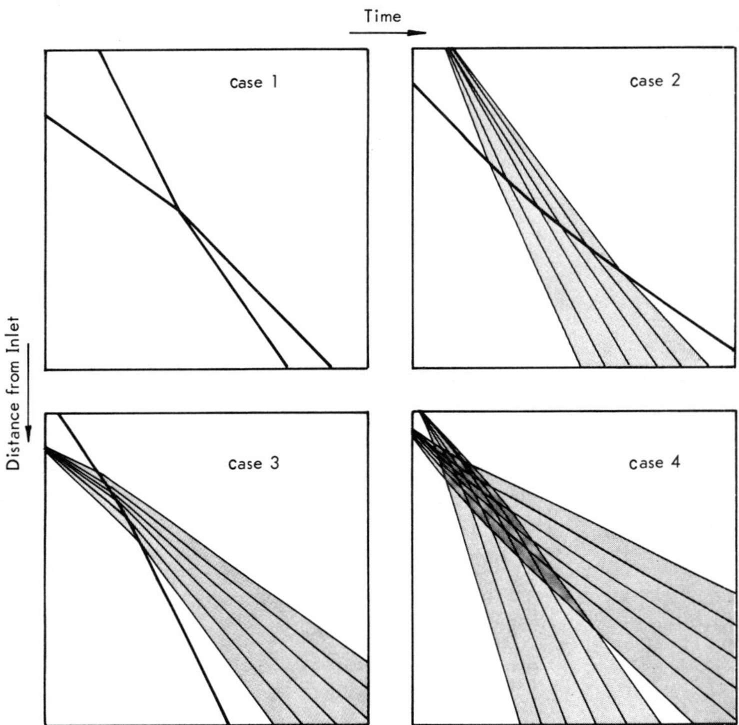


**Fig. 3.39.** Root and step trajectories for crossover of two coherent boundaries with various combinations of sharpening characteristics. Sharp boundaries shown as heavy lines; diffuse boundaries, shaded; region of noncoherence, darkly shaded. (Calculated for three-component system with $\alpha_{12} = 2$, $\alpha_{13} = 4$ and compositions as shown in Fig. 3.40.)

a step, because the variation of the other root is gradual in the first case and abrupt in the second.

When crossing the flanks of a pulse, which have opposite sharpening characteristics, a given root value or step alternatingly accelerates and decelerates. A comparison of criteria (3.74) and (3.86) shows that crossing a transient pulse produces a net velocity change in the same direction as does crossing a self-sharpening boundary, and crossing a permanent pulse produces a change in the same direction as does crossing a nonsharpening boundary.

### 4. *Composition Routes and Transient Behavior*

Since all roots other than $h_j$ and $h_k$ remain uniform and constant before, during, and after the crossover, the composition route remains confined at all times to one and the same common plane of paths with variable $h_j$ and $h_k$. The initial, transient, and final routes of two crossing boundaries or pulses thus are readily depicted in two-dimensional common-plane diagrams.

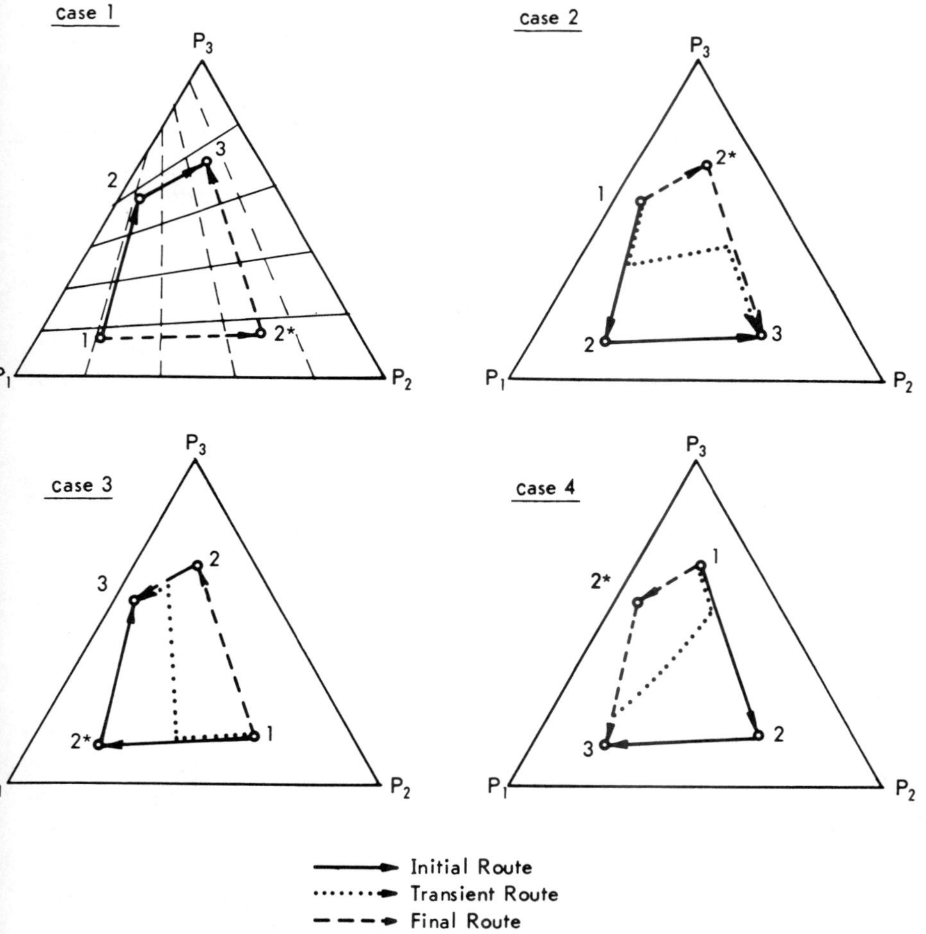


**Fig. 3.40.** Initial, transient, and final composition profile routes for crossover of two coherent boundaries in three-component system. Cases correspond to those in Fig. 3.39.

Profile routes corresponding to the four root-trajectory diagrams in Fig. 3.39 are shown in Fig. 3.40. Three-component systems have been chosen for this illustration to facilitate mental translation of the routes into concentration profiles. In all four cases, the initial and final routes run exclusively along composition paths. If both boundaries are ideally sharp and self-sharpening, as in case 1, the change from the initial to the final route is instantaneous. Here, a noncoherent step 1–3 exists only momentarily and is immediately resolved (for proof of immediate resolution of an ideally sharp noncoherent step, see Section V.E.1). On the other hand, if one boundary is sharp and the other is diffuse, as in cases 2 and 3, crossover requires a finite time but, nevertheless, all transient routes still run exclusively along composition paths because the self-sharpening step remains sharp and coherent while crossing the diffuse boundary. However, if both boundaries are diffuse, as in case 4, the transient routes are truly noncoherent in that they comprise portions which do not follow any composition paths.

### B. Conditions for Crossover. Other Interferences

For simplicity, we have so far considered only cases in which the interfering boundaries or pulses have different variable roots, and have presumed that regular crossovers, as shown schematically in Fig. 3.38, do occur. In fact, however, the interfering boundaries or pulses may have bracketing affinity cuts and involve variations of more than one root, or may have one or several variable roots in common. Interference of boundaries or pulses with bracketing affinity cuts may nevertheless result in a crossover. On the other hand, a crossover does not necessarily occur even if the interfering boundaries or pulses have different variable roots. We shall now establish under which conditions a normal crossover takes place, and examine other possible types of interferences.

#### 1. *Boundaries or Pulses with Different Variable Roots*

The discussion of noncoherence in Section V has shown that any arbitrary initial variations of the *H*-function roots, upon development, sort themselves out into separate variations of individual roots in the sequence of their index numbers, except if watershed compositions are involved. This applies also in the special case of interference of coherent boundaries or pulses having different variable roots, so that, here, a normal crossover takes place in all but the following exceptional situations involving watersheds.

There is only one possible cause that can indefinitely delay the complete resolution of trajectory bundles of two different roots: A composition must persist or be approached for which the velocities of the two roots are equal. This, in turn, can happen only if the two roots have adjacent index numbers (see Section V.F.). Thus, interference of boundaries or pulses having different variable roots can deviate from normal crossover only if the roots have adjacent index numbers, say, $k-1$ and $k$, and will do so if compositions on the watershed $h_{k-1} = h_k = \alpha_{1k}$, entailing $u_{h_{k-1}} = u_{h_k}$, persist or are approached.

The four possible situations in which two boundaries with variable $h_{k-1}$ and $h_k$ interfere without crossing in a normal manner are illustrated in Fig. 3.41. In cases a and b, the composition point of zone 1 or 3 is on the watershed. The noncoherent boundary arising from interference of the initial boundaries then has a watershed composition on its upstream or downstream side. As shown in Section V.F.1, such a boundary is completely resolved into coherent boundaries with variable $h_{k-1}$ and $h_k$, but no intermediate plateau zone is formed (see Fig. 3.35a and b). Thus, the two initial coherent boundaries cross but remain immediately adjacent rather than becoming separated by a plateau zone.

In Fig. 3.41c, the composition point of the initial intermediate plateau zone 2 falls on the watershed. Although having different variable roots, both initial boundaries then have the same (bracketing) affinity cut. Crossover does not occur. Rather, the two initial boundaries merge into a single coherent boundary between zones 1 and 3.

In Fig. 3.41d, species $k$ is present in zone 2 but is absent from zones 1 and 3, and both initial boundaries are nonsharpening. The noncoherent boundary formed when the initial boundaries merge is of the type discussed in Section V.F.2 (see Fig. 3.35d); resolution remains incomplete and a small portion of the final pattern remains permanently noncoherent.

The same considerations apply if the interfering root variations constitute pulses rather than boundaries. The root variations of one or both interfering boundaries or pulses, if they have bracketing affinity cuts, may also extend over more than one root, say, over $h_i, \ldots, h_j$ and $h_k, \ldots, h_l$ ($i \leqq j < k \leqq l$; for cases with $j \geqq k$, see p. 141). Although seemingly more complex in that variations of additional roots are involved, the resulting interferences are of exactly the same type as those of variations of $h_j$ and $h_k$ alone. Regardless of bracketing affinity cuts, the composition routes of the interfering boundaries or pulses are on a common plane, as in the cases discussed previously. The variation of more than two roots on this common plane results merely from the fact that the plane is on a border or sub-

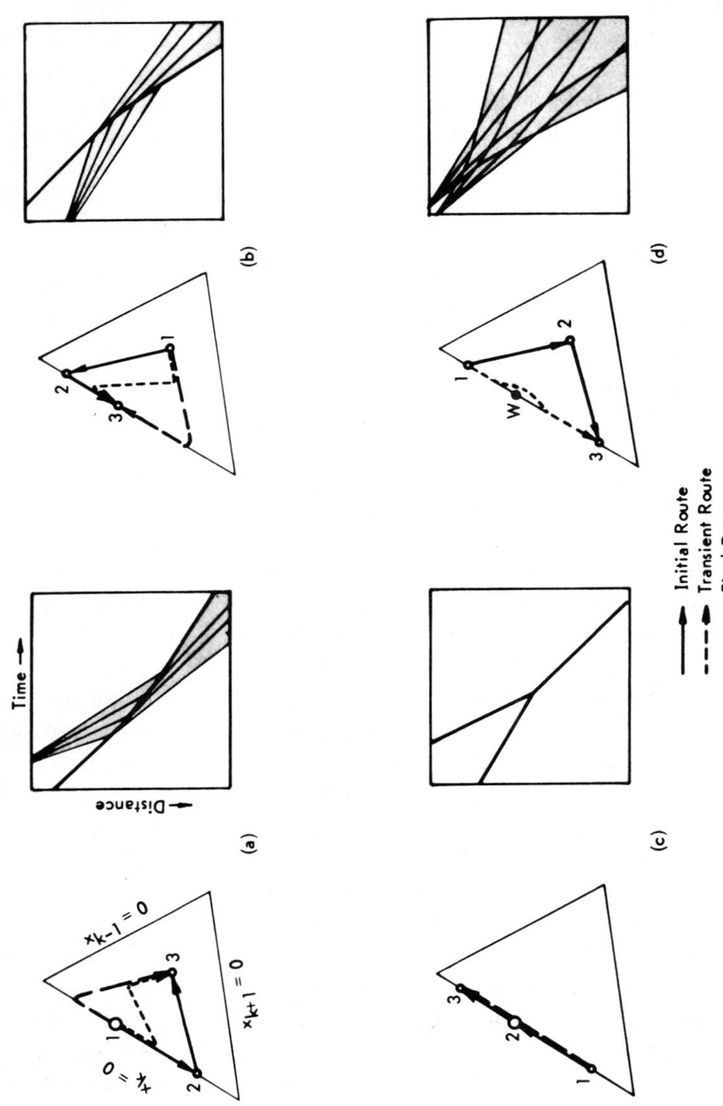


**Fig. 3.41.** Interference not resulting in crossover. Composition profile routes on common plane and root trajectories for interference of boundaries with variable $h_{k-1}$ and $h_k$ in cases involving composition on or near watershed of $x_k = 0$. Sharp boundaries shown as heavy lines; diffuse boundaries, shaded. (Schematic.)

ordinate border with one or several species $m$ absent and thus is "folded" by the $h$ transformation at its intersections with the watersheds $h_{m-1} = h_m = \alpha_{1m}$. These watersheds have no effect on the interference behavior since the species $m$, to whom they owe their existence, are entirely absent from the interfering boundaries or pulses. Thus, interference of boundaries or pulses with variable $h_i, \ldots, h_j$ and $h_k, \ldots, h_l$, $(j < k)$ results in a normal crossover, except that the situations illustrated in Fig. 3.41 can arise if $j = k-1$. The complete proof is straightforward, but involves cumbersome bookkeeping of root index numbers and will not be reproduced here.

### 2. *Boundaries or Pulses with Common Variable Roots*

Two coherent boundaries or pulses having the same variable root merge with one another if the intermediate plateau zone shrinks and disappears, as it does if either or both boundaries or pulse flanks bounding this zone are self-sharpening (see Section IV.F). Crossover does not occur. Rather, since the two initial boundaries or pulses necessarily have the same affinity cut, their merger produces a single coherent boundary or pulse of this cut. For example, two self-sharpening boundaries merge into one, and a self-sharpening and a nonsharpening boundary merge to form a pulse.

Still to be considered are cases in which either or both interfering boundaries or pulses involve root variations extending over several roots, of which one or more are common to both (variable roots $h_i, \ldots, h_j$ and $h_k, \ldots, h_l$, with $i \leqq j \geqq k \leqq l$). The boundaries or pulses then may or may not have the same bracketing affinity cut. If they do, they merge to form a single coherent boundary or pulse. If they do not, a more complex rearrangement resembling neither a crossover nor a merger takes place, as will now be shown.

The simplest cases of this type are interferences of two coherent boundaries, one with root variation extending over $h_{k-1}$ and $h_k$, and the other with either $h_{k-1}$ or $h_k$ as variable root and with a different affinity cut. The four possible situations of this type are illustrated in Fig. 3.42, which shows the composition routes on the common planes and their $h$ transforms. In all four cases, the initial intermediate plateau zone 2 shrinks and disappears, as is readily established with the rules derived in Section IV.F. The noncoherent boundary formed when the two initial boundaries merge involves variation of $h_{k-1}$ and $h_k$ and is resolved into separate variations of these two roots. The profile of the variable root common to both initial boundaries—$h_{k-1}$ in Fig. 3.42a and b, and $h_k$ in c and d—has a maximum or minimum, whereas that of the other root is monotonic.

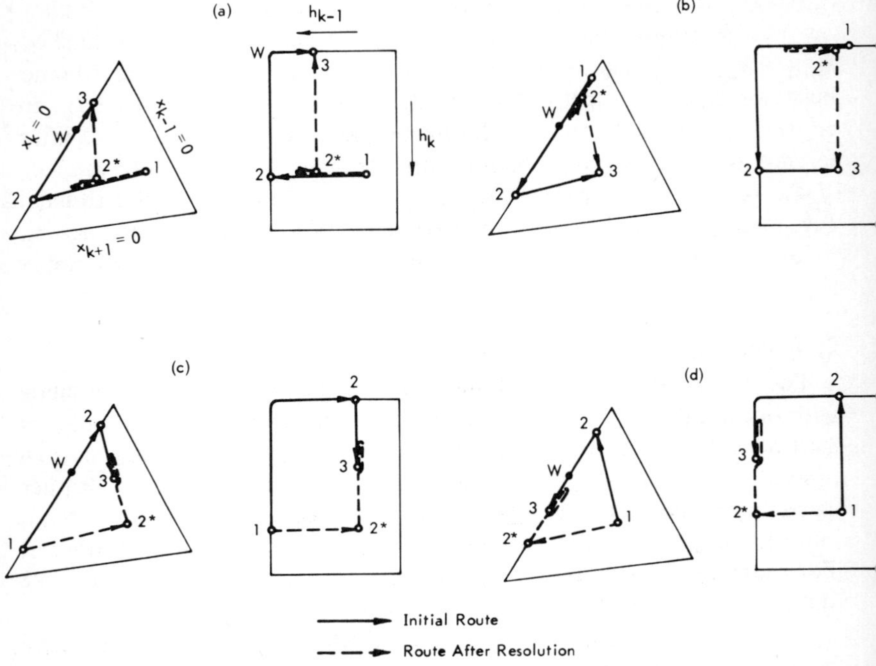


**Fig. 3.42.** Interference of boundaries with common variable root but different affinity cuts. Composition profile routes on common plane and its $h$ transform for four typical cases (schematic).

Resolution of the noncoherent boundary thus yields a coherent pulse and a coherent boundary. The pulse may be transient or permanent, depending on whether the value of the common variable root is higher or lower in zone 1 than in zone 3.

Essentially the same considerations apply if the interfering root variations initially constitute pulses rather than boundaries.

The interference behavior is the same if variations of more than two roots are involved. As indicated earlier for crossovers, such additional variations result merely from "folding" of the respective common plane by the $h$ transformation and do not affect the interference behavior. As a moderately complex example of this type, Fig. 3.43 shows the composition routes for interference of a boundary with 1 | 4 affinity cut (variable $h_1$, $h_2$, and $h_3$, initially upstream) and a pulse with 1 | 2 affinity cut (variable $h_1$, initially downstream) in a four-component system.

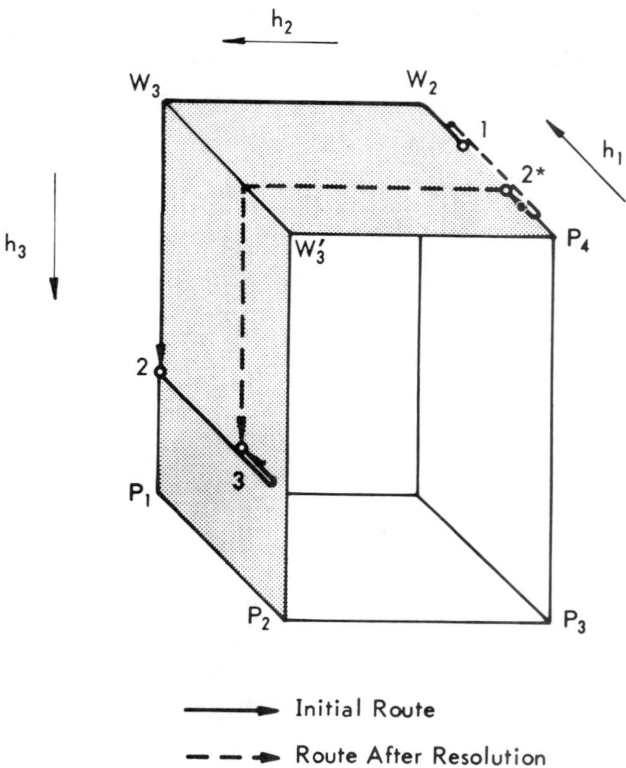


**Fig. 3.43.** Composition profile routes for interference of pulse with variable $h_1$ and boundary with variable $h_1$, $h_2$, and $h_3$ in $h$ space of four-component system (schematic).

## C. Multiple Interferences

Three or more coherent boundaries or pulses may interfere with one another simultaneously or successively. More detailed calculations are then required to establish the sequence of interferences and the transient behavior. However, the final route and plateau-zone compositions are readily determined by the simple methods given for the general case of noncoherence (see Section V.D). Since variation of a given root does not induce variations of others, the type of interference of any two root variations is the same regardless of whether or not other roots also vary. Simultaneous interference or any sequence of successive interferences of a given set of boundaries or pulses thus leads to the same final pattern.

As an example, Fig. 3.44 shows schematic root trajectories for simultaneous and two different sequential crossovers of three coherent boundaries or pulses with variable $h_j$, $h_k$, and $h_l$ $(j < k < l)$. The values of these roots in the various plateau zones are indicated. The final pattern and plateau-zone compositions are the same in all three cases.

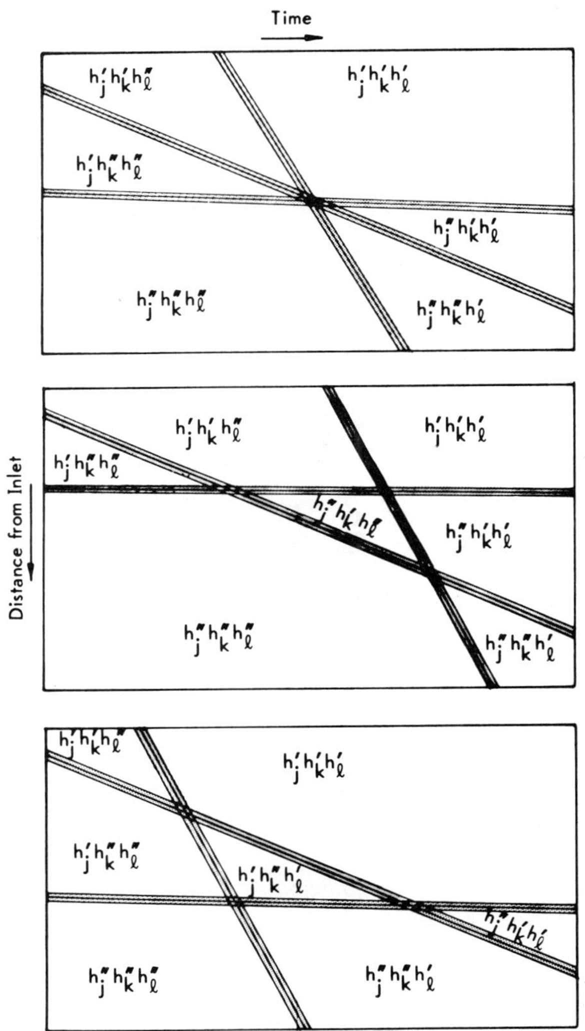


**Fig. 3.44.** Root trajectories for simultaneous and sequential crossovers of three coherent boundaries with variable $h_j$, $h_k$, and $h_l$ $(j < k < l)$. (Schematic.)

## D. Summary

The most important rules for interference of coherent boundaries or pulses can be summarized as follows:

In principle, the root variations of two interfering coherent boundaries or pulses may extend over one or several roots $h_i, \ldots, h_j$ and $h_k, \ldots, h_l$ ($i < j, k \leqq l$), of which one or more may or may not be common to both ($j \geqq k$ or $j < k$).

With the exception noted below, boundaries or pulses having no common variable root cross one another, that is, the "faster" boundary or pulse overtakes the "slower" one. Both retain the index numbers and upstream and downstream values of their variable roots as well as their sharpening characteristics, permanent or transient character, and multiplicity. No additional boundaries or pulses are generated. The velocity of a boundary or pulse is increased by crossover with a slower nonsharpening boundary or permanent pulse or with a faster self-sharpening boundary or transient pulse, and is decreased by crossover with a slower self-sharpening boundary or transient pulse or with a faster nonsharpening boundary or permanent pulse.

In exceptional cases a regular crossover does not occur even though the interfering boundaries or pulses have no common variable root. The exceptions, illustrated in Fig. 3.41, arise if watershed compositions with $h_j = h_k = \alpha_{1k}$ persist or are approached. The prerequisite is that the variable roots have adjacent index numbers (i.e., $j = k - 1$).

Two interfering boundaries or pulses having one or more common variable roots may or may not have the same affinity cut. If they do, they merge to form a single coherent boundary or pulse of the same cut. If they do not, they rearrange into a boundary and a pulse, or two pulses, without common variable roots, as illustrated in Figs 3.42 and 3.43.

Three or more coherent boundaries or pulses may interfere with one another simultaneously. The resulting final pattern is the same as would be produced by any sequence of successive interferences of the same set of initial boundaries or pulses.

The rules derived in this section will be particularly useful when responses to multiple influent composition changes are considered (see Chapter 4, Section IV).

To this point, we have examined elementary phenomena in columns without reference to particular initial and boundary conditions. Having completed this study we can now turn to the more practical problem of deducing how a column will respond under specified conditions.

## REFERENCES

1. J. N. Wilson, *J. Am. Chem. Soc.*, **62**, 1583 (1940).
2. D. DeVault, *J. Am. Chem. Soc.*, **65**, 532 (1943).
3. J. Weiss, *J. Chem. Soc.*, **1943**, 297.
4. J. E. Walter, *J. Chem. Phys.*, **13**, 229 (1945).
5. E. Glückauf, *Nature*, **156**, 748 (1945).
6. L. G. Sillén and E. Ekedahl, *Arkiv Kemi Mineral. Geol.*, **A 22**, No. 15 and No. 16 (1946).
7. T. Vermeulen, in *Advances in Chemical Engineering* (T. B. Drew and J. W. Hoopes, Jr., eds.), Vol. 2, Academic Press, New York, 1958, p. 147.
8. F. Helfferich, *Ion Exchange*, McGraw-Hill, New York, 1962, Chap. 9; *Angew. Chem. Intern. Ed. Engl.*, **1**, 440 (1962); *Chem. Ing.-Tech.*, **34**, 269 (1962).
9. N. K. Hiester, T. Vermeulen, and G. Klein, in *Chemical Engineers' Handbook* (R. H. Perry, C. H. Chilton, and S. D. Kirkpatrick, eds.), 4th ed., McGraw-Hill, New York, 1963, Sec. 16.
10. J. C. Giddings, *Dynamics of Chromatography*, Part I, Marcel Dekker, New York, 1965.
11. H. C. Thomas, *J. Am. Chem. Soc.*, **66**, 1664 (1944).
12. S. Goldstein, *Proc. Roy. Soc. (London)*, **A 219**, 151, 171 (1953).
13. E. Glueckauf, in *Ion Exchange and its Applications*, Society of Chemical Industry, London, 1955, p. 34.
14. A. L. LeRosen, *J. Am. Chem. Soc.*, **67**, 1683 (1945).
15. G. S. Bohart and E. Q. Adams, *J. Am. Chem. Soc.*, **42**, 523 (1920).
16. E. Wicke, *Kolloid-Z.*, **86**, 295 (1939).
17. R. H. Beaton and C. C. Furnas, *Ind. Eng. Chem.*, **33**, 1500 (1941).
18. E. Glückauf, *Proc. Roy. Soc. (London)*, **A 186**, 35 (1946).
19. E. Glueckauf, *Discussions Faraday Soc.*, **7**, 12 (1949).
20. L. G. Sillén, *Arkiv Kemi*, **2**, 477 (1950).
21. G. G. Baylé and A. Klinkenberg, *Rec. Trav. Chim.*, **73**, 1037 (1954).
22. A. C. Offord and J. Weiss, *Nature*, **155**, 725 (1945); *Discussions Faraday Soc.*, **7**, 26 (1949).
23. F. I. Stalkup and H. A. Deans, *A. I. Ch. E. J.*, **9**, 106 (1963).
24. D. O. Cooney and E. N. Lightfoot, *Ind. Eng. Chem. Process Design Develop.*, **5**, 25 (1966).
25. P. C. Mangelsdorf, Jr., *Anal. Chem.*, **38**, 1540 (1966).
26. G. Klein, D. Tondeur, and T. Vermeulen, *Ind. Eng. Chem. Fundamentals*, **6**, 339 (1967).
27. D. Tondeur and G. Klein, *Ind. Eng. Chem. Fundamentals*, **6**, 351 (1967).
28. L. D. Landau and E. M. Lifshitz, *Fluid Mechanics*, Pergamon Press, London, 1959, p. 367.
29. M. J. E. Golay, in *Gas Chromatography 1964* (A. Goldup, ed.), Institute of Petroleum, London, 1965, p. 143.
30. L. P. Eisenhart, *A Treatise on the Differential Geometry of Curves and Surfaces*, Constable, London, 1909, Chap. VII; reprinted by Dover, New York, 1960.
31. T. Muir, *The Theory of Determinants in the Historical Order of Development*, Vol. 2, Macmillan, London, 1911, p. 156.

32. N. R. Amundson, *Mathematical Methods in Chemical Engineering*, Prentice-Hall, Englewood Cliffs, 1966, Sec. 5.13.
33. H. Pleijel, *Z. Physik. Chem.*, **72**, 1 (1910).
34. R. Schlögl, *Z. Physik. Chem. (Frankfurt)*, **1**, 305 (1954).
35. V. P. Dole, *J. Am. Chem. Soc.*, **67**, 1119 (1945).
36. A. J. V. Underwood, *J. Inst. Petrol.*, **32**, 614 (1946); *Chem. Eng. Progr.*, **44**, 603 (1948).
37. A. A. Zhukhovitskii, M. L. Sazonov, A. F. Shlyakhov, and A. I. Karymova, *Zavodsk. Lab.*, **31**, 1048 (1965).
38. A. A. Zhukhovitskii, M. L. Sazonov, A. F. Shlyakhov, and V. P. Shvartsman, *Zh. Fiz. Khim.*, **41**, 2640 (1967).
39. J. P. Johnston and A. G. Ogston, *Trans. Faraday Soc.*, **42**, 789 (1946).
40. F. G. Helfferich, *Ind. Eng. Chem. Fundamentals*, **6**, 362 (1967).
41. F. Helfferich, in *Adsorption From Aqueous Solution* (Advances in Chemistry Series No. 79, R. F. Gould, ed.), American Chemical Society, 1968, p. 30.
42. Ref. 28, pp. 368–370.
43. F. Helfferich and G. Klein, to be published.

# 4

# COLUMN RESPONSE

The previous chapter has provided the tools with which column behavior under arbitrary conditions can be predicted. In particular, the methods presented in its Section V can be used to derive composition profiles and histories with no restrictions other than the premises stated at the outset. In many practical applications, however, the operating conditions are of a relatively simple type and allow simpler methods to be used. The present chapter will be devoted to such cases, with the aim of illustrating the application of principles derived earlier, of providing solutions and rules for special cases of practical interest, and of comparing the results of the present approach with existing theories of chromatographic behavior. No new concepts will be introduced, so that the reader may take up the various parts of this chapter selectively as the occasion arises.

The cases to be treated are of the following type. After uniform presaturation of the column with an influent of constant composition, the influent composition is varied. Single and multiple abrupt changes, gradual changes, and pulses will be considered. The familiar chromatographic techniques of frontal analysis, displacement development, elution development, and vacancy chromatography will appear as special cases.

The extension of the treatment to other types of operating conditions will be outlined in the last section.

## I. Single Abrupt Influent Composition Changes

The simplest, but for practical applications by no means unimportant, case is that of a single abrupt influent change from a presaturant of constant composition, with which the entire column has been equilibrated, to a different influent, also of constant composition. Many plant-scale and preparative sorption and ion-exchange columns operate under conditions approximating this situation when they receive, after practically complete regeneration, an influent of constant composition.

### A. Origin and Type of Response. The Proportionate Pattern

An influent composition change, whether abrupt or gradual, amounts to a variable composition history at the column inlet. In the type of operation discussed here, the variation is between two constant compositions, to be called *presaturant* and *influent*, and the influent composition history thus consists of two plateaus separated by a boundary. This is but a special case of the more general situation of an arbitrarily located initial boundary between plateau zones, discussed in detail in Chapter 3, Section V.

As a rule, the presaturant and influent compositions are not on the same composition path, so that the boundary generated at the column inlet is noncoherent. As shown in Chapter 3, Section V.B, such a boundary is resolved upon development into a set of coherent ones, each with a different variable $H$-function root. Thus, in general, a single composition change at the column inlet gives rise to a multiplicity of coherent boundaries.

Multiple responses to single influent composition changes are a general feature of chromatographic behavior rather than one caused by mutual interference of species. In systems with noninterfering species, multiple boundaries or pulses arise as the initially simultaneous concentration changes of the various species are propagated independently and at different rates and thus are resolved from one another. The complication caused by interference is that variations of $H$-function roots rather than of species concentrations are resolved from one another. As a result, each response boundary involves variations of the concentrations of *all* species present locally, including even those which might have been held constant at the inlet, and the number of response boundaries depends on the number of $H$-function roots, not of species concentrations, varied in the course of the composition change at the inlet.

These considerations apply regardless of whether the composition change at the column inlet is executed gradually or abruptly. The faster this change, the sharper is the initial noncoherent boundary it generates, and therefore, as shown in Chapter 3, Section V.E, the shorter are the develop-

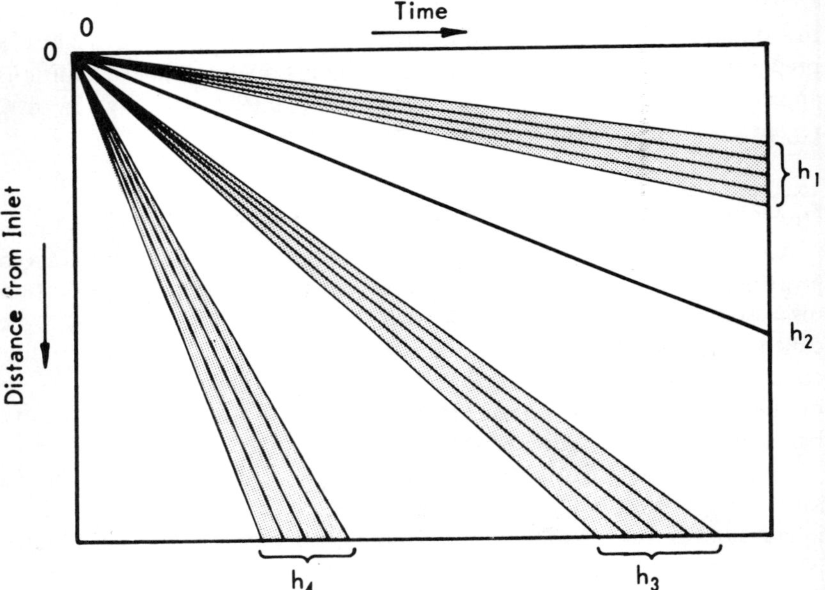


**Fig. 4.1.** Typical distance-time diagram for response to single abrupt influent composition change. Five-component system with one self-sharpening and three nonsharpening response boundaries.

ment time and distance required for resolution into coherent boundaries. In the limiting case of a step change, the initial boundary is ideally sharp and its resolution is instantaneous, so that the response pattern is coherent from the outset. This behavior is illustrated in Fig. 4.1, which shows composition or root trajectories for a five-component system. All trajectories originate from the same point $z = 0$, $t = 0$ (column inlet, and time of composition change). Furthermore, all trajectories are linear because the compositions and steps advance, from the outset, at their constant composition-velocity eigenvalues and step velocities. Because of

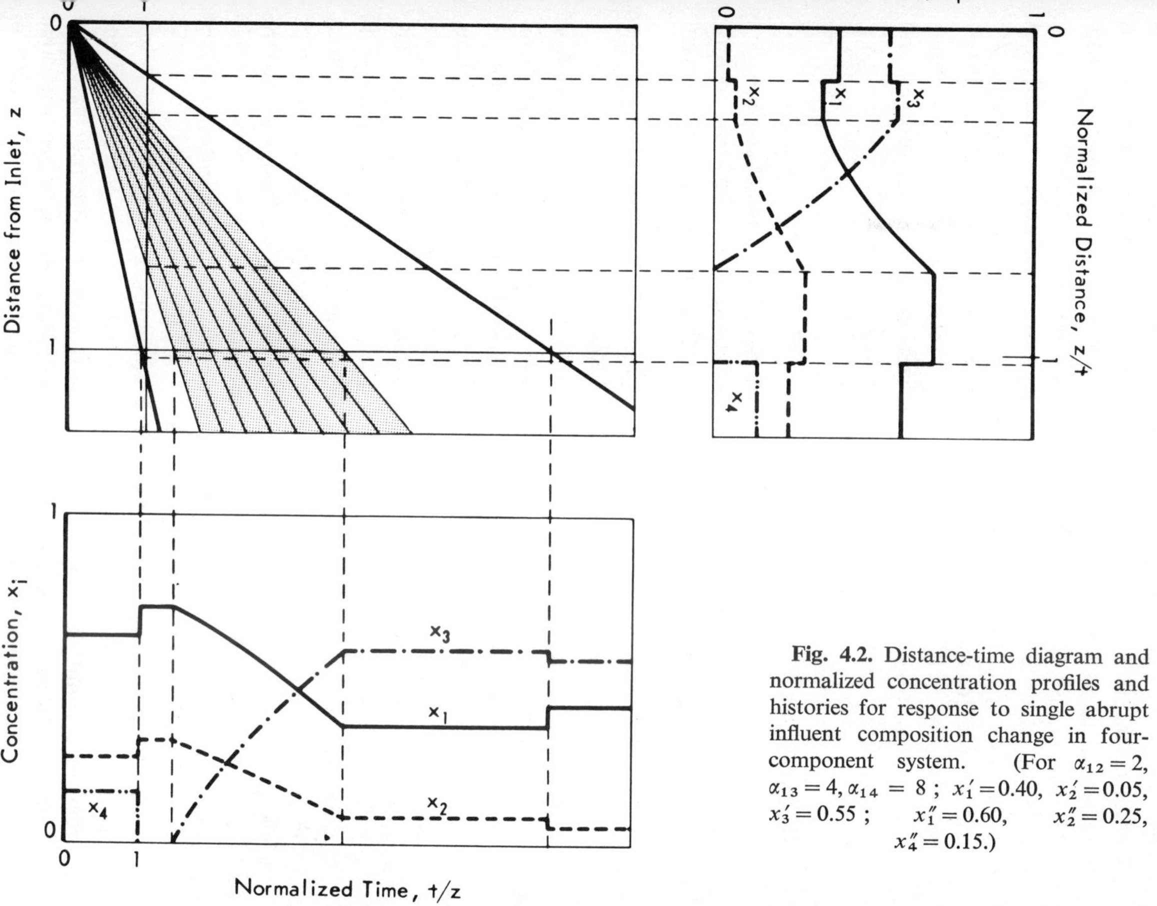


**Fig. 4.2.** Distance-time diagram and normalized concentration profiles and histories for response to single abrupt influent composition change in four-component system. (For $\alpha_{12}=2$, $\alpha_{13}=4, \alpha_{14}=8$; $x_1'=0.40$, $x_2'=0.05$, $x_3'=0.55$; $x_1''=0.60$, $x_2''=0.25$, $x_4''=0.15$.)

these properties, the spread of the pattern, as measured by the widths of the plateau zones and boundaries, increases in proportion to the traveled distance or elapsed time [1–4]. For such behavior, the term *proportionate pattern* has been coined [1]. For each composition or step in a proportionate pattern, the traveled distance $z$ is proportional to the elapsed time $t$, so that $z/t$ has a constant value (equaling the composition or step velocity). Thus, composition profiles for different times $t$ coincide if plotted versus a normalized distance $z/t$, and so do composition histories at different distances $z$ if plotted versus a normalized time $t/z$. These properties are well-known for systems with noninterfering species. As shown here, they are not impaired by mutual interference of the species. As an example, the composition route and the normalized concentration profiles and histories for a four-component case are shown in Fig. 4.2.

## B. Response Patterns

The response patterns resulting from single abrupt influent composition changes obey a number of simple qualitative rules, which are readily derived by application of the more general treatment of noncoherent boundaries to this special case.

### 1. *Number, Type, and Sequence of Boundaries*

It has been shown that noncoherent boundaries in general are resolved into separate variations of the $H$-function roots, which arrange themselves in numerical sequence of their index numbers (see Chapter 3, Section V). In principle, each such root variation may, after resolution, constitute a single or several separate coherent boundaries or pulses, depending on the manner in which the root varied initially (see Fig. 3.29). In the special case of an ideally sharp, noncoherent initial boundary, as generated by an abrupt composition change at the inlet, all initial root variations are monotonic and without intermediate plateaus and thus each give rise to a single coherent boundary. Hence, the response pattern consists exclusively of coherent boundaries, each with a different variable root, and arranged in numerical sequence of the index numbers of their variable roots (for singular cases, see Section I.B.4).

The maximum number of response boundaries is $n-1$, since there are $n-1$ roots, which may all vary. Fewer boundaries are generated if the presaturant and influent compositions have one or more root values in common. In the extreme case, both these compositions are on the same composition path, and only one response boundary appears.

Each response boundary may be self-sharpening or nonsharpening, depending on the presaturant and influent compositions. Since the response pattern is a "proportionate" one, all self-sharpening boundaries are sharp from the outset and all nonsharpening ones start to spread immediately. Also, all plateau zones lengthen as they travel.

### 2. *H-Function Roots and Sharpening Characteristics*

Rules and criteria concerning the sharpening characteristics of boundaries, and the absence of species from portions of the pattern, generated by an abrupt influent composition change are conveniently derived from the sets of $H$-function roots of the various plateau zones. To conform with the notation used in Chapter 3, Section V, we use primes and double primes to refer to the influent and presaturant, respectively, and number the plateau zones in the direction of flow in such a way that the boundary with variable $h_k$ is between zones $k$ and $k+1$; zones 1 and $n$ then are the portions of the column in equilibrium with the influent $\mathbf{x}'$ and presaturant $\mathbf{x}''$, respectively. With this notation, the sets of roots are

$$\begin{array}{ll} \text{Zone } 1: & h_1', h_2', \ldots, h_{k-1}', h_k', \ldots, h_{n-2}', h_{n-1}' \\ \text{Zone } 2: & h_1'', h_2', \ldots, h_{k-1}', h_k', \ldots, h_{n-2}', h_{n-1}' \\ \quad\vdots & \qquad\qquad\qquad\vdots \\ \text{Zone } k: & h_1'', h_2'', \ldots, h_{k-1}'', h_k', \ldots, h_{n-2}', h_{n-1}' \\ \quad\vdots & \qquad\qquad\qquad\vdots \\ \text{Zone } n-1: & h_1'', h_2'', \ldots, h_{k-1}'', h_k'', \ldots, h_{n-2}'', h_{n-1}' \\ \text{Zone } n: & h_1'', h_2'', \ldots, h_{k-1}'', h_k'', \ldots, h_{n-2}'', h_{n-1}'' \end{array} \tag{3.92}$$

as was derived in Chapter 3, Section V.D.2.

The sharpening characteristics of the boundaries can now be established with the root sets (3.92) and condition (3.74). Since its variable root is $h_k$, the $k$th boundary (between the zones $k$ and $k+1$) is

$$\begin{array}{lll} \text{self-sharpening} & \text{if} & h_k' > h_k'' \\ \text{nonsharpening} & \text{if} & h_k' < h_k'' \end{array} \tag{4.1}$$

If the root value is the same in the influent and presaturant, the response pattern does not contain a variation of the root, and the respective boundary is absent. Thus, the $k$th boundary is

$$\text{nonexistent} \quad \text{if} \quad h_k' = h_k'' \tag{4.2}$$

The plateau zones $k$ and $k+1$, normally separated by this boundary, then have identical compositions and appear as a single zone.

3. *Absence of Species from Portions of the Pattern*

The root sets (3.92) can also be used to establish which species, if any, are absent from portions of the pattern generated by an abrupt influent composition change, as will now be shown.

In terms of *H*-function roots, the necessary and sufficient condition for the absence of an arbitrary species $k$ is

$$\begin{aligned} &k_{k-1} = \alpha_{1k} \quad \text{or} \quad h_k = \alpha_{1k} && \text{if} \quad x_k = 0 \qquad (k \neq 1, n) \\ &h_1 = \alpha_{11} \equiv 1 && \text{if} \quad x_1 = 0 \\ &h_{n-1} = \alpha_{1n} && \text{if} \quad x_n = 0 \end{aligned} \tag{3.55}$$

(see Chapter 3, Section IV.D.2). Hence, if a species $k$ is present in both the influent and the presaturant, none of the root values $h'_{k-1}$, $h'_k$, $h''_{k-1}$, and $h''_k$ equals $\alpha_{1k}$. According to the root sets (3.92), no other than these values of $h_{k-1}$ and $h_k$ appear in any of the plateau zones of the response pattern, so that condition (3.55), necessary for absence of species $k$, is not met in any of these zones either. Thus, a species present in both the influent and presaturant is also present in all zones of the response pattern. The negative parallel is also true: Obviously, a species absent from both the influent and presaturant is entirely absent from the system.

For the case of a species $k$ being absent from the influent but present in the presaturant, condition (3.55) shows that $h'_{k-1}$ or $h'_k$ equals $\alpha_{1k}$ whereas $h''_{k-1}$ and $h''_k$ do not. According to the root sets (3.92) and condition (3.55), species $k$ then is absent from the first $k-1$ or $k$ plateau zones and is present in all others. Similarly, if species $k$ is present in the influent but absent from the presaturant, one finds this species to be present in the first $k-1$ or $k$ plateau zones and absent from all others. In either case, species $k$ is present in all zones on one side, and absent from all zones on the other side, of the $(k-1)$th or $k$th boundary. Thus, pattern portions from which a species is absent cannot alternate with others in which this species is present.

We shall adopt the convention of saying a species "appears" at a boundary if it is absent from the upstream side but is present on the downstream side, and a species "disappears" if the reverse is true; thus, the terms in their literal sense apply to composition profiles viewed in the direction of flow, not to composition or effluent histories viewed in the direction of progressing time.

The considerations above show that no species can appear or disappear at more than one place in any given pattern, and that a species $k$ absent from the influent or presaturant appears or disappears at either of two boundaries, the $(k-1)$th or the $k$th, depending on whether $h_{k-1}$ or $h_k$

equals $\alpha_{1k}$ where species $k$ is absent. One can distinguish between these alternatives without calculating $H$-function roots, since the root equalling $\alpha_{1k}$ can be identified by means of condition (3.64). Thus, a species $k$ absent from the influent appears

$$\begin{aligned} &\text{at the } (k-1)\text{th boundary} \qquad \text{if } \sum_{i\neq k} \frac{x'_i}{1-\alpha_{ki}} > 0 \\ &\text{at the } k\text{th boundary} \qquad \text{if } \sum_{i\neq k} \frac{x'_i}{1-\alpha_{ki}} \leqq 0 \end{aligned} \tag{4.3}$$

since condition (3.64) shows the root equalling $\alpha_{1k}$ to be $h'_{k-1}$ in the first case, and $h'_k$ in the second. [This criterion includes the case $\Sigma_{i\neq k}$ $[x'_i/(1-\alpha_{ki})]=0$, for which condition (3.63) gives $h'_{k-1}=h'_k=\alpha_{1k}$; species $k$ then remains absent until both $h_{k-1}$ and $h_k$ have varied from their influent values, and thus appears at the $k$th rather than the $(k-1)$th boundary.] Similarly, a species $k$ absent from the presaturant disappears

$$\begin{aligned} &\text{at the } (k-1)\text{th boundary} \qquad \text{if } \sum_{i\neq k} \frac{x''_i}{1-\alpha_{ki}} \geqq 0 \\ &\text{at the } k\text{th boundary} \qquad \text{if } \sum_{i\neq k} \frac{x''_i}{1-\alpha_{ki}} < 0 \end{aligned} \tag{4.4}$$

Species 1 and $n$ can appear and disappear only at the first and last boundary, respectively, since there is no $(k-1)$th boundary for $k=1$ and no $k$th boundary for $k=n$.

A consequence of these rules is that the $k$th boundary, whose variable root is $h_k$, may involve the appearance or disappearance of species $k$ or $k+1$ if these are absent from the influent or presaturant. Species $k$ may be absent from one side of this boundary and species $k+1$ from the other, since $h_k$ may vary from $\alpha_{1k}$ to $\alpha_{1,k+1}$ across the boundary. However, the two species may not both be absent from the same side while being present on the other side, because this would require $h_k$ to equal $\alpha_{1k}$ as well as $\alpha_{1,k+1}$ where both species are absent. These conclusions are in agreement with the general rules for coherent boundaries (see Chapter 3, Section IV.B). For cases with several species absent from the influent or presaturant, it follows that these species appear or disappear at different boundaries and in the sequence of decreasing affinity (counted in the direction of flow).

The physical cause for this sequence of appearance or disappearance of species is plain: of a group of species jointly introduced into a column presaturated with others, those with lower affinities have higher species velocities at any point in the column, and their front boundaries should therefore advance more rapidly, i.e., be farther downstream in the pattern. Similarly, of a group of species jointly being eluted from the column, those with lower affinities and thus higher species velocities should have their rear boundaries, again, farther downstream. This reasoning is applicable regardless of whether or not the species being introduced or eluted are accompanied by others present in both the presaturant and the influent.

#### 4. *Singular Cases Involving Compositions on Watersheds*

It has so far been taken for granted that the initial noncoherent boundary is completely resolved into separate variations of the different $H$-function roots. This requires proof, however, since the discussion of the general case of development of a diffuse noncoherent boundary has shown that deviations from the normal resolution behavior can be caused by compositions on or near watersheds (see Chapter 3, Section V.F). The proof is supplied by the following examination, which shows that the response pattern generated by an abrupt composition change cannot involve compositions on a watershed unless the composition of the influent or presaturant is on this watershed, and that the root variations are completely resolved even in such singular cases.

For a composition point on the watershed of the arbitrary border $x_k = 0$ $(k \neq 1, n)$, condition (3.62) requires both $h_{k-1}$ and $h_k$ to equal $\alpha_{1k}$. In a response pattern produced by an abrupt composition change and thus with monotonic variations of the roots, $h_{k-1}$ and $h_k$ cannot equal $\alpha_{1k}$ unless they do so in the influent or presaturant, since $\alpha_{1k}$ is the highest, or lowest, value which these roots can assume. For compositions with $h'_{k-1} = h''_k = \alpha_{1k}$ or $h''_{k-1} = h'_k = \alpha_{1k}$, condition (3.61) would require species $k$ to be absent from both the influent and presaturant and thus from the entire system. Hence, provided entirely absent species are not counted as components, compositions on the watershed of $x_k = 0$ can only occur if

$$h'_{k-1} = h'_k = \alpha_{1k} \qquad \text{or} \qquad h''_{k-1} = h''_k = \alpha_{1k} \tag{4.5}$$

In these singular cases, the influent or presaturant composition is on the watershed, and so is the composition route to or from the point at which one of the roots varies, that is, from the influent point to the $(k-1)$th boundary or from the $k$th boundary to the presaturant point. Even here, in the only cases involving watersheds at all, the watershed is not crossed by the composition route.

In the cases (4.5) the initial noncoherent boundary has a watershed composition on one or the other side. As the examination of such cases in Chapter 3, Section V.F.1 has shown, complete resolution of the root variations takes place, although the variations of $h_{k-1}$ and $h_k$ remain directly adjacent to one another rather than becoming separated by a plateau zone (see Figs. 3.35 and 3.36, a and b). Thus, the response pattern does not contain a plateau zone $k$, since the boundaries with variable $h_{k-1}$ and $h_k$, normally separated by this zone, remain adjacent. These two boundaries have opposite sharpening characteristics, because $h_{k-1}$ and $h_k$ vary in opposite directions, $\alpha_{1k}$ being the maximum value for one and the minimum value for the other root. In consequence, the concentrations of all species other than $k$ have maxima or minima between the two boundaries. The profile portion consisting of the two boundaries thus somewhat resembles a coherent pulse, except that the pulse amplitude does not decline and that one species, $k$, is present in only one flank. As examples, composition routes and profiles of two three-component cases are shown in Fig. 4.3.

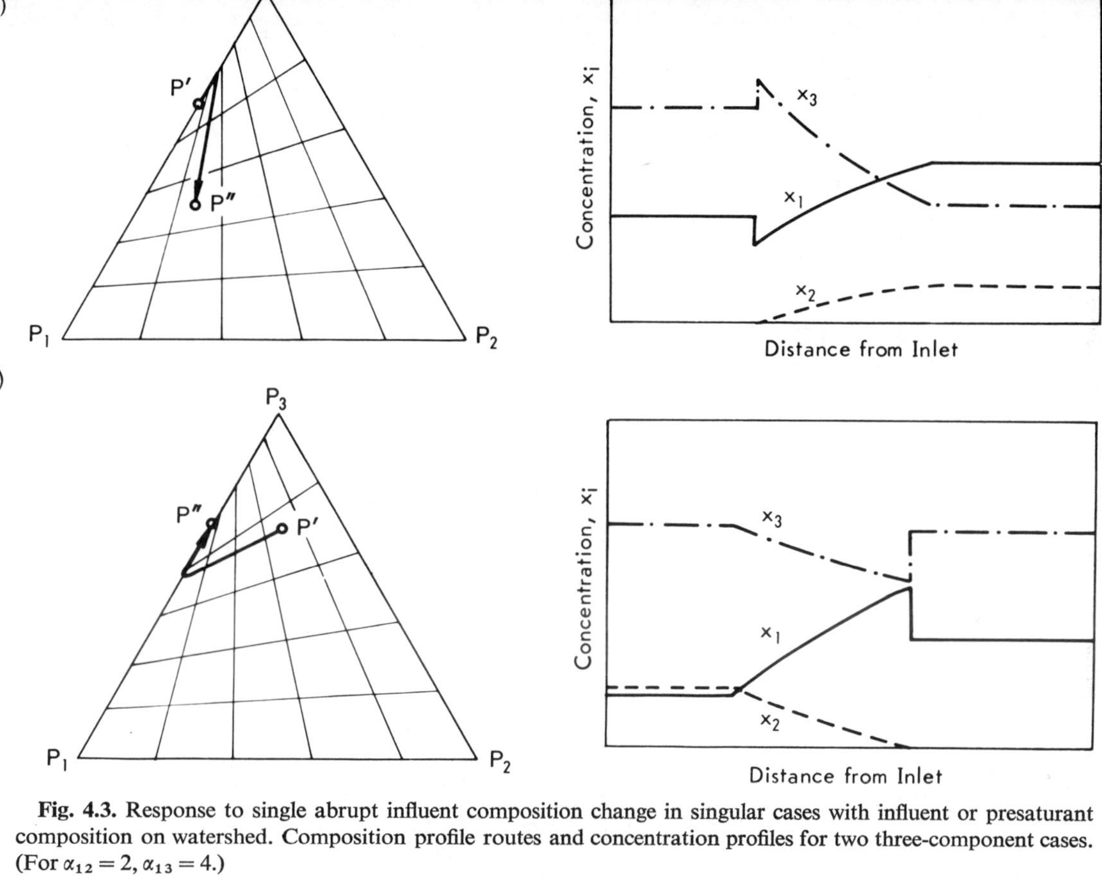


**Fig. 4.3.** Response to single abrupt influent composition change in singular cases with influent or presaturant composition on watershed. Composition profile routes and concentration profiles for two three-component cases. (For $\alpha_{12} = 2, \alpha_{13} = 4$.)

No $H$-function roots need be calculated to establish whether the composition of the influent or presaturant falls on a watershed. According to condition (3.63), the watershed on the arbitrary border $x_k = 0$ is given by

$$\sum_{i \neq k} \frac{x_i}{1-\alpha_{ki}} = 0, \qquad x_k = 0 \tag{4.6}$$

The influent or presaturant composition thus falls on this watershed if it obeys this relation.

It is now easy to show that complete resolution into coherent boundaries with variation of only one root each does indeed take place. Incomplete resolution of two boundaries (permanent noncoherence, see Chapter 3, Section V.F.2) can be ruled out because it would require a species to be present within the initial boundary but absent from both its sides, a requirement which a sharp initial boundary obviously cannot meet. Complete resolution into coherent boundaries thus is assured. This alone would not prevent the pattern from containing a coherent boundary whose root variation extends over more than one root, say, over $h_{k-1}$ and $h_k$, as could be the case if the affinity cut were bracketing (see Table 3.1 in Chapter 3, Section IV.D.5). However, since the roots vary monotonically, this would require $h'_{k-1} = h''_k = \alpha_{1k}$ or $h''_{k-1} = h'_k = \alpha_{1k}$, and in either case species $k$ would be entirely absent from the system. Hence, again provided that absent species are not counted as components, the pattern cannot involve a boundary with variation of more than one root. Moreover, with the exception of the singular cases (4.5) discussed earlier, all boundaries become separated from one another by plateau zones; this follows from the fact that watershed compositions, which alone can cause a different behavior, are absent from the pattern in all other cases.

### 5. *Affinity Cuts*

The affinity cuts of the boundaries in the response to an abrupt influent composition change are readily established once it is known which species, if any, are absent from the various plateau zones.

As shown earlier, the cut of a boundary with variable $h_k$ is $k \mid k+1$, with the exception that species $k$ or $k+1$ or both, or a group of species with successive index numbers including $k$ or $k+1$ or both, are bracketed by the cut if they are absent from the boundary and both its sides (see Chapter 3, Section IV.D.5). Thus, if all species are present in all plateau zones, all boundaries have nonbracketing cuts; if, for each boundary, the species present are listed and the cut is marked by a vertical line, the pattern looks as shown in (4.7).

If any species are absent from portions of the pattern, bracketing cuts may, but need not, occur. Which boundaries, if any, have bracketing cuts becomes apparent if all species absent at the respective boundaries are omitted from the pattern (4.7). As an example, a possible pattern for a six-component case with species 1, 2, and 5 as the influent and species 2, 3, 4, and 6 as the presaturant is shown in (4.8).

| Influent | 1 | 2 | 3 | ... | $k$ | $k+1$ | ... | $n-1$ | $n$ |
|---|---|---|---|---|---|---|---|---|---|
| 1st boundary | 1 \| | 2 | 3 | ... | $k$ | $k+1$ | ... | $n-1$ | $n$ |
| 2nd boundary | 1 | 2 \| | 3 | ... | $k$ | $k+1$ | ... | $n-1$ | $n$ |
| ⋮ | | | | | ⋮ | | | | |
| $k$th boundary | 1 | 2 | 3 | ... | $k$ \| | $k+1$ | ... | $n-1$ | $n$ |
| ⋮ | | | | | ⋮ | | | | |
| $(n-1)$th boundary | 1 | 2 | 3 | ... | $k$ | $k+1$ | ... | $n-1$ \| | $n$ |
| Presaturant | 1 | 2 | 3 | ... | $k$ | $k+1$ | ... | $n-1$ | $n$ |

(4.7)

| Influent | 1 | 2 | | | 5 | |
|---|---|---|---|---|---|---|
| 1st boundary | 1 \| | 2 | | | 5 | |
| 2nd boundary | | 2 \| | | | 5 | |
| 3rd boundary | | 2 | 3 \| | | 5 | |
| 4th boundary | | 2 | 3 | 4 \| | 5 | |
| 5th boundary | | 2 | 3 | 4 | 5 \| | 6 |
| Presaturant | | 2 | 3 | 4 | | 6 |

(4.8)

In this example, the second and third boundaries are seen to have bracketing cuts 2 | 5 and 3 | 5, respectively, since the species bracketed by these cuts are absent.

### 6. *Composition Routes*

For cases with not too many components, composition routes can provide a readily perceived overall picture of the column response and, furthermore, are a convenient means of obtaining graphical solutions.

Routes for responses to abrupt influent composition changes are readily constructed. Since a final coherent pattern is attained instantaneously (see Section I.A), the route runs exclusively along composition paths and does not vary with time. The sequence of paths along the route is that of the boundaries in the pattern, i.e., of increasing index numbers of the variable roots (counted in the direction of flow). The orthogonality of the paths in the $h$ space makes it obvious that there is one and only one such route connecting any pair of given influent and presaturant composition points.

One may represent the composition change at the column inlet by a vector pointing from the influent to the presaturant composition; development can then be said to resolve this vector into its components in the directions of the $h$ coordinates, the composition route being the head-to-tail sequence of the component vectors in the order of increasing root index numbers. The route does not cross any watershed, as the earlier discussion of singular cases has shown.

Examples of two- and three-dimensional representations of composition routes appear in Figs. 4.3, 4.5, and 4.7, in which the $x$ rather than $h$ space

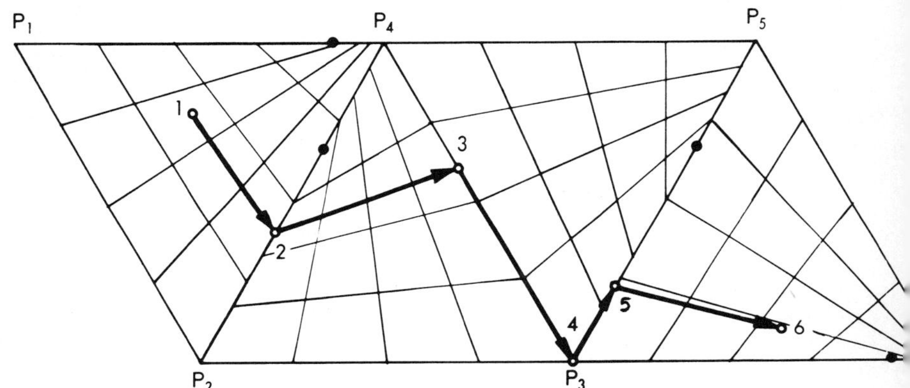


**Fig. 4.4.** Two-dimensional composition-route diagram for response to single abrupt influent composition change in six-component system with species 1, 2, and 4 in influent and 3, 5, and 6 in presaturant. (For $\alpha_{12} = 2, \alpha_{13} = 4, \alpha_{14} = 8, \alpha_{15} = 16, \alpha_{16} = 32$.)

has been chosen for greater ease of translation into concentration profiles and histories. In some cases with up to six components, the route can be displayed in simple two-dimensional diagrams, namely, if no more than three species coexist at any point along the route. As an example, the route of a six-component case with species 1, 2, and 4 in the influent and species 3, 5, and 6 in the presaturant is shown in Fig. 4.4, in which the contiguous triangles constitute the composition diagrams of the respective subordinate three-component systems. More complex systems can be handled with analogous, somewhat more elaborate constructions involving common planes.

The previously stated rules that the composition route follows the paths in the sequence of increasing index numbers of the variable roots and does not cross any watershed are subject to the provision that absent species are not counted as components. For example, the route for a case with only two species, 1 and 3, if plotted in the path grid of a three-component system $\{1, 2, 3\}$, may cross this system's watershed point, which owes its existence merely to the absent species 2. Such a case is shown in Fig. 4.5a; it is true that the three-component path grid provides a route 1 1′ 2 which bypasses the watershed point and has the prescribed sequence of root variations, but this route involving species 2 can obviously not be taken by a system from which species 2 is absent.

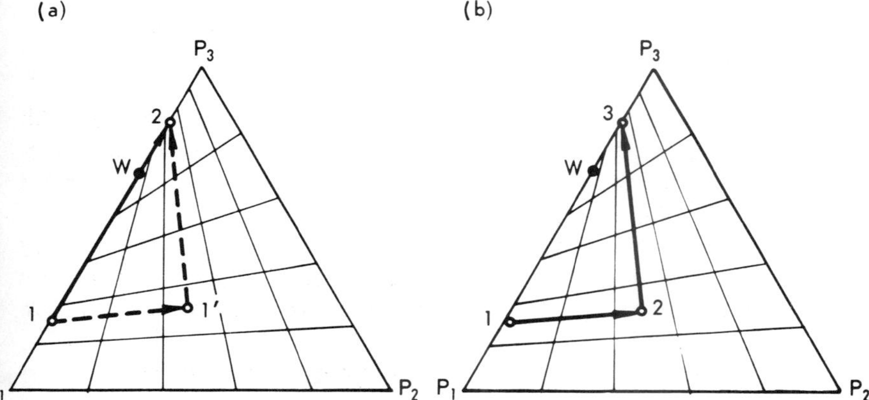


**Fig. 4.5.** Effect of trace species on composition route. Profile routes in composition simplex of three-component system. (a) Route in absence of species 2; (b) route resulting if traces of species 2 are added to influent and presaturant.

The route shown in Fig. 4.5a is altered drastically if even only a trace of species 2 is added to the influent, the presaturant, or both. As illustrated in Fig. 4.5b, the route then bypasses the watershed point, and an intermediate plateau zone rich in species 2 is formed. This effect is surprising but real and serves to show that, under suitable conditions, a trace species can be highly accumulated in an intermediate zone. In the case shown in Fig. 4.5b, conditions are such that species 2 in zone 1 has a velocity higher than that of the sharp front boundary of this zone, and is therefore accumulated at this boundary. In keeping with the very low concentration of species 2 in the influent, the zone rich in species 2 is very narrow (see Section I.E). Except for the generation of this narrow zone, the addition of a trace of species 2 has not significantly changed the concentration profiles.

### 7. *Summary*

The rules so far derived for responses to single abrupt influent composition changes can be consolidated as follows:

1. An abrupt composition change at the column inlet generates a response pattern which is completely coherent from the outset and is "proportionate," i.e., the widths of all boundaries and plateau zones increase in proportion to the traveled distance or elapsed time.
2. The pattern consists exclusively of boundaries and plateau zones, i.e., does not contain composition pulses.
3. The usual, and maximum, number of boundaries is $n-1$. Fewer boundaries arise if the influent and presaturant have one or more values of $H$-function roots in common.
4. For each boundary, the root variation extends over only one root. All boundaries have different variable roots. The sequence of the boundaries is that of increasing index numbers of their variable roots (counted in the direction of flow).
5. Successive boundaries are separated by plateau zones, with only this exception: If the composition of the influent or presaturant falls on the watershed of $x_k = 0$, no plateau zone forms between the $(k-1)$th and $k$th boundaries.
6. Species present in both the influent and presaturant are present everywhere in the response pattern. Any species $k$ absent from the influent or presaturant is absent from all zones on one side, and present in all zones on the other side, of the $(k-1)$th or $k$th boundary. Thus, pattern portions from which a species is absent cannot alternate with others in which this species is present. Also, two species cannot simultaneously be absent from the same side of a boundary while being present on the other side. Thus, if several species are absent from the influent or presaturant, they appear or disappear at different boundaries and in the sequence of decreasing affinity (counted in the direction of flow).
7. Each boundary involves concentration changes of all species, including those whose concentrations may have been held constant at the inlet.

The last rule does not apply in trace-component systems, as will be discussed in Section I.G.

## C. Prediction of Response Patterns

For any given compositions of the presaturant and influent, the qualitative features of the response pattern (number, sharpening characteristics, and affinity cuts of the boundaries, and species present in or absent from the various plateau zones) generated by an abrupt influent composition change are readily predicted with the rules derived in the previous section, as will now be reviewed.

## 1. *Shorthand Notation*

For convenience, a shorthand notation for characterizing cases and response patterns will be introduced.

To denote the basic compositions of the influent and presaturant, the species constituting the influent are written on the left of an arrow pointing toward those constituting the presaturant. For example, the case leading to the pattern (4.8) would appear as

$$1, 2, 5 \rightarrow 2, 3, 4, 6$$

Response patterns are written as follows: Each boundary is represented by a letter, "S" for self-sharpening, or "N" for nonsharpening; any species disappearing at the boundary (i.e., present upstream but absent downstream) is indicated as superscript, and any species appearing (absent upstream but present downstream), as a subscript; the boundaries are stated in sequence, counted in the direction of flow, with any "nonexistent" ones (i.e., boundaries missing from the pattern because the respective *H*-function root is constant) marked by dashes. For example, the pattern (4.8) could be

$$\mathrm{S^1S\ N_3N_4S_6^5}, \qquad \mathrm{S^1N\ N_3N_4S_6^5}, \qquad \text{or} \qquad \mathrm{S^1\text{–}\ N_3N_4S_6^5}$$

depending on whether its second boundary is self-sharpening, nonsharpening, or nonexistent. Even this brief notation is not free of possible redundance, since the sharpening characteristics of boundaries at which species appear or disappear can be inferred from the general rules for coherent boundaries (see Chapter 3, Section IV.B), from which it follows that a boundary is self-sharpening if it involves the disappearance of a high-affinity species or appearance of a low-affinity species, and is nonsharpening in the opposite cases.

## 2. "*Unique*" *Patterns*

In a few cases, all one needs to know to predict the qualitative pattern is which species constitute the influent and presaturant; the species concentrations need not be given since they do not affect the pattern. The cases with such "unique" (i.e., concentration-independent) patterns are of the type

$$1, \ldots, j \quad \rightarrow \quad j+1, \ldots, n \qquad (1 \leqq j \leqq n) \tag{4.9}$$

or

$$j+1, \ldots, n \quad \rightarrow \quad 1, \ldots, j \qquad (1 \leqq j \leqq n) \tag{4.10}$$

that is, the influent and presaturant have no species in common, and the influent species have all higher affinities, or all lower affinities, than the presaturant species. Regardless of the species concentrations, the response pattern for the case (4.9) is

$$S^1 \ldots S^{j-1} S^j_{j+1} S_{j+2} \ldots S_n \tag{4.11}$$

and for the case (4.10),

$$N_1 \ldots N_{j-1} N_j^{j+1} N^{j+2} \ldots N^n \tag{4.12}$$

This is readily shown with the rules that any species $k$ can appear or disappear only at the $(k-1)$th or $k$th boundaries, and that at most one species can appear, or can disappear, at any boundary (see Section I.B.3). In the case (4.9), species 1 through $j$, being present in the influent but absent from the presaturant, must disappear along the profile. Species 1 can do so only at the first boundary. This precludes the disappearance of any other species at this boundary, so that species 2 can disappear only at the second boundary. By the same token, any further species $i \leqq j$ can disappear only at the $i$th boundary. A similar argument pertains to the appearing species $j+1$ through $n$; each such species $i > j$ can appear only at the $(i-1)$th boundary. All boundaries are self-sharpening because each involves disappearance of a high-affinity species or appearance of a low-affinity species. The resulting pattern thus is as stated in (4.11). Analogous considerations for the case (4.10) lead to the pattern (4.12).

The affinity cuts of the boundaries are readily found if the patterns are written in their longhand forms [i.e., as in (4.7) and (4.8)]. In the case (4.9), all boundaries have nonbracketing cuts $i \mid i+1$ $(1 \leqq i < n)$, whereas in the case (4.10) the cuts are $i \mid j+1$ $(1 \leqq i \leqq j)$ and $j \mid i$ $(j < i \leqq n)$.

A special case of the type (4.9), namely, frontal analysis, is of practical importance and will be discussed in more detail in Section I.F.

### 3. *Composition-Dependent Patterns*

In all cases other than those examined above, the response patterns are not "unique" but, rather, depend on the relative species concentrations (or values of the $H$-function roots) in the influent and presaturant. Consider, for example, the simple case

$$1, 3 \rightarrow 2, 4 \tag{4.13}$$

While here, again, species 1 and $n$ can appear and disappear only at the first and last boundary, respectively, species 2 may appear at the first or second boundary, and species 3 may disappear at the second or third. Furthermore, the second boundary, if not involving appearance or disappearance of a species, may be self-sharpening, nonsharpening, or nonexistent. Six patterns thus are possible:

$$\mathrm{S}_2^1\mathrm{N}^3\mathrm{S}_4,\quad \mathrm{S}_2^1\mathrm{S}\,\mathrm{S}_4^3,\quad \mathrm{S}_2^1\mathrm{N}\,\mathrm{S}_4^3,\quad \mathrm{S}_2^1\text{–}\,\mathrm{S}_4^3,\quad \mathrm{S}^1\mathrm{N}_2^3\mathrm{S}_4,\quad \mathrm{S}^1\mathrm{N}_2\mathrm{S}_4^3 \tag{4.14}$$

For the slightly more complex case 1, 3, 4 → 1, 2, 4, the number of possible patterns is sixteen. As calculated by Tondeur and Klein [5], the total number of possible patterns for all cases with $n$ components is $8(6)^{n-2}$, e.g., 1728 for five components and over $10^7$ for ten components.

The first step in predicting the response pattern from given influent and presaturant compositions is to establish at which boundaries any species absent from the influent or presaturant appear or disappear. This can be achieved with the criteria (4.3) and (4.4), for which $H$-function roots need not be calculated. If desired, the affinity cuts of all boundaries can then be determined; for this purpose it is convenient to write the pattern in its longhand form, as shown in the example (4.8). The sharpening characteristics of the boundaries at which species appear or disappear can be inferred from the general rules for coherent boundaries: A boundary is self-sharpening if it involves appearance of a low-affinity species or disappearance of a high-affinity species, and is nonsharpening in the opposite cases. For a complete description of the pattern, the sharpening characteristics of the other boundaries still remain to be determined. Conditions (4.1) and (4.2) can be used for this purpose. Alternatively, the following criteria can be applied, which require calculation of only one set of $H$-function roots, that of either the influent or presaturant: The $k$th boundary is

$$\begin{array}{lllll}
\text{self-sharpening} & \text{if } \sum_i x_i''/(h_k'-\alpha_{1i}) > 0 & \text{or} & \sum_i x_i'/(h_k''-\alpha_{1i}) < 0 \\
\text{nonsharpening} & \text{if } \sum_i x_i''/(h_k'-\alpha_{1i}) < 0 & \text{or} & \sum_i x_i'/(h_k''-\alpha_{1i}) > 0 \\
\text{nonexistent} & \text{if } \sum_i x_i''/(h_k'-\alpha_{1i}) = 0 & \text{or} & \sum_i x_i'/(h_k''-\alpha_{1i}) = 0
\end{array} \tag{4.15}$$

(for proof, see Appendix I).

## D. Quantitative Calculations of Composition Profiles and Histories

As shown at the outset, an abrupt composition change at the column inlet produces a "proportionate" response pattern, which allows normalized composition profiles and histories to be given as plots of concentrations versus velocity and versus reciprocal velocity, respectively (see Section I.A). How compositions and velocities can be calculated from the $H$-function roots of the influent and presaturant and be used for construction of normalized profiles and histories will now be examined.

1. *Plateau-Zone Compositions*

Equations stating the concentrations of all species in the various plateau zones resulting from undisturbed development of an initial noncoherent boundary have been derived in Chapter 3, Section V.D.2. They are applicable here, too, since the same conventions as to numbering of zones and use of primes and double primes have been adhered to (see also Section I.B.2). Thus, the mobile- and stationary-phase concentrations of the arbitrary species $j$ in the arbitrary $k$th plateau zone are

$$x_{jk} = \prod_{i=1}^{k-1} (h''_i - \alpha_{1j}) \prod_{i=k}^{n-1} (h'_i - \alpha_{1j}) \Big/ \prod_{i \neq j} (\alpha_{1i} - \alpha_{1j}) \tag{3.93}$$

$$y_{jk} = \prod_{i=1}^{k-1} \left( \frac{1}{h''_i} - \alpha_{j1} \right) \prod_{i=k}^{n-1} \left( \frac{1}{h'_i} - \alpha_{j1} \right) \Big/ \prod_{i \neq j} (\alpha_{i1} - \alpha_{j1}) \tag{3.94}$$

2. *Adjusted Velocities*

Equations for the adjusted velocities of composition steps and of compositions in diffuse boundaries in terms of the local $H$-function roots have been derived in Chapter 3, Section IV.D. 6–7. These equations can be used here after the appropriate root values have been substituted. As shown earlier, the boundary with variable $h_k$ is the $k$th in the pattern, and the other roots at this boundary have the values

$$h_i = h''_i \quad \text{for all } i < k, \qquad h_i = h'_i \quad \text{for all } i > k \tag{4.16}$$

(see Section I.B. 1–2). Inserting these values into Eqs. (3.71) and (3.68) one obtains the following relations: If the $k$th boundary is self-sharpening, its adjusted step velocity is

$$(u_{\Delta})_k = h'_k h''_k P_k \tag{4.17}$$

where

$$P_k \equiv \prod_{i=1}^{k-1} h''_i \prod_{i=k+1}^{n-1} h'_i \prod_i \alpha_{i1} \tag{4.18}$$

If the $k$th boundary is nonsharpening and thus diffuse, the adjusted composition velocities of any composition with root value $h_k$ in this boundary is

$$(u)_k = h_k^2 P_k \tag{4.19}$$

At the upstream and downstream end points, at which the nonsharpening $k$th boundary is bounded by the adjacent plateau zones, the values of $h_k$ are $h_k'$ and $h_k''$, respectively, and the adjusted velocities of these end points thus are

$$(u')_k = h_k'^2 P_k \qquad \text{and} \qquad (u'')_k = h_k''^2 P_k \tag{4.20}$$

For a diffuse $k$th boundary, it will be more convenient to transform Eq. (4.19) into relations which give the variable concentrations $x_j$ and $y_j$ of an arbitrary species $j$ within the boundary as functions of the velocity

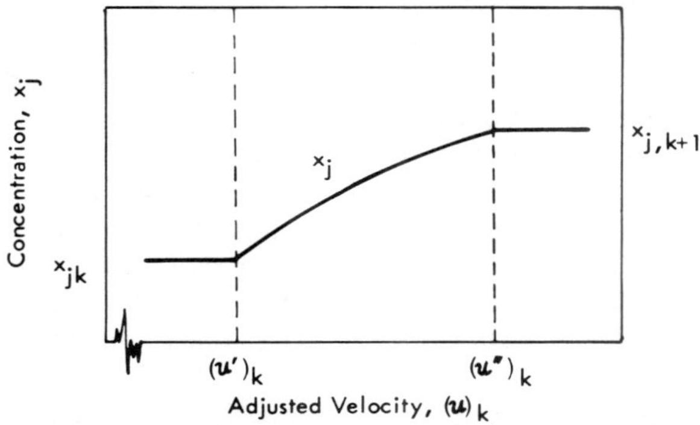


**Fig. 4.6.** Mobile-phase concentration of species $j$ as function of adjusted composition velocity in nonsharpening $k$th boundary (schematic).

$(u)_k$. This is achieved as follows: In view of the fixed values (4.16) of the roots other than $h_k$, the concentrations $x_j$ and $y_j$ at an arbitrary point in the $k$th boundary are given by equations which differ from Eqs. (3.93) and (3.94) only in that the respective value $h_k$ appears instead of $h_k'$. Thus,

$$\frac{x_j}{x_{jk}} = \frac{h_k - \alpha_{1j}}{h_k' - \alpha_{1j}} \qquad \text{and} \qquad \frac{y_j}{y_{jk}} = \frac{1/h_k - \alpha_{j1}}{1/h_k' - \alpha_{j1}} \tag{4.21}$$

Substituting $h_k$ by means of Eq. (4.19) and solving for $x_j$ and $y_j$ one obtains the desired relations giving the concentrations as functions of the velocity:

$$x_j = \frac{[(u)_k/P_k]^{1/2} - \alpha_{1j}}{h_k' - \alpha_{1j}} x_{jk} \tag{4.22}$$

$$y_j = \frac{[P_k/(u)_k]^{1/2} - \alpha_{j1}}{1/h_k' - \alpha_{j1}} y_{jk} \tag{4.23}$$

When Eqs. (4.22) and (4.23) are applied to species $k$, the right-hand sides can become indeterminate. This happens if the species "appears" at this boundary, so that $x_{kk} = y_{kk} = 0$ and $h_k' = \alpha_{1k}$. In this case, the following relations, derived in an analogous manner, can be used:

$$x_k = \frac{[(\mathscr{u})_k/P_k]^{1/2} - \alpha_{1k}}{h_k'' - \alpha_{1k}} x_{k,k+1} \tag{4.24}$$

$$y_k = \frac{[P_k/(\mathscr{u})_k]^{1/2} - \alpha_{k1}}{1/h_k'' - a_{k1}} y_{k,k+1} \tag{4.25}$$

3. *Construction of Composition Histories and Profiles*

Once the plateau-zone compositions and the adjusted step and composition velocities have been calculated, composition histories and profiles are readily constructed as follows:

Since all compositions and steps originate at $z = 0$, $t = 0$ and travel at constant velocities, a plot of concentrations versus reciprocal velocity gives the concentrations as a function of time at unit distance from the column inlet and thus constitutes a normalized composition history. Histories at any other distances, the effluent history at the column exit included, can be obtained from the normalized one by multiplication of the abscissa values by the respective distance $z$. Similarly, a plot of concentrations versus velocity gives the concentrations as a function of distance at unit time and thus constitutes a normalized composition profile, from which profiles at any other times can be obtained by multiplication of the abscissa values by the respective time $t$. After it has been determined with conditions (4.1) and (4.2) which boundaries are self-sharpening, nonsharpening, or nonexistent, the plots versus velocity or reciprocal velocity are readily constructed with Eqs. (4.17) and (4.20), which give the velocities of the bounds between the various plateau zones and boundaries, and with Eqs. (3.93), (3.94), and (4.22) to (4.25), which give the concentrations in the plateau zones and nonsharpening boundaries.

For this construction of composition histories and profiles, true rather than adjusted composition and step velocities are needed. The adjusted velocities, used in Eqs. (4.17) to (4.25) because the respective expressions take simpler forms, are readily converted to true velocities by means of the relations in Chapter 2, Section IV.B. For composition histories the conversion is particularly simple; the adjusted and true reciprocal velocities are linearly related:

$$\frac{1}{\mathscr{u}} = \frac{Cu_0}{\bar{C}}\left(\frac{1}{u} - \frac{1}{u_0}\right) \tag{2.23}$$

and the concentrations can thus be plotted versus adjusted velocity and the composition history then be obtained by a mere change in the scale and origin of the abscissa. For composition profiles the conversion to the true-velocity scale is nonlinear and must precede the plotting; in a plot versus adjusted velocity the "faster" plateau zones and boundaries would appear too wide relative to the "slower" ones. This distortion, however, is negligible if even the highest composition and step velocities are much lower than the velocity $u_0$ of the bulk mobile phase, a requirement met if $C \ll \bar{C}$.

## E. Relative Widths of Boundaries and Plateau Zones

While all self-sharpening boundaries in the response to an abrupt influent composition change, under the simplifying assumptions stated at the outset, remain ideally sharp throughout, the various nonsharpening boundaries and plateau zones spread to different degrees as they travel. As a convenient measure of the rate of spreading, and thus of the relative width of a plateau zone or boundary, the difference between the adjusted velocities of its downstream and upstream end points may be used. In particular, if this difference is almost nil, virtually no spreading will take place and the boundary remains sharp or the plateau zone infinitesimally narrow.

Summarized below are the step and end-point velocities from which the end-point velocity differences for the $k$th boundary and $k$th plateau zone can be obtained:

$$(u_\Delta)_k = h_k' h_k'' P_k \tag{4.17}$$

$$(u_\Delta)_{k-1} = \prod_{i=1}^{k-1} h_i'' \prod_{i=k-1}^{n-1} h_i' \prod_i \alpha_{i1} = h_{k-1}'' h_k' P_k \tag{4.26}$$

$$(u')_k = h_k'^2 P_k \quad \text{and} \quad (u'')_k = h_k''^2 P_k \tag{4.20}$$

$$(u'')_{k-1} = h_{k-1}''^2 \prod_{i=1}^{k-2} h_i'' \prod_{i=k}^{n-1} h_i' \prod_i \alpha_{i1} = h_{k-1}'' h_k' P_k \tag{4.27}$$

where $P_k$ is defined as in Eq. (4.18).

For the arbitrary nonsharpening $k$th boundary, the difference between the adjusted velocities of the end points is, with Eqs. (4.20),

$$\Delta u = (u'')_k - (u')_k = (h_k''^2 - h_k'^2) P_k \tag{4.28}$$

For the arbitrary plateau zone $k$, flanked by the $(k-1)$th and $k$th boundaries, the form of the expression for the difference between the end-point velocities depends on the sharpening characteristics of the boundaries, as shown in Table 4.1.

We can now establish under which conditions a nonsharpening boundary spreads very little or a plateau zone remains very narrow. Equation (4.28) shows that

$$h_k' \simeq h_k'' \tag{4.29}$$

is required for the spreading of the arbitrary nonsharpening $k$th boundary to remain insignificant. This requirement is met only if the composition change across the boundary is very small, or if the species $k$ and $k+1$ have almost identical affinities (i.e.,

**Table 4.1**

DIFFERENCE BETWEEN ADJUSTED END-POINT VELOCITIES OF PLATEAU ZONE $k$

| Sharpening characteristics of | | End-point velocity difference |
|---|---|---|
| $k$th boundary | $(k-1)$th boundary | |
| Self-sharpening | Self-sharpening | $\Delta u = (u_\Delta)_k - (u_\Delta)_{k-1} = (h_k'' - h_{k-1}')h_k'P_k$ |
| Self-sharpening | Nonsharpening | $\Delta u = (u_\Delta)_k - (u'')_{k-1} = (h_k'' - h_{k-1}'')h_k'P_k$ |
| Nonsharpening | Self-sharpening | $\Delta u = (u')_k - (u_\Delta)_{k-1} = (h_k' - h_{k-1}')h_k'P_k$ |
| Nonsharpening | Nonsharpening | $\Delta u = (u')_k - (u'')_{k-1} = (h_k' - h_{k-1}'')h_k'P_k$ |

$\alpha_{1k} \simeq \alpha_{1,k+1}$) so that the interval $\alpha_{1k} \leqq h_k \leqq \alpha_{1,k+1}$, to which all values of $h_k$ including $h_k'$ and $h_k''$ are confined, is very small.

As Table 4.1 shows, the arbitrary plateau zone $k$ remains very narrow in the four cases

$$h_{k-1}' \simeq h_k'' \simeq \alpha_{1k}, \qquad h_{k-1}'' \simeq h_k'' \simeq \alpha_{1k},$$
$$h_{k-1}' \simeq h_k' \simeq \alpha_{1k}, \qquad \text{and} \qquad h_{k-1}'' \simeq h_k' \simeq \alpha_{1k} \tag{4.30}$$

(Since $h_{k-1} \leqq \alpha_{1k} \leqq h_k$, any two approximately equal values of $h_{k-1}$ and $h_k$ must approximately equal $\alpha_{1k}$.) The second and third of these cases involve presaturant or influent composition points near the watershed of the border $x_k = 0$ and, in their limiting forms, were discussed earlier in Section I.B.4 [see conditions (4.5)]. An example of the first of the cases (4.30), with a very narrow zone $k$ between self-sharpening boundaries, was shown earlier in Fig. 4.5b. The fourth case, involving a zone $k$ whose composition point is near the watershed of $x_k = 0$, is of lesser interest since species $k$ then is a trace component in all zones and has little effect on the pattern. In addition, zone $k$ remains very narrow regardless of the influent and presaturant compositions if the species $k-1$, $k$, and $k+1$ have almost the same affinities, since all values of $h_{k-1}$ and $h_k$ then are confined to a very narrow interval $\alpha_{1,k-1} \leqq h_{k-1}, h_k \leqq \alpha_{1,k+1}$. On the other hand, it can be noted that zone $k$ spreads at a finite rate if species $k$ is present in significant concentrations in both the influent and presaturant, provided that species $k-1$, $k$, and $k+1$ differ significantly in their affinities.

## F. FRONTAL ANALYSIS

A particular type of operation with a single abrupt influent composition change is well known from chromatographic experience. In *frontal analysis* (as applied to ion exchange or exchange sorption), a column presaturated with a single species $n$ of low affinity receives as influent a mixture of species $1, \ldots, n-1$ of higher affinities.*

* Frontal analysis is more commonly used in sorption chromatography, as originally suggested by Tiselius [6], with a column containing no solute initially.

As one of the cases in which all species in the influent have higher affinities than those in the presaturant, frontal analysis has a unique (i.e., composition-independent) response pattern (see Section I.C.2). Along the composition route in the direction of flow, the influent species 1, ..., $n-1$ disappear in this sequence, one at each boundary, and the presaturant species $n$ appears at the last boundary. All $n-1$ boundaries are self-sharpening and have nonbracketing affinity cuts. Thus, regardless of the relative concentrations in the influent, the pattern is $S^1 \ldots S^{n-2}S_n^{n-1}$. Written to show the species present at the various boundaries and with affinity cuts marked by vertical lines, the pattern is

$$
\begin{array}{llllllll}
\text{Influent} & 1 \quad 2 \quad 3 & \ldots & k & k+1 & \ldots & n-1 & \\
\hline
\text{1st boundary} & 1 \mid 2 \quad 3 & \ldots & k & k+1 & \ldots & n-1 & \\
\text{2nd boundary} & \quad\; 2 \mid 3 & \ldots & k & k+1 & \ldots & n-1 & \\
\vdots & & & & \vdots & & & \\
k\text{th boundary} & & & k \mid & k+1 & \ldots & n-1 & \\
\vdots & & & & \vdots & & & \\
(n-1)\text{th boundary} & & & & & & n-1 \mid & n \\
\hline
\text{Presaturant} & & & & & & & n
\end{array}
\tag{4.31}
$$

As an example, the composition route and profile of a four-component frontal analysis are given in Fig. 4.7.

As can be seen in the pattern (4.31), each influent species is a low-affinity species at all boundaries upstream of that at which it disappears. (This rule, of course, does not apply to species 1, for which there are no such boundaries.) Since all boundaries are self-sharpening and thus have their low-affinity species present in higher concentrations on their downstream than on their upstream sides (see Chapter 3, Section IV.E.2), it follows that the concentrations of all influent species increase from zone to zone from the inlet down to the boundaries at which the species disappear. The example in Fig. 4.7 displays such behavior.

Quantitative calculations of composition profiles and effluent histories are particularly simple for frontal analysis. The absence of species $n$ from the influent, and of all other species from the presaturant, fixes the following $H$-function root values:

$$h'_{n-1} = \alpha_{1n}, \qquad h''_n = \alpha_{1i} \quad (i = 1, \ldots, n-1) \tag{4.32}$$

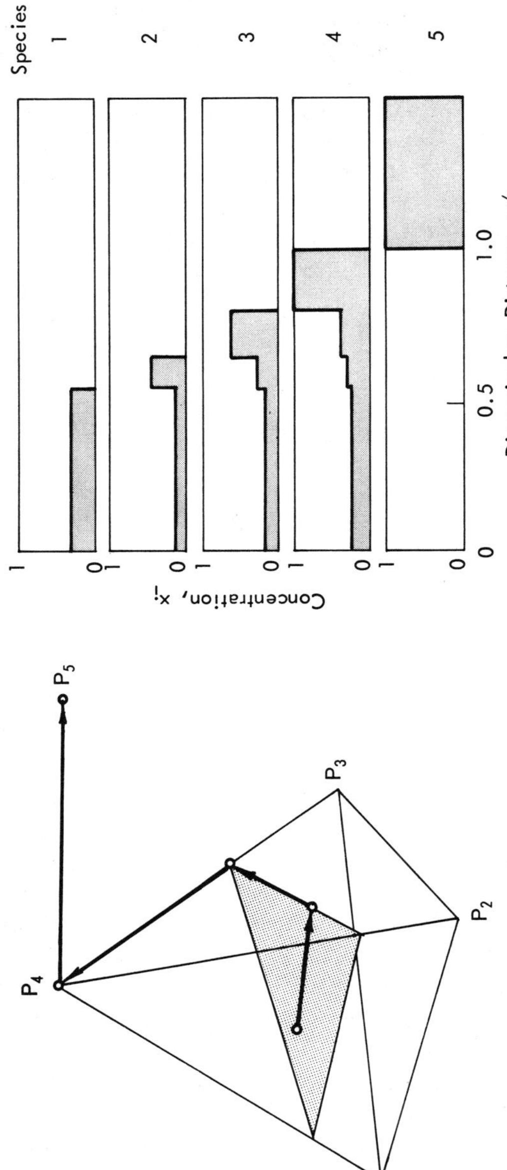


**Fig. 4.7.** Frontal analysis of four-component mixture. Composition profile route and concentration profiles. 1 | 2–2 | 3 Common plane of influent composition point is shown for orientation (shaded). (For $\alpha_{12} = 2$, $\alpha_{13} = 4$, $\alpha_{14} = 8$; $x_1' = 0.40$, $x_2' = 0.14$, $x_3' = 0.19$, $x_4' = 0.27$.)

[trivial roots, see condition (3.55)]. With these root values, the general relations (3.93) and (3.94) for the mobile- and stationary-phase concentrations of species $j$ in the $k$th plateau zone reduce to

$$x_{jk} = \prod_{i=k}^{n-2} (h_i' - \alpha_{1j}) \Bigg/ \prod_{\substack{i=k \\ i \neq j}}^{n-1} (\alpha_{1i} - \alpha_{1j}) \qquad (4.33)$$

$$(j \geqq k)$$

$$y_{jk} = \prod_{i=k}^{n-2} \left( \frac{1}{h_i'} - \alpha_{j1} \right) \Bigg/ \prod_{\substack{i=k \\ i \neq j}}^{n-1} (\alpha_{i1} - \alpha_{j1}) \qquad (4.34)$$

(For $j < k$, one obtains $x_{jk} = y_{jk} = 0$.) With the root sets (3.92) and Eqs. (4.32), Eq. (4.17) for the adjusted velocity of the $k$th boundary reduces to

$$(u_\Delta)_k = \prod_{i=k}^{n-2} (h_i' \alpha_{i+1,1}) \qquad (k < n-1)$$

$$(u_\Delta)_{n-1} = 1 \qquad (4.35)$$

Composition profiles and histories can be constructed from the plateau-zone compositions and boundary velocities as shown in Section I.D.

The qualitative properties of the theoretical frontal-analysis pattern are in complete agreement with chromatographic experience [6–16], and the quantitative results are equivalent to theories derived much earlier and with the same premises by Claesson [8, 9] and Sillén [7]. Regarding the evaluation of experimental frontal-analysis patterns for analytical purposes, the present treatment does not go beyond this earlier work, except for offering greater convenience through the use of $H$-function roots.

## G. Infinitesimal Composition Changes. Trace-Component Systems

Two special cases of single abrupt influent composition changes, namely, that of an infinitesimal composition change and that with all but one species at trace levels in both the influent and presaturant, will later be of interest in connection with practical applications and for establishing relations between the present approach and conventional theories of chromatography. For these special cases, some of the rules stated previously must be qualified, as will now be shown.

### 1. *Infinitesimal Composition Changes*

We consider first the response to an infinitesimal abrupt composition change at the inlet. Since the $H$-function roots are continuous functions of composition, the difference between the sets of roots for the influent and presaturant is infinitesimal if that between the influent and presaturant compositions is infinitesimal:

$$\mathbf{h}' \simeq \mathbf{h}'' \qquad \text{if} \quad \mathbf{x}' \simeq \mathbf{x}'' \tag{4.36}$$

Thus, all root variations are infinitesimal:

$$h_i \simeq h_i' \simeq h_i'' \qquad \text{for all } i \tag{4.37}$$

and all response boundaries involve but infinitesimal composition changes; no plateau zone can have a composition grossly different from those of the influent and presaturant.

Since the composition variations across the boundaries are infinitesimal, the composition velocity also is practically constant across each boundary. Self-sharpening and nonsharpening tendencies of the boundaries, having their origin in composition-velocity variations (see Chapter 3, Section IV.E), therefore are virtually absent. Thus, granted the simplifying premises stated in Chapter 3, Section I, all response boundaries remain sharp as they travel (see also Section I.E). (In practice, boundary spreading through effects neglected here—finite sorption and desorption rates, axial diffusion and dispersion, etc.—will take place uncombated by self-sharpening, and unaided by nonsharpening, tendencies arising from composition-velocity variations and will affect all response boundaries to a similar degree; see Chapter 5, Section V.A.3.)

Since the response boundaries remain sharp, their velocities are well-defined and are given by the respective, virtually constant, composition velocities. The velocity $(u)_k$ of the arbitrary $k$th boundary thus is the $k$th velocity eigenvalue of the virtually constant composition of the system:

$$(u)_k = h_k \prod_i h_i \prod_i \alpha_{i1} \tag{4.38}$$

[see Eq. (3.68) for the velocity eigenvalues]. This result is readily verified: with condition (4.37), Eqs. (4.17) and (4.20) for the step or end-point velocities of a self-sharpening or nonsharpening $k$th boundary reduce to Eq. (4.38).

It is interesting to note that the boundary velocities are independent of the signs and relative magnitudes of the influent concentration changes of the various species, provided these changes are infinitesimal. Thus, with a presaturant of given composition, any arbitrary infinitesimal influent composition change produces a response pattern with the same set of (adjusted) boundary velocities, given by the velocity eigenvalues of the presaturant composition. Experiments with very small influent composition changes can be used to determine the composition-velocity eigenvalues, from which information about the isotherm surface of the system can be derived. Usually, however, small influent composition pulses rather than step changes are employed for such purposes (see concentration-pulse chromatography, Section III.E).

In all other respects, the rules derived for finite influent composition changes apply to infinitesimal changes also, with exceptions only for trace components. In particular, for infinitesimal as for finite influent composition changes, each response boundary involves concentration variations of *all* species present at that boundary, and the condition for absence of a boundary from the response pattern is that an *H*-function root, not a species concentration, has the same value in the influent and presaturant. These rules do not, however, extend to trace-component systems, as will be shown in the following section.

### 2. *Trace-Component Systems*

The special case of trace-component systems, i.e., systems with all but one species at trace levels in both the influent and presaturant, is of interest in that it provides the link between the present approach and theories for systems without interference. As discussed earlier, the large excess of the bulk species suppresses mutual interference of the trace species, so that the latter advance independently (see Chapter 3, Sections II and III.B.1). Thus, as in theories neglecting interference, the behavior of each trace species can be calculated separately and the response pattern be obtained by superposition of the separate trace-species responses.

The influent composition change in trace-component systems is necessarily infinitesimal. In consequence, as shown previously, all response boundaries involve but infinitesimal composition changes, and self-sharpening and nonsharpening tendencies are virtually eliminated so that all boundaries remain sharp. In the following respects, however, trace-component systems differ from others with infinitesimal composition changes; because interference between the trace species is eliminated, each response boundary involves a concentration variation of only one trace species (and of the bulk species), and the condition for absence of a boundary from the pattern is that a trace species has the same concentration in the influent and presaturant. Furthermore, since for a trace species the concentration velocity equals the species velocity (see Chapter 3, Sections III.B.1 and IV.D.6), each boundary advances at a velocity equal to that of the trace species whose concentration varies.

The equations for the plateau-zone compositions and boundary velocities reduce, in the limiting case of trace-component systems, to simple expressions in accordance with the behavior stated above. This is readily shown as follows: For a composition with all species except $k$ at trace levels, the $H$-function roots in the limit $x_k \to 1$ assume the values of the trivial roots:

$$\begin{aligned} \lim_{x_k \to 1} h_i &= \alpha_{1i} \qquad &&\text{for all } i < k \\ \lim_{x_k \to 1} h_i &= \alpha_{1,i+1} \qquad &&\text{for all } i \geqq k \end{aligned} \tag{4.39}$$

(see Chapter 3, Section IV.D.2). For the concentrations of the trace species $j$ in the $i$th plateau zone, Eqs. (3.93) and (3.94) then reduce to

$$\left.\begin{aligned} x_{ji} &= x_j'', \quad y_{ji} = y_j'' \qquad \text{for all } j < i \\ x_{ji} &= x_j', \quad y_{ji} = y_j' \qquad \text{for all } j \geqq i \end{aligned}\right\} \quad \text{if} \quad j < k$$
$$\left.\begin{aligned} x_{ji} &= x_j'', \quad y_{ji} = y_j'' \qquad \text{for all } j \leqq i \\ x_{ji} &= x_j', \quad y_{ji} = y_j' \qquad \text{for all } j > i \end{aligned}\right\} \quad \text{if} \quad j > k \tag{4.40}$$

For the velocity of the $j$th boundary, Eq. (4.38) reduces to

$$\begin{aligned} (u)_j &= \alpha_{kj} \qquad &&\text{if} \quad j < k \\ (u)_j &= \alpha_{k,j+1} \qquad &&\text{if} \quad j \geqq k \end{aligned} \tag{4.41}$$

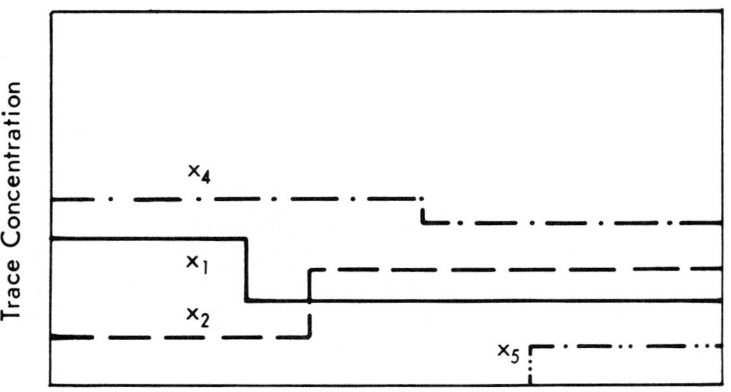


**Fig. 4.8.** Response of trace-component system to single abrupt influent composition change. Concentration profiles of trace components in typical five-component case with species 3 as bulk species.

Equations (4.40) show that each trace species varies its concentration across only one boundary; namely, species $j$ at the $j$th boundary if $j < k$, and at the $(j-1)$th boundary if $j \geqq k$. This behavior is illustrated in Fig. 4.8, which shows the composition profile of a trace-component system. The striking fact that coherent boundaries, normally involving concentration variations of *all* species, in trace-component systems degenerate to concentration variations of only one trace species and the bulk species has been discussed earlier (see Chapter 3, Sections IV.B and C.4). Returning to Eqs. (4.41) one finds that the adjusted velocity of the boundary involving the concentration variation of the trace species $j$ equals $\alpha_{kj}$. This is also the species velocity of species $j$ [see Eq. (3.69)], in accordance with the rule that trace-concentration variations advance at the respective trace-species velocities (see Chapter 3, Sections III.B.1 and IV.D.6).

## H. Comparison with Other Theories

The response of a uniformly presaturated column to an influent of constant (i.e., time-independent) composition, as treated in the present section, has been the subject of other theories for interfering solutes [5, 7, 16–23]. In fact, lacking the concept of coherence, most of these theories have confined themselves to this simple case.

There is a basic difference in approach between the present and other theories. The latter are exclusively based on the stated or implied postulate that all concentrations, in general functions of the two separate variables time and distance (or their equivalents), can instead be expressed as functions of a single variable.* With this premise, which appears in various guises and amounts to postulating coherence, solutions are derived which satisfy all pertinent conditions and thus describe possible states of the systems. However, since the basic postulate remains hypothetical, there is no proof that the solutions are unique and, thus, that the systems in fact attain the respective states. In the present treatment, no such postulate is used, and the uniqueness of the solution is not in question. Operation with uniform presaturation and constant influent composition appears as a limiting case of a general treatment of arbitrary, gradual influent composition changes. Eventual attainment of coherence is proved for the general case, and is shown to be instantaneous in the limiting case. The present treatment thus provides a proof for the validity of the postulate on which the other theories are based.

---

* For a detailed critical discussion, see [21]. For one exception [16], see end of this section.

Most of the other theories operate with simplifying assumptions essentially equivalent to those in the present approach. In particular, as to equilibria, the most detailed treatments, presented by Glueckauf [19, 20], Sillén [7], and Tondeur and Klein [5], postulate either stoichiometric exchange with constant separation factors, as has been done here, or mathematically equivalent compound Langmuir sorption isotherms (see also Chapter 5, Section III.B). With identical premises, the various approaches lead to equivalent results. Different sets of independent variables are used by different authors, but are readily shown to be equivalent to time and distance from the column inlet as used here (see Chapter 5, Section I). However, since the other theories do not make use of the $H$ function, their equations take much more complicated forms. In fact, closed-form expressions for systems with more than three components have previously been given only for simple special cases such as frontal analysis [7, 8, 16]. On the other hand, several other theories have gone to considerable length to account for effects, e.g., of changes in total influent concentration [19], deviations from local equilibrium [22], and variable separation factors [23], which are not restricted to interfering solutes and will receive only cursory treatment in this book (see Chapter 5).

The adjusted velocities used in the present approach are related in a simple manner to the independent variable $\psi$ introduced by Sillén [7] and called "throughput parameter" **T** by Vermeulen and co-workers ([24, 1, 4, 5, 23]; symbols $t/s$ or **Z** used in the earlier publications). The throughput parameter of a composition equals the reciprocal adjusted composition velocity:

$$\psi = \mathbf{T} = 1/u \tag{4.42}$$

[see also Eq. 2.26)]. An analogous relation holds for composition steps. However, both $\psi$ and **T** are restricted to cases with uniform presaturation and constant influent and can at best be generalized for other entirely coherent systems, whereas the adjusted velocities used here are universally applicable.

One curious exception to the common, uncritical acceptance of coherent behavior is the approach taken by Rachinskii [16].* This author, like others before him, derives a velocity for the set of concentrations constituting a composition at a given location and time, under premises which imply that this velocity is the same for all species. He then finds that the equation for this velocity has $n$ roots. On the strength of the argument that the species have different distribution ratios and should therefore have different

---

* The authors are indebted to Dr. D. Tondeur for having brought this approach and its error to their attention.

velocities, he assigns each root to a different species and concludes that all concentrations advance at different rates, which is equivalent to saying that diffuse coherent boundaries cannot exist. The paradox that Rachinskii's conclusion contradicts his premise is readily resolved as follows. The roots are the composition-velocity eigenvalues, i.e., the velocities which the composition *may* assume under coherent conditions. ($n$ rather than $n-1$ roots appear because stoichiometric exchange is not postulated; see Chapter 5, Section III.) Rachinskii's assignment of the roots to different species is an error, brought about by his confusing the species velocity, which depends on the distribution ratio, with the concentration velocity, which does not.

## II. Gradual Influent Composition Changes

The response to an abrupt influent composition change, treated in detail in the preceding section, constitutes but a limiting case of the more general situation in which the change from a constant presaturant $\mathbf{x}''$ to a different influent $\mathbf{x}'$ is allowed to extend over a finite period of time. The rules and criteria derived in the previous section are readily generalized to account for gradual as well as abrupt influent composition changes, although the quantitative calculation of composition profiles and histories in the general case requires numerical methods.

As pointed out in Section I.A, an influent composition change, whether abrupt or gradual, generates at the column inlet a boundary which usually is noncoherent and is resolved into separate variations of the different $H$-function roots, i.e., into separate coherent boundaries. What distinguishes gradual from abrupt influent composition changes is that the boundary generated at the inlet is diffuse and requires a finite development time and distance for resolution. This behavior is illustrated in Fig. 4.9 by a typical distance-time diagram. The root trajectories do not radiate out linearly from a common point of origin, as they do in response to an abrupt influent change (compare Fig. 4.1), and the overall pattern thus is not a "proportionate" one; a proportionate pattern may, however, be approached at long development times, as will become apparent.

The simplest case of a gradual influent composition change is that along a *linear influent composition route*. This situation arises, for example, if a composition step in the influent, before entering the actual bed, looses its sharpness through back-mixing. The response is very similar to that generated by an abrupt influent composition change; since along the linear route all variations of $H$-function roots are monotonic, the rules derived in the previous section apply. In particular, the final pattern consists exclusively of coherent boundaries which are separated by growing plateau zones and each have a different variable root, the plateau-zone

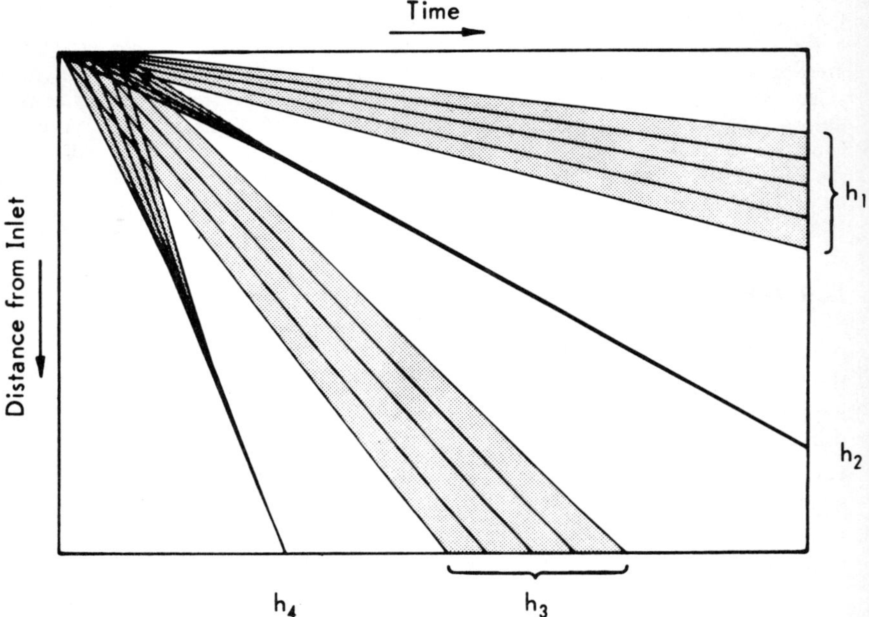


**Fig. 4.9.** Typical distance-time diagram for response to gradual influent composition change. Five-component system with two self-sharpening and two nonsharpening boundaries.

compositions are given by Eqs. (3.93) and (3.94), and the criteria (4.1) and (4.2) for sharpening characteristics and absence of boundaries are valid. The self-sharpening boundaries become completely sharp, and, since the composition trajectories after resolution are linear, the widths of the resolved nonsharpening boundaries increase linearly with traveled distance or elapsed time; in consequence, a proportionate pattern is approached as the development time becomes long compared with the time required for resolution (provided the column is long enough to retain the response boundaries for such lengths of time). Thus, for given influent and presaturant compositions, the response differs from that to an abrupt influent composition change merely in that resolution into coherent boundaries and attainment of complete sharpness of the self-sharpening boundaries require finite times and distances, and that the nonsharpening boundaries at any given time or distance are somewhat broader because the diffuseness of the influent composition change contributes to their widths. As development progresses, these differences between the responses to gradual and abrupt influent composition changes become less and less significant.

To produce a response as described above, the influent composition change need not follow a linear route. The weaker condition of *monotonic variations of all H-function roots* in the influent history is sufficient. (Note that the restriction to monotonic variations of the species concentrations in the influent is neither necessary nor sufficient for producing the described response; compare Fig. 3.31).

A more complex response may arise if *no restrictions on the influent history* are imposed. Since the diffuse noncoherent boundary generated at the column inlet may then involve any arbitrary intermediate compositions, the general rules for development of noncoherent boundaries with arbitrary routes now apply. As shown in Chapter 3, Section V, the final coherent pattern may contain composition pulses as well as additional plateau zones between boundaries having the same variable root. With the rules derived in that section, the response is readily deduced from the root variations constituting the influent composition change. For example, if a root has a maximum or minimum in its influent history, the variation of this root in the response pattern takes the form of a composition pulse, which may be transient or permanent. A compilation of typical responses to influent root variations is given in Fig. 4.10. (This figure is the exact counterpart to Fig. 3.29, with influent root histories instead of initial root profiles shown in the graphs; thus, in the graphs in Fig. 4.10, the change of the variable root from its presaturant to its influent value is from left to right, rather than from right to left as in Fig. 3.29.) Since the final pattern may contain pulses with diminishing amplitudes as well as plateau zones of constant lengths, a proportionate pattern is not necessarily approached even upon prolonged development. Moreover, under the exceptional conditions discussed in Chapter 3, Section V.F.2, a portion of the response pattern may remain permanently noncoherent. Equations (3.93) and (3.94) for the compositions of the plateau zones between variations of different $H$-function roots remain valid but, as can be seen in Fig. 4.10, the criteria (4.1) and (4.2) for the sharpening characteristics and absence, respectively, of boundaries are no longer adequate for response patterns that may contain pulses and multiple boundaries having the same variable root. Specifically, condition (4.2) must be replaced by the following condition: Response boundaries and pulses with variable $h_k$ are

$$\text{nonexistent} \qquad \text{if} \quad \partial h_k/\partial t\big|_{z=0} = 0 \tag{4.43}$$

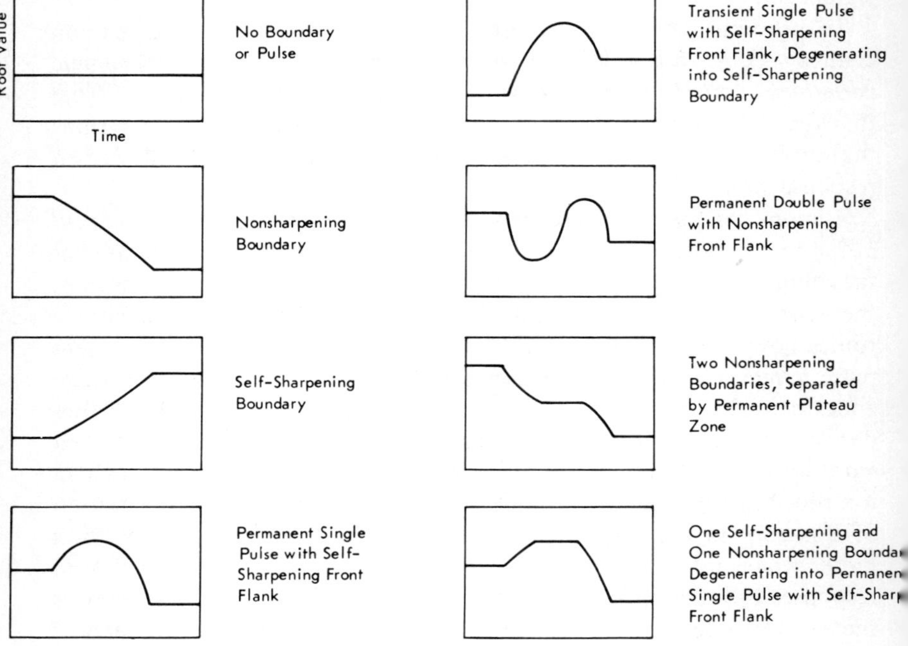


**Fig. 4.10.** Various types of influent histories of a root in noncoherent gradual influent composition change, and resulting responses.

(The weaker condition $h_k' = h_k''$ is not sufficient because it does not rule out the existence of a response pulse with maximum or minimum of $h_k$.) Similarly, the criterion (4.1) is applicable only to monotonic variations of $h_k$.

For a more complete and detailed discussion of the patterns arising from development of arbitrary initial boundaries, as generated by influent composition changes with arbitrary history routes, the reader is referred to Chapter 3, Section V.

The quantitative calculation of composition profiles and histories requires integration of the set of simultaneous differential equations (3.87). For systems in which noncoherence persists over finite periods of time and finite distances, as is the case here, no analytical solutions have been obtained, and numerical or graphical solution methods must be used.

## III. Influent Composition Pulses

In many applications of chromatography, particularly for analytical purposes, the response to the injection of a pulse into an influent of otherwise constant composition is observed. Foremost in this category is conventional chromatography by so-called elution development, in which a sample of the mixture to be analyzed is injected into a carrier gas or solvent or, in ion-exchange or exchange-sorption systems, into a stream of a species of low affinity. A more recent counterpart to elution development is "vacancy chromatography," in which a carrier gas, solvent, or species of low affinity is injected into an otherwise continuous stream of the mixture to be analyzed. Still other applications are the recent concentration-pulse and tracer-pulse techniques, in which information on phase equilibria is obtained from responses to small influent pulses.

In the present section, we shall examine responses to influent composition pulses, i.e., systems in which an influent of otherwise constant composition ("base influent," with composition $\mathbf{x}'$ and a set of roots $\mathbf{h}'$) is temporarily replaced by one of a different composition ("influent pulse," with composition $\mathbf{x}^0$ and a set of roots $\mathbf{h}^0$). The discussion is not restricted to instantaneous pulses; rather, these will appear as a limiting case of pulses of finite duration. Also, the base influent as well as the influent pulse will be allowed to consist of arbitrary mixtures; elution development and vacancy chromatography, in which the base influent or influent pulse, respectively, consists of only one species, will be discussed as special cases. Furthermore, no limits are imposed on the magnitude of the influent composition change; the special case of a very small influent pulse, as used in the concentration-pulse technique, will also be discussed separately.

For the sake of simplicity, only square-wave influent pulses will be considered, i.e., it will be assumed that the composition changes constituting the front and rear flanks of the influent pulse are abrupt. This idealization does not introduce a serious limitation, because a diffuseness caused by back-mixing upstream of the column (e.g., imperfect injection, dead volumes) does not alter the qualitative features of the response; back-mixing leaves the composition history routes of the influent-pulse flanks linear, and as long as its route is linear a gradual influent composition change generates the same response pattern as does an abrupt one (see Section II). Even for quantitative calculations, a slight diffuseness of the influent-pulse flanks has little effect because, soon, the self-sharpening boundaries sharpen completely and the nonsharpening ones spread so much that the width increment stemming from a slight initial diffuseness becomes negligible.

## A. Arbitrary Influent Pulses

We consider first the response to an arbitrary influent pulse of finite duration, injected into a base influent of arbitrary composition. In this general case, no restrictions are imposed on the presence of species in the base influent and the influent pulse.

### 1. *Development Behavior and Reponse Pattern*

The general qualitative features of the developing response pattern are readily deduced and illustrated with the aid of a schematic distance-time diagram as shown in Fig. 4.11. At the column inlet, the flanks of the influent pulse constitute two consecutive abrupt influent composition changes, each of which generates $n-1$ coherent boundaries with different velocities and different variable roots in the sequence of increasing index numbers, counted in the direction of flow (see Section I.A for response to abrupt influent composition changes). Soon, the fastest boundaries of the

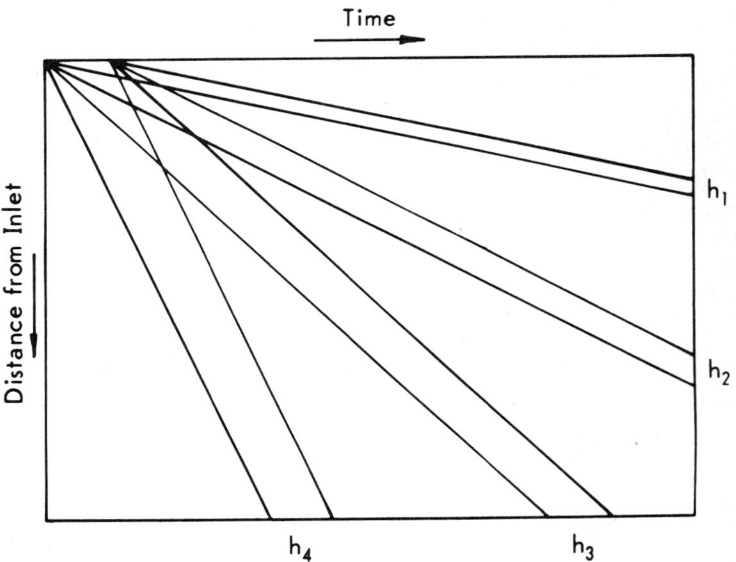


**Fig. 4.11.** Boundary trajectories for response to two successive abrupt influent composition changes in five-component system (schematic).

set generated later catch up with the slowest ones of the set generated earlier, so that the two sets begin to interfere. In such interferences of coherent boundaries, examined in detail in Chapter 3, Section VI, each boundary eventually overtakes any other boundaries whose variable roots have lower index numbers and which thus are slower (e.g., see Fig. 3.37). Accordingly, in the present case, the boundaries sort themselves out in such a way that the final pattern consists of pairs of boundaries having the same variable root, arranged in numerical sequence of the index

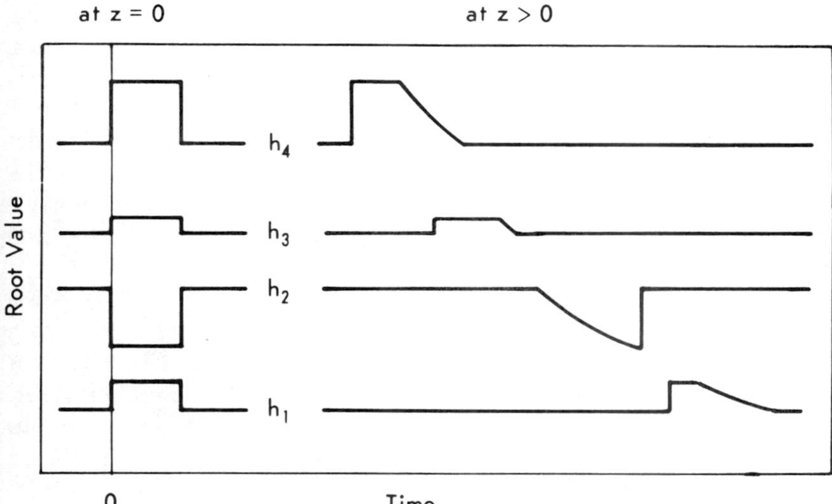


**Fig. 4.12.** Resolution of influent pulse. Root histories at inlet ($z = 0$) and point downstream ($z > 0$) in typical five-component case (schematic).

numbers of the variable roots. This is shown schematically in Fig. 4.11 where spreading of nonsharpening boundaries and velocity variations upon crossovers have been disregarded for the sake of clarity.

The foregoing considerations apply to systems with two consecutive influent composition changes in general. For the particular case of an influent pulse, where the second change restores the initial influent composition, the resolution of the variations of the $H$-function roots is illustrated in Fig. 4.12. In the influent history, the root variations are simultaneous square-wave pulses. Development separates the pulses of the various roots and lets the nonsharpening pulse flanks become diffuse. In the influent history as well as in the response pattern, each root deviates from its base-influent value only within its pulse, so that all plateau zones that develop between the pulses of the different roots have the same root values and, thus, the same composition as the base influent. One may say that the noncoherent influent pulse is resolved into coherent response pulses which travel on a background of the base-influent composition. (Anomalous resolution behavior in singular cases involving compositions on watersheds will be discussed later in this section.)

Each response pulse, corresponding to a trajectory pair in Fig. 4.11, consists of two boundaries with the same variable root but with opposite directions of the root variation and thus with opposite sharpening characteristics. In the early stages of development, the two boundaries, generated at different times, are separate, so that the profile (or history) of the variable root has a "flat top." Later, this flat top disappears since the two boundaries merge (e.g., see Chapter 3, Section VI.B.2). The flat top of the root profile may or may not survive the resolution of the pulse from the others; if it does, the resolved response pulse has a transient flat-top zone, i.e., a zone of uniform composition between its front and rear flanks.

As a simple example illustrating the development of the response pattern, a three-component case is shown in Fig. 4.13. In the composition diagrams, points A and B are the composition points of the base influent and influent pulse, respectively, so that the influent history route is ABA. The two influent composition changes AB and BA are each resolved into two coherent boundaries with 1 | 2 and 2 | 3 affinity cuts, yielding a transient composition route ACBDA. Zone B, having the influent-pulse composition, shrinks and disappears as the faster upstream boundary with 2 | 3 cut catches up with the slower downstream boundary with 1 | 2 cut. Crossover of these two boundaries changes the route to ACADA, and resolution into two coherent pulses, both still with flat tops, is achieved. As the pulse flanks merge, the flat-top zones disappear. The pulse amplitudes then decrease, so that the turning points on the route move from C and D toward A.

One may regard the influent pulse as an $h$-space vector pointing from the base-influent to the influent-pulse composition; the action of the column then is to resolve this vector into its components along the composition paths, each component corresponding to a response pulse. This description, however, glosses over the fact that decrease of the response-pulse amplitudes may precede resolution, so that the response-pulse vectors may in reality no longer have their full lengths when resolution is complete. Nevertheless, for the sake of simplicity, arrows corresponding

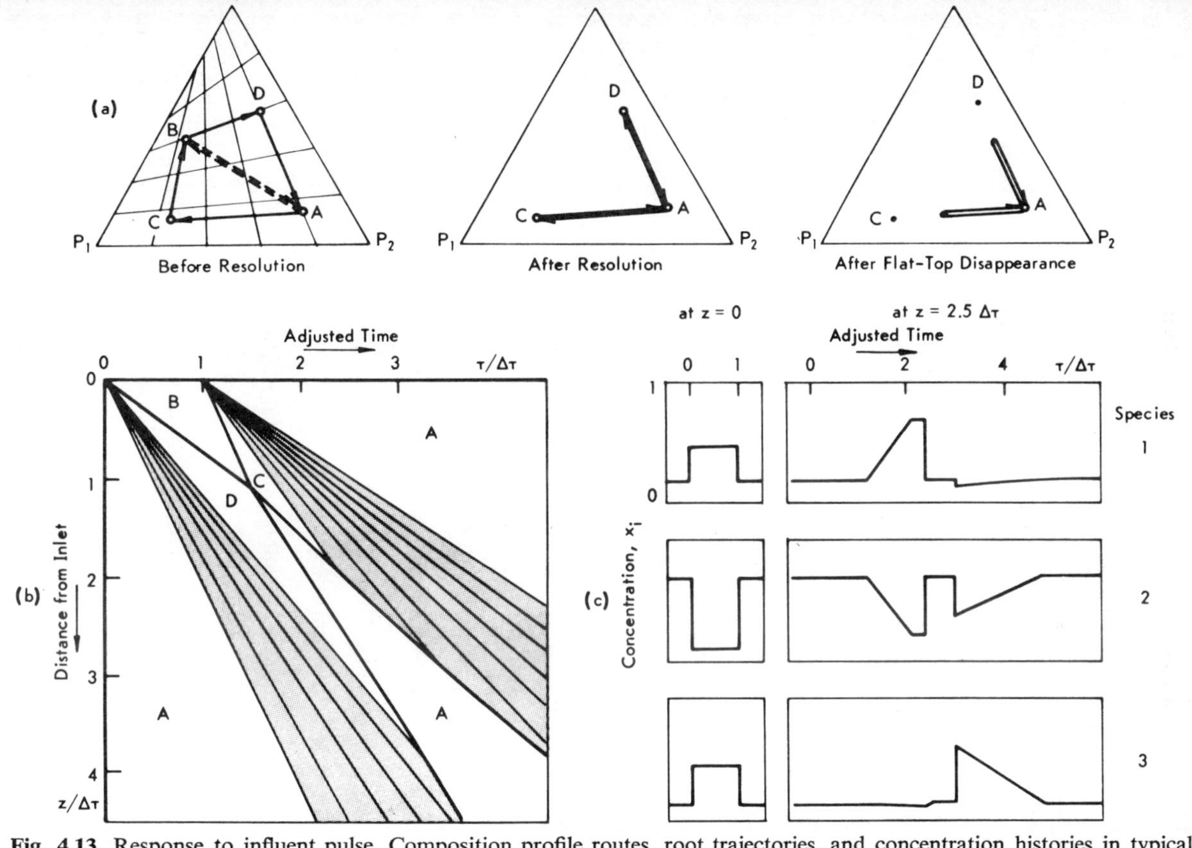


**Fig. 4.13.** Response to influent pulse. Composition profile routes, root trajectories, and concentration histories in typical three-component case. (For $\alpha_{12} = 2$, $\alpha_{13} = 4$; $x_1' = 0.18$, $x_2' = 0.68$, $x_3' = 0.14$; $x_1^0 = 0.45$, $x_2^0 = 0.10$, $x_3^0 = 0.45$.)

to this vector description will be used to indicate influent and response pulses in later composition diagrams.

2. *Properties of Response Pulses*

Being coherent composition pulses, the response pulses in the final pattern generated by an influent pulse obey the rules derived in Chapter 3, Section IV.G. All are "single" pulses, i.e., with only one locus of concentration maxima and minima. Each pulse involves variations of all species present (except for trace components; see Section III.B.3) and has either concentration maxima of the high-affinity species and concentration minima of the low-affinity species or vice versa, depending on whether the variable root has a maximum or minimum in the pulse. In the $k$th pulse, with variable $h_k$, the high- and low-affinity species are $1, \ldots, k$ and $k+1, \ldots, n$, respectively (absent species excluded). Also, each pulse may have either of two shapes, namely, a sharp front and diffuse rear flank or vice versa, again depending on whether the variable root has a maximum or minimum in the pulse. Since the variable root in a response pulse varies in the same direction as it does in the influent pulse, the directions of the concentration variations and the shapes of the response pulses are readily predicted from the base-influent and influent-pulse values of the roots. Thus, with the rules and criteria derived in Chapter 3, Sections IV.E and G one finds that the $k$th pulse

$$\begin{array}{ll} \text{has} & \left.\begin{array}{l}\text{sharp front, diffuse rear flank,}\\ \text{concentration maxima of } 1, \ldots, k,\\ \text{concentration minima of } k+1, \ldots, n\end{array}\right\} \quad \text{if} \quad h_k^0 > h_k' \\ \text{has} & \left.\begin{array}{l}\text{diffuse front, sharp rear flank,}\\ \text{concentration minima of } 1, \ldots, k,\\ \text{concentration maxima of } k+1, \ldots, n\end{array}\right\} \quad \text{if} \quad h_k^0 < h_k' \\ \multicolumn{2}{l}{\text{and is nonexistent} \qquad\qquad\qquad\qquad\qquad \text{if} \quad h_k^0 = h_k'} \end{array} \tag{4.44}$$

(The statements as to concentration maxima and minima do not apply to species that are absent from the pulse and the plateau zones on both its sides.)

3. *Compositions of Flat-Top Zones*

The compositions of the plateau zones constituting the flat tops, if any, of the fully resolved response pulses are readily calculated from the

$H$-function roots of the base influent and influent pulse. In the flat-top zone, the variable root of the pulse has still retained its influent-pulse value while all other roots, after resolution, are at their base-influent values (e.g., see Fig. 4.12). Inserting these root values into the general expressions (3.57) and (3.58) relating concentrations to roots, one obtains for the mobile- and stationary-phase concentrations of the arbitrary species $j$ in the flat-top zone of the arbitrary $k$th pulse:

$$x_{jk} = (h_k^0 - \alpha_{1j}) \prod_{i \neq k} (h_i' - \alpha_{1j}) \Big/ \prod_{i \neq j} (\alpha_{1i} - \alpha_{1j}) \tag{4.45}$$

$$y_{jk} = \left(\frac{1}{h_k^0} - \alpha_{j1}\right) \prod_{i \neq k} \left(\frac{1}{h_i'} - \alpha_{j1}\right) \Big/ \prod_{i \neq j} (\alpha_{i1} - \alpha_{j1}) \tag{4.46}$$

These equations differ from those for the base-influent concentrations $x_j'$ and $y_j'$ only in that $h_k^0$ appears instead of $h_k'$. Combination with the latter equations gives the following more convenient expressions:

$$x_{jk} = \frac{h_k^0 - \alpha_{1j}}{h_k' - \alpha_{1j}} x_j' \tag{4.47}$$

$$y_{jk} = \frac{1/h_k^0 - \alpha_{j1}}{1/h_k' - \alpha_{j1}} y_j' \tag{4.48}$$

Equations (4.47) and (4.48) become indeterminate if $h_k' = \alpha_{1j}$, as is the case for either $j = k$ or $j = k+1$ if species $j$ is absent from the base influent [see condition (3.55)]. The concentrations $x_{jk}$ and $y_{jk}$ can then be obtained from those of the other species with $\Sigma_i x_{ik} = 1$ and $\Sigma_i y_{ik} = 1$.

It is not unusual for the flat-top zone of a resolved pulse to contain one or several species in concentrations higher or lower than in both the base influent and the influent pulse. For example, in the case shown in Fig. 4.13, the concentration of species 1 is higher in zone C, and lower in zone D, than in both the base influent and the influent pulse.

### 4. *Absence of Species from Response Pulses*

We shall now establish under which conditions any species is absent from portions of the response pattern generated by an influent pulse.

First, it is readily shown that any species present in both the base influent and the influent pulse is also present everywhere in the response pattern. According to condition (3.55), a species $j$ can only be absent from a composition if one of the $H$-function roots of the latter is a trivial root $\alpha_{1j}$. Hence, if species $j$ is present in both the base influent and the influent pulse, none of the roots $h_i'$ and $h_i^0$ equals $\alpha_{1j}$. In the response pattern, $h_i'$

and $h_i^0$ are the extremes between which any root $h_i$ varies. If neither extreme equals $\alpha_{1j}$, no intermediate value of $h_i$ can do so because of the general condition $\alpha_{1i} \leqq h_i \leqq \alpha_{1,i+1}$ [see condition (3.50)]. Thus, since no root equals $\alpha_{1j}$ anywhere in the response pattern, species $j$ is present everywhere.

On the other hand, if a species $j$ is present in the base influent but absent from the influent pulse, one root $h_i^0$ equals $\alpha_{1j}$, and species $j$ then is absent from the flat top of one response pulse. The pulse is either the $(j-1)$th or the $j$th, depending on whether the root equalling $\alpha_{1j}$ is $h_{j-1}^0$ or $h_j^0$. When the flat top disappears and the pulse amplitude decreases, the root variation no longer extends to the extreme value $\alpha_{1j}$, and species $j$ is from then on present everywhere in the pattern.

Conversely, if a species $j$ is present in the influent pulse but absent from the base influent, one root $h_i'$ equals $\alpha_{1j}$. Except in the pulse across which this root varies, the root value then is $\alpha_{1j}$ everywhere in the pattern. Species $j$ thus is present in only one response pulse. The pulse containing species $j$ is the $(j-1)$th or the $j$th, depending on whether the root equalling $\alpha_{1j}$ is $h_{j-1}'$ or $h_j'$.

For ease of reference, the results are summarized in Table 4.2. Special cases with single-component base influents or influent pulses will be further discussed in Sections III.C and D.

**Table 4.2**

RULES FOR PRESENCE AND ABSENCE OF AN ARBITRARY SPECIES $k$ FROM PORTIONS OF THE RESPONSE PATTERN GENERATED BY AN INFLUENT PULSE

| Influent | | | Response |
|---|---|---|---|
| Base influent | Influent pulse | Trivial root | |
| $x_k' > 0$ | $x_k^0 > 0$ | — | $x_k > 0$ in all pulses and plateau zones |
| $x_k' > 0$ | $x_k^0 = 0$ | $h_{k-1}^0 = \alpha_{1k}$ | $x_k > 0$ except in flat top of $(k-1)$th pulse |
| | | $h_k^0 = \alpha_{1k}$ | $x_k > 0$ except in flat top of $k$th pulse |
| $x_k' = 0$ | $x_k^0 > 0$ | $h_{k-1}' = \alpha_{1k}$ | $x_k = 0$ except in $(k-1)$th pulse |
| | | $h_k' = \alpha_{1k}$ | $x_k = 0$ except in $k$th pulse |

### 5. *Crest Velocities of Response Pulses*

Once having lost their flat tops, the response pulses have well defined crests (e.g., loci of concentration maxima and minima), for which velocities can be stated. The velocity of the crest is the same as that of the sharp flank of the pulse, since crest and sharp flank ideally occupy the same location at any given time. Applying the general relation (3.71) for step velocities to the present case one finds the adjusted velocity of the sharp flank, and thus of the crest, of the arbitrary $k$th pulse to be

$$(u)_k = h_k^* \prod_i h_i' \prod_i \alpha_{i1} \tag{4.49}$$

where $h_k^*$ is the value of the variable root $h_k$ at the crest.

At the moment at which the flat top of the pulse disappears, $h_k^*$ still has the influent-pulse value $h_k^0$. As the pulse amplitude shrinks, $h_k^*$ monotonically approaches the base-influent value $h_k'$. Hence, the crest velocity is not in general constant. Rather, as development progresses, the crest velocities of pulses with sharp front flanks [and thus with $h_k^0 > h_k'$, see criterion (4.44)] decrease while those of pulses with sharp rear flanks (and thus with $h_k^0 < h_k'$) increase. Only if the variation of the variable root is very small (i.e., $h_k^0 \simeq h_k'$) does the crest velocity of a pulse remain virtually constant; this happens if the composition variation is very small or if $\alpha_{1k}$ and $\alpha_{1,k+1}$, between which both $h_k^0$ and $h_k'$ are confined, are nearly equal.

### 6. *Distances and Times Required for Resolution of Pulses and Disappearance of Flat Tops*

The resolution of an influent pulse of finite duration as well as the disappearance of the flat tops of the resulting response pulses require finite development distances and times, since boundaries generated at different times must cross or merge. The distance-time diagrams show this at a glance.

Quantitative calculations of the distances and times needed for resolution and flat-top disappearance usually require numerical methods, except for three-component systems, for which explicit solutions will be given in Table 4.3. In a qualitative manner, however, the decisive factors are readily understood, as will now be shown.

The distance and adjusted time required for resolution in a three-component system are chiefly determined by three factors. First, the distance and time are proportional to the duration of the influent pulse, since this duration constitutes the initial time lag between the boundaries which must cross one another. Second, early resolution is favored if the boundaries which must cross differ greatly in their velocities. The velocity difference depends on composition (e.g., see Fig. 3.34), being particularly small near the watershed. For given compositions, the velocity difference is large if the species differ widely in their affinities. Third, early resolution is favored if the boundaries which must cross either remain sharp or, if

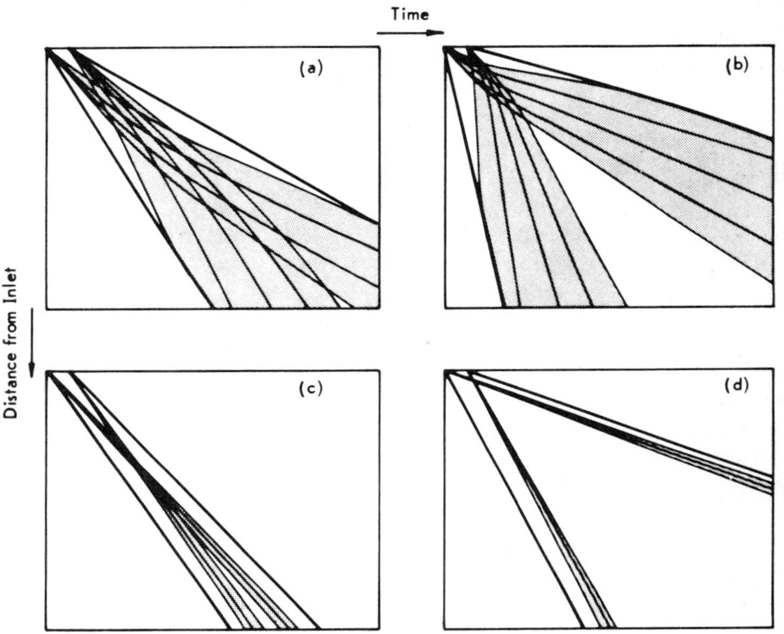


**Fig. 4.14.** Effects of composition variation and velocity difference on response to influent pulse. Root and step trajectories for four typical three-component cases (schematic): (a) Composition variation large, velocity difference between response pulses small; (b) composition variation large, velocity difference large; (c) composition variation small, velocity difference small; (d) composition variation small, velocity difference large.

they are nonsharpening, spread slowly, as is the case if the pulses involve but small composition variations (see Section III.B.2). The effects of the influent-pulse amplitude and the velocity difference of the boundaries are illustrated in Fig. 4.14 by distance-time diagrams for four characteristic cases.

The preceding considerations refer to the adjusted rather than the true time required for resolution. The true resolution time (but not the resolution distance) depends in addition on the flow rate $u_0$ and the concentration ratio $C/\overline{C}$. High values of these parameters result in high boundary velocities and thus in short resolution times.

In systems with more than three components there are, in general, more than two response pulses. Accordingly, a set of resolution distances and times can be defined, each referring to resolution of a different pair of pulses from one another. For each of these resolution distances and times the same considerations as in three-component systems apply. The "faster" pulses with variable roots of higher index numbers tend to require longer distances, but shorter times, for resolution. It is therefore not unusual that the longest resolution distance and the longest resolution time are not for the same pair of pulses. For example, in the case schematically shown in Fig. 4.11, resolution of the third from the fourth pulse requires the longest distance, whereas resolution of the second from the third pulse requires the longest time.

The main factors controlling the distances and times required for disappearance of the flat tops of the response pulses are also readily understood. First, as for resolution and for the same reason, the distances and times for flat-top disappearance are proportional to the duration of the influent pulse. Second, a flat-top tends to be long-lived if the amplitude of the respective response pulse is low; the merger of the pulse flanks then requires a long distance and time because the nonsharpening flank spreads but slowly. Thus, whether or not a flat top survives resolution of the pulse from the others is independent of the duration of the influent pulse, because prolonging this pulse delays resolution as much as it delays flat-top disappearance. On the other hand, an influent pulse of low amplitude is particularly likely to yield resolved flat-top pulses, because the slow spread of the nonsharpening pulse flanks favors not only early resolution, but also late flat-top disappearance (e.g., see Fig. 4.14d).

For three-component systems, explicit relations for the distances and adjusted times for resolution and flat-top disappearance in terms of the $H$-function roots of the base influent and influent pulse can be derived. The respective values for the four possible combinations of response-pulse shapes are stated in Table 4.3. The derivation is given in Appendix III. The relations bear out the qualitative rules stated earlier; in particular, they show that all distances and adjusted times for resolution as well as for flat-top disappearance are proportional to the duration $\Delta\tau$ (in units of adjusted time) of the influent pulse. That the equations giving the distance and adjusted time for flat-top disappearance of a pulse of given shape are independent of the shape of the other pulse is an unforeseen result of the

**Table 4.3**

Development Distances and Adjusted Development Times Required for Resolution and Flat-Top Disappearance in Three-Component Systems. (Zero Time is Taken as Moment of Entry of Front of Influent Pulse; $\Delta\tau$ = Duration of Influent Pulse.)

| Case | Pulse shapes | | | | Relative root values | Distance and adjusted time for | | |
|---|---|---|---|---|---|---|---|---|
| | 1st Pulse | | 2nd Pulse | | | Resolution | Disappearance of flat top | |
| | Rear | Front | Rear | Front | | | 1st Pulse | 2nd Pulse |
| 1 | Diffuse | Sharp | Sharp | Diffuse | $h_1^0 > h_1'$<br>$h_2^0 < h_2'$ | $z_{1/2} = \frac{h_1'h_1^0h_2'h_2^0}{(h_2'-h_1')\alpha_{12}\alpha_{13}}\Delta\tau$<br>$\tau_{1/2} = \frac{h_2'}{h_2'-h_1'}\Delta\tau$ | $z_1 = \frac{h_1'h_1^{0\,2}h_2'(h_2^0-h_1')}{(h_2'-h_1')(h_1^0-h_1')\alpha_{12}\alpha_{13}}\Delta\tau$<br>$\tau_1 = \left(1+\frac{h_1'(h_2^0-h_1')}{(h_2'-h_1')(h_1^0-h_1')}\right)\Delta\tau$ | $z_2 = \frac{h_1'h_2'h_2^{0\,2}(h_2'-h_1^0)}{(h_2'-h_1')(h_2'-h_2^0)\alpha_{12}\alpha_{13}}\Delta\tau$<br>$\tau_2 = \frac{h_2'(h_2'-h_1^0)}{(h_2'-h_1')(h_2'-h_2^0)}\Delta\tau$ |
| 2 | Diffuse | Sharp | Diffuse | Sharp | $h_1^0 > h_1'$<br>$h_2^0 > h_2'$ | $z_{1/2} = \frac{h_1'h_1^0h_2'^2(h_2^0-h_1')}{(h_2'-h_1')^2\alpha_{12}\alpha_{13}}\Delta\tau$<br>$\tau_{1/2} = \left(1+\frac{h_1'(h_2^0-h_1')}{(h_2'-h_1')^2}\right)\Delta\tau$ | Same as case 1 | $z_2 = \frac{h_1'h_2'h_2^{0\,2}(h_2^0-h_1^0)}{(h_2^0-h_1')(h_2^0-h_2')\alpha_{12}\alpha_{13}}\Delta\tau$<br>$\tau_2 = \frac{h_2^0(h_2^0-h_1^0)}{(h_2^0-h_1')(h_2^0-h_2')}\Delta\tau$ |
| 3 | Sharp | Diffuse | Sharp | Diffuse | $h_1^0 < h_1'$<br>$h_2^0 < h_2'$ | $z_{1/2} = \frac{h_1'^2h_2'h_2^0(h_2'-h_1^0)}{(h_2'-h_1')^2\alpha_{12}\alpha_{13}}\Delta\tau$<br>$\tau_{1/2} = \frac{h_2'(h_2'-h_1^0)}{(h_2'-h_1')^2}\Delta\tau$ | $z_1 = \frac{h_1'h_1^{0\,2}h_2'(h_2^0-h_1^0)}{(h_2'-h_1^0)(h_1'-h_1^0)\alpha_{12}\alpha_{13}}\Delta\tau$<br>$\tau_1 = \left(1+\frac{h_1^0(h_2^0-h_1^0)}{(h_2'-h_1^0)(h_1'-h_1^0)}\right)\Delta\tau$ | Same as case 1 |
| 4 | Sharp | Diffuse | Diffuse | Sharp | $h_1^0 < h_1'$<br>$h_2^0 > h_2'$ | No analytical solution available | Same as case 3 | Same as case 2 |

calculations and is difficult to prove in any other way; the numerical values of the distance and time for flat-top disappearance are nevertheless affected by the shape and amplitude of the other pulse since they depend on the values of the latter's variable root.

### 7. *Quantitative Calculations*

For three-component systems, with only one exception, the algebraic calculation of the response to an influent pulse can be carried at least to the disappearance of the flat tops of the response pulses. A convenient approach is to calculate the root and step trajectories as shown in Appendix III, and then convert to concentrations with Eqs. (3.57) and (3.58).

The mentioned exception is the fourth case in Table 4.3, also shown in the examples in Fig. 4.14. Here, the boundaries which must cross are both diffuse, so that a finite region in the distance-time diagram is noncoherent. Within this region, the simultaneous nonlinear differential equations (3.87) must be integrated, for which no analytical solution has been found. However, the numerical or graphical methods for computing the resolution behavior of arbitrary noncoherent boundaries (see Chapter 3, Section V.E) can be applied without difficulty.

If more than two response pulses arise, as is the rule in systems with more than three components, parts of the response are no longer covered by the available explicit analytical solutions, even if crossovers of diffuse flanks with one another do not occur. This is because no explicit solution exists for trajectories having crossed more than one diffuse flank (see Appendix III). However, here and for cases involving crossovers of diffuse flanks, the numerical methods developed for noncoherent boundaries in general are applicable.

### 8. *Singular Cases Involving Watershed Compositions*

Complete resolution of the response pulses from one another has so far been taken for granted. This premise is justified if no compositions on watersheds are involved, since the higher root velocities of the roots with higher index numbers then guarantee regular crossover and resolution behavior (see Chapter 3, Sections V.B and VI.B.1). For compositions on watersheds, however, the velocities of two roots of adjacent index numbers are equal and resolution therefore is impaired. The resulting deviations will now be examined.

We establish first under which conditions a watershed composition can occur in the response pattern. For any composition on the watershed of an arbitrary border $x_k = 0$:

$$h_{k-1} = h_k = \alpha_{1k} \qquad (k \neq 1, n) \tag{3.62}$$

As shown earlier in this section, no root in the response pattern can equal a separation factor unless it does so in the influent pulse or base influent. Hence, a composition on the watershed of $x_k = 0$ can occur only if

$$h^0_{k-1} = h^0_k = \alpha_{1k} \tag{4.50}$$

or

$$h'_{k-1} = h'_k = \alpha_{1k} \tag{4.51}$$

that is, only if the composition of the influent pulse or base influent is on the watershed. (The cases $h'_{k-1} = h^0_k = \alpha_{1k}$ and $h^0_{k-1} = h'_k = \alpha_{1k}$ are irrelevant since species $k$ then is absent from both the base influent and influent pulse and therefore is not to be counted as a component of the system.) The further examination can thus be restricted to the two singular cases (4.50) and (4.51).

If the influent-pulse composition is on the watershed of a border $x_k = 0$ [case (4.50)], the qualitative development behavior of the roots $h_{k-1}$ and $h_k$ is as shown schematically in Fig. 3.36c (see Chapter 3, Section V.F). As the amplitudes of the root variations decrease, these variations no longer extend to their extreme value $\alpha_{1k}$; from then on, the root and step velocities of $h_k$ are higher than those of $h_{k-1}$ at any point in the pattern, so that the two root variations separate and a plateau zone appears between them. Thus, resolution into coherent response pulses separated by plateau zones eventually takes place, the only deviation from the normal behavior being that the appearance of the plateau zone between the $(k-1)$th and $k$th pulse is delayed until the

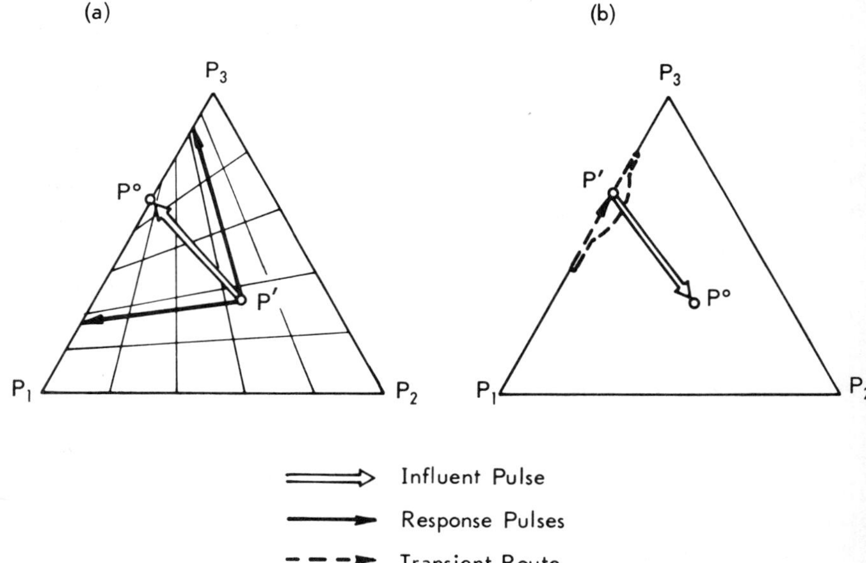


**Fig. 4.15.** Response to influent pulse involving composition on watershed. Influent- and response-pulse vectors for three-component system with (a) influent-pulse composition on watershed, (b) base-influent composition on watershed.

amplitude of either pulse begins to decrease, i.e., until either flat top has disappeared. As a simple example, the composition-route diagram of a three-component case with the influent-pulse composition on the watershed is shown in Fig. 4.15a.

On the other hand, if the base-influent composition is on the watershed of a border $x_k = 0$ [case (4.51)], the variations of $h_{k-1}$ and $h_k$ keep extending to the extreme value $\alpha_{1k}$ even when their amplitudes decrease. In consequence, these root variations are not completely resolved, so that a portion of the final response pattern remains permanently noncoherent. As a simple example, the composition-route diagram of a three-component case is shown in Fig. 4.15b. As in the similar case discussed in Chapter 3, Section V.F.2, and shown in Figs. 3.35d and 3.36d, resolution of the two nonsharpening boundaries with variable $h_{k-1}$ and $h_k$ remains incomplete.

These considerations show that eventual resolution of the influent pulse into coherent response pulses separated by plateau zones fails to occur in only one singular case, namely, if the composition of the base influent is on a watershed.

### 9. *Summary of Rules*

The chief rules for responses to square-wave influent composition pulses can be summarized as follows:

1. The final response pattern consists exclusively of single, coherent, permanent response pulses on a background of the base-influent composition (singular cases excepted). Each response pulse involves concentration changes of all species present.
2. The response pulses are of either of two types, with sharp front and diffuse rear flanks or vice versa. Initially they have flat tops, which later disappear as the front and rear flanks merge. The pulse amplitudes then diminish.
3. The usual, and maximum, number of response pulses is $n-1$. Fewer response pulses arise if the influent pulse and base influent have one or more $H$-function roots in common.
4. The sequence of the response pulses (counted in the direction of flow) is that of increasing index numbers of their variable roots.
5. Any species present in both the base influent and influent pulse is also present everywhere in the response pattern. A species $j$ absent from the influent pulse is absent from the flat top of the $(j-1)$th or $j$th response pulse. A species $j$ absent from the base influent is present only in the $(j-1)$th or $j$th response pulse.
6. The crest velocities of the response pulses are variable. Those of the pulses with sharp front flanks decrease, and those of the pulses with diffuse front flanks increase, as development progresses. The set of crest velocities eventually approaches that of the composition-velocity eigenvalues of the base influent.

7. The resolution of adjacent response pulses from one another requires finite development distances and times. The resolution distances and times are proportional to the duration of the influent pulse. Furthermore, rapid resolution is favored by a low amplitude of the influent pulse, large separation factors, and compositions far removed from watersheds. High flow rates and high concentration ratios $C/\bar{C}$ shorten the (true) resolution times but not the resolution distances.
8. The flat tops of the response pulses may or may not survive resolution. The development distances and times required for disappearance of the flat tops are proportional to the duration of the influent pulse. Survival of the flat tops is particularly favored by a low amplitude of the influent pulse.

The subsequent sections will show that some of these rules must be qualified for infinitesimal influent pulses and trace-component systems.

## B. Instantaneous and Infinitesimal Influent Pulses. Trace-Component Systems

Three interrelated special cases deserve further consideration, namely, influent pulses of extremely short duration, of extremely small amplitude, and in systems with all but one species at trace levels.

### 1. *Instantaneous Influent Pulses*

In the preceding section, the resolution distances and times were shown to be proportional to the duration of the influent pulse. Accordingly, in the limiting case of an instantaneous influent pulse (i.e., one of infinitesimal duration), resolution into coherent response pulses is instantaneous, granted the simplifying premises of the present treatment. In practice, of course, resolution is delayed by disturbances such as deviations from local equilibrium and axial diffusion and dispersion, which have so far been disregarded here and will be briefly discussed in Chapter 5, Section V.

The duration of the influent pulse being very short, the total amount injected is necessarily very small. Therefore, as the response pulses stretch out over finite widths, their amplitudes become very small. Thus, the behavior of the response pulses quickly approaches that found in responses to infinitesimal influent pulses, to be discussed next. Specifically, the peak velocities rapidly approach the composition-velocity eigenvalues of the base influent, and the nonsharpening pulse flanks soon cease to spread.

## 2. *Infinitesimal Influent Pulses*

A limiting case of some practical interest is that of an influent pulse of infinitesimal amplitude. Here, all concentration changes in the influent and thus all root variations remain infinitesimal:

$$h_i \simeq h_i' \simeq h_i^0 \qquad \text{for all } i \tag{4.52}$$

and, therefore, so do the amplitudes of all response pulses [compare condition (4.37) and its deduction].

The behavior of response boundaries generated by infinitesimal influent composition changes was examined in detail in Section I.G.1. The same behavior is displayed here by the flanks of the response pulses, particularly with respect to retention of sharpness and constancy of velocities. Since all pulse flanks remain sharp, all response pulses retain their widths and square-wave shapes indefinitely. (In practice, spreading caused by disturbances neglected here affects both pulse flanks to virtually the same degree and thus produces attenuating, symmetrical response pulses of a general shape well known from experience and theory of conventional chromatography.) The velocities of the pulse flanks, given by the composition-velocity eigenvalues of the system, remain virtually constant regardless of crossovers; for the flanks of the $k$th pulse:

$$(u)_k = h_k \prod_i h_i \prod_i \alpha_{i1} \tag{4.38}$$

Because of the retention of sharpness and the constancy of the velocities of the pulse flanks, the appearance of the distance-time diagram is quantitatively as shown in Fig. 4.11, i.e., the pattern consists of pairs of parallel, linear trajectories.

The response pulses furthermore share the following properties with the response boundaries generated by infinitesimal influent composition changes: The velocities of the pulse flanks, given by the composition-velocity eigenvalues, are independent of the direction of the influent-pulse vector, and with exceptions only for trace components, each response pulse involves concentration changes of all species present (see Section I.G.1).

The development distances and times required for resolution of any two response pulses from one another are readily calculated. The distance $z_{j/k}$ and adjusted time $\tau_{j/k}$ required for resolution of the $j$th from the $k$th pulse ($j < k$) are

$$z_{j/k} = \frac{h_j h_k}{h_k - h_j} \prod_i h_i \prod_i \alpha_{i1} \Delta\tau \tag{4.53}$$

$$\tau_{j/k} = \frac{h_k}{h_k - h_j} \Delta\tau \tag{4.54}$$

where $\Delta\tau$ is the duration of the influent pulse.

*Derivation.* Resolution requires crossover of the front flank of the $j$th pulse and the rear flank of the (faster) $k$th pulse. Being generated at $z=0$ at the times $\tau=0$ and $\tau=\Delta\tau$, these two pulses cover the distance $z_{j/k}$ in the times $\tau_{j/k}$ and $\tau_{j/k}-\Delta\tau$, respectively (see Fig. 4.16). Thus, with Eq. (4.38):

$$\frac{z_{j/k}}{\tau_{j/k}} = h_j \prod_i h_i \prod_i \alpha_{i1} \quad \text{and} \quad \frac{z_{j/k}}{\tau_{j/k}-\Delta\tau} = h_k \prod_i h_i \prod_i \alpha_{i1} \tag{4.55}$$

Solving for $z_{j/k}$ and $\tau_{j/k}$ one obtains Eqs. (4.53) and (4.54).

Infinitesimal influent composition pulses can be used for determining composition-velocity eigenvalues, from which information about equilibrium properties can be deduced. This so-called *concentration-pulse chromatography* will be briefly discussed in Section III.E.

### 3. *Trace-Component Systems*

The particular properties of response boundaries in systems with all but one species at trace levels were discussed in detail in Section I.G.2. The behavior of response pulses in such systems is analogous. A brief survey may therefore suffice.

The response pulses combine the properties of infinitesimal pulses

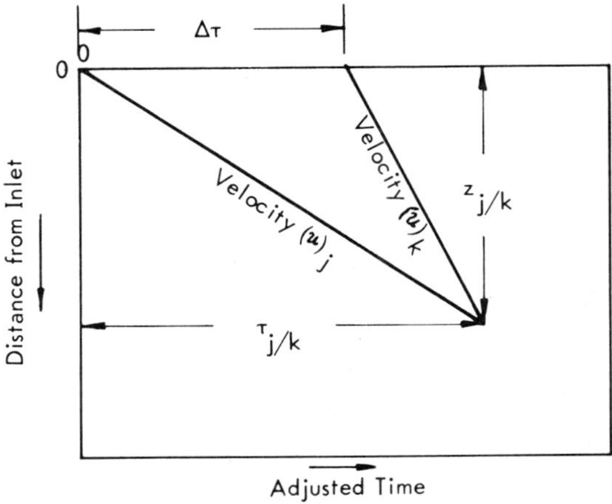


**Fig. 4.16.** Intersecting trajectories of sharp boundaries with variable $h_j$ and $h_k$, originating from influent composition changes at $\tau=0$ and $\tau=\tau\Delta$.

with those of trace-component systems. As infinitesimal pulses, they retain their widths and square-wave shapes indefinitely (granted the simplifying premises as to absence of disturbances) and advance at constant rates that are independent of the direction of the influent-pulse vector, as discussed earlier in this section. The characteristic property of trace-component systems is that the predominance of the bulk species suppresses mutual

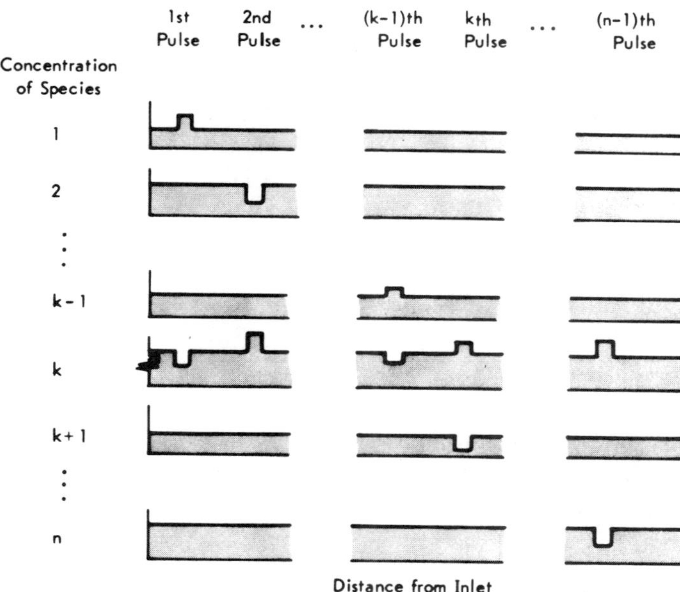


**Fig. 4.17.** Typical response pattern generated by influent pulse in trace-component system with bulk species $k$ (schematic).

interference of the trace species. Each trace species thus has its own response pulse, across which, after resolution, the concentrations of all other trace species are constant and which advances independently at the respective trace-species velocity. For lack of interference and attenuation, each trace species preserves its influent-pulse concentration in its response pulse. A characteristic response pattern is shown in Fig. 4.17.

## C. Pulses Injected into Single-Component Influents. Elution Development

Much of the remaining discussion of influent pulses will be devoted to various special cases involving single-component base influents or influent pulses. For ease of reference, the cases will be characterized in the shorthand notation adopted earlier, which lists the species in the successive influents and indicates the sequence by arrows (compare Section I.C.1). Thus,

$$k \quad \rightarrow \quad 1, \ldots, n \quad \rightarrow \quad k \tag{4.56}$$

signifies an influent pulse of all species 1 to $n$ in a base influent consisting only of species $k$.

The case (4.56) is particularly important since most chromatographic techniques employ such a procedure, in which an agent $k$ serves in the dual role of presaturant and development agent. Usually, the base-influent species $k$ is that of lowest affinity (i.e., $k = n$) or, in sorption chromatography, is the solvent or carrier gas, and the technique is called *elution development*. Detailed theories for elution development have been available for many years and are found in any text on chromatography, in some to the exclusion of any other modes of operation. Nevertheless, the results of the present treatment are of interest for comparison.

### 1. *Finite Pulses*

We examine first the response to influent pulses containing all their species at finite concentrations. The response to pulses with all species other than $k$ at trace levels will be discussed separately.

Pulse injection into a single-component base influent [case (4.56)] is one of the few cases which give a "unique" final response pattern, that is, the sequence and shapes of the response pulses and the distribution of the species over the various portions of the pattern are independent of the relative concentrations in the influent. The pattern is shown schematically in Fig. 4.18. The species other than $k$ are completely resolved from one another, each having its separate response pulse. The $n-1$ response pulses, on a background of the base-influent species $k$, are in the sequence of decreasing affinity of the respective species (counted in the direction of flow). Qualitatively, this behavior is in full agreement with chromatographic experience, the response pulses corresponding to what are usually called "peaks".

The properties of the response pulses are readily derived as follows: With $k$ as the only species in the base influent, the $H$-function roots of the latter are all trivial:

$$\begin{aligned} h_i' &= \alpha_{1i} && \text{for all } i < k, \\ h_i' &= \alpha_{1,i+1} && \text{for all } i \geqq k \end{aligned} \tag{4.57}$$

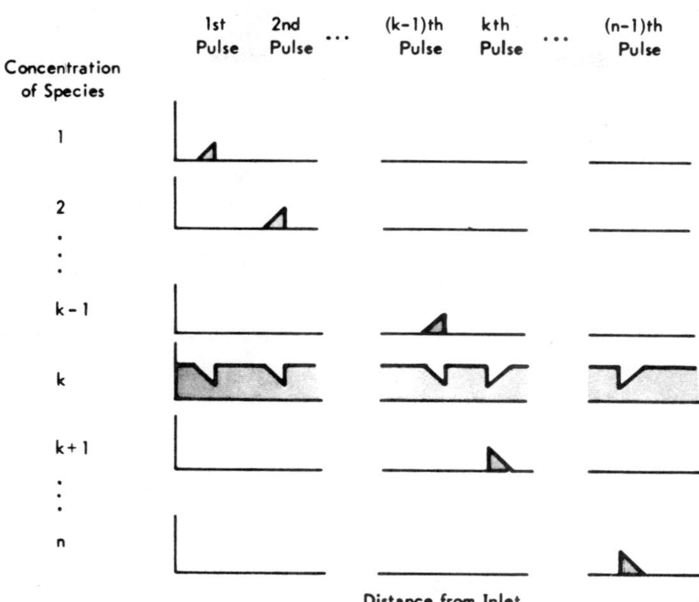


**Fig. 4.18.** Response pattern generated by multicomponent influent pulse in single-component base influent of species $k$ (schematic).

[see condition (3.55)]. As shown in Section III.A.4, any species $j$ absent from the base influent is present in only one response pulse, namely, in the $(j-1)$th pulse if $h'_{j-1} = \alpha_{1j}$, and in the $j$th pulse if $h'_j = \alpha_{1j}$. In the present case, Eqs. (4.57) decide between the alternatives and thus uniquely determine the distribution of the species over the response pulses in the manner shown in Fig. 4.18. Furthermore, since no species is absent from the influent pulse, no root $h_i^0$ equals a separation factor $\alpha_{1i}$ or $\alpha_{1,i+1}$, so that, with condition (3.50),

$$\alpha_{1i} < h_i^0 < \alpha_{1,i+1} \tag{4.58}$$

Therefore, with Eqs. (4.57):

$$\begin{aligned} h_i^0 &> h_i' \qquad \text{for all } i < k \\ h_i^0 &< h_i' \qquad \text{for all } i \geqq k \end{aligned} \tag{4.59}$$

Invoking the criterion (4.44), one then finds that the first through $(k-1)$th pulses have sharp front and diffuse rear flanks while the $k$th through

$(n-1)$th have the opposite behavior, as indicated in Fig. 4.18. One also finds that none of the $n-1$ response pulses can be missing from the pattern, since this would require $h_i^0 = h_i'$ for at least one root.

The factors controlling the distances and times required for resolution of the response pulses from one another and for disappearance of the flat tops are the same as in the general case of multicomponent base influents (see Section III.A.6). The concentration of a species in the flat-top zone, if still existing, of its fully resolved response pulse is readily calculated as follows: The arbitrary influent-pulse species $j \neq k$ occurs only in the $j$th response pulse if $j < k$, and only in the $(j-1)$th if $j > k$. With the root values (4.57), Eqs. (4.45) and (4.46) for the flat-top concentrations of species $j$ reduce to

$$\begin{aligned} x_{jl} &= \frac{h_l^0 - \alpha_{1j}}{\alpha_{1k} - \alpha_{1j}} \qquad l = j \qquad \text{if} \quad j < k \\ y_{jl} &= \frac{1/h_l^0 - \alpha_{j1}}{\alpha_{k1} - \alpha_{j1}} \qquad l = j-1 \qquad \text{if} \quad j > k \end{aligned} \tag{4.60}$$

(For all other values of $l$ one finds $x_{jl} = y_{jl} = 0$.)

As in the general case of multicomponent base influents, a species may have a higher concentration in the flat-top zone of its fully resolved response pulse than in the influent pulse. Chromatographic separation of a species from others by development with the base influent can thus be accompanied by an enrichment, i.e., the effluent concentration of a species may be higher than the influent concentration. Such an enrichment has indeed been observed in practice, for example, for glycerol in its separation from NaCl by "ion exclusion" [25–27], but has not been adequately explained by earlier theories. It is shown below that, with $k$ as the base-influent species, enrichment invariably results for species $k-1$ and $k+1$, provided their flat-top zones survive resolution. Other species may or may not be enriched, depending on the influent-pulse composition and the separation factors.

*Derivation.* The concentration of the arbitrary species $j$ in its resolved flat-top zone is given by Eq. (4.60), and that in the influent pulse is given by Eq. (3.57) with root values $h_i^0$ substituted for the $h_i$. For the enrichment factor $R_j$, defined as the ratio of the two concentrations, these equations give

$$R_j \equiv \frac{x_{jl}}{x_j^0} = \frac{\prod\limits_{i \neq j,k} (\alpha_{1i} - \alpha_{1j})}{\prod\limits_{i \neq l} (h_i^0 - \alpha_{1j})} \qquad \begin{aligned} l &= j \qquad \text{if} \quad j < k \\ l &= j-1 \qquad \text{if} \quad j > k \end{aligned} \tag{4.61}$$

Replacement of $l$ by the appropriate value $j$ or $j-1$ and rearrangement of the products yields

$$R_j = \prod_{i=1}^{j-1}\left(\frac{\alpha_{1i}-\alpha_{1j}}{h_i^{\text{o}}-\alpha_{1j}}\right)\prod_{i=j+1}^{k-1}\left(\frac{\alpha_{1i}-\alpha_{1j}}{h_i^{\text{o}}-\alpha_{1j}}\right)\prod_{i=k+1}^{n}\left(\frac{\alpha_{1i}-\alpha_{1j}}{h_{i-1}^{\text{o}}-\alpha_{1j}}\right) \quad \text{if} \quad j < k \tag{4.62}$$

$$R_j = \prod_{i=1}^{k-1}\left(\frac{\alpha_{1i}-\alpha_{1j}}{h_i^{\text{o}}-\alpha_{1j}}\right)\prod_{i=k+1}^{j-1}\left(\frac{\alpha_{1i}-\alpha_{1j}}{h_{i-1}^{\text{o}}-\alpha_{1j}}\right)\prod_{i=j+1}^{n}\left(\frac{\alpha_{1i}-\alpha_{1j}}{h_{i-1}^{\text{o}}-\alpha_{1j}}\right) \quad \text{if} \quad j > k \tag{4.63}$$

With the condition

$$\alpha_{1i} < h_i^{\text{o}} < \alpha_{1,i+1} \tag{4.58}$$

one finds that in both Eqs. (4.62) and (4.63) all ratios under the first and third products are larger than unity whereas all ratios under the second product, while positive, are smaller than unity. Accordingly, $R_j$ is larger than unity if the index numbers are such that the second product does not appear. This is the case in Eq. (4.62) if $j = k-1$, and in Eq. (4.63) if $j = k+1$. Hence

$$R_{k-1} > 1 \qquad \text{and} \qquad R_{k+1} > 1 \tag{4.64}$$

that is, species $k-1$ and $k+1$ are enriched.

By the same token, $R_j$ is smaller than unity if the index numbers are such that the first and third products in Eq. (4.62) or (4.63) do not appear. This is the case in Eq. (4.62) if $j = 1, k = n$, and in Eq. (4.63) if $k = 1, j = n$. Accordingly,

$$R_1 < 1 \quad \text{if} \quad k = n \qquad \text{and} \qquad R_n < 1 \quad \text{if} \quad k = 1 \tag{4.65}$$

that is, regardless of the influent-pulse composition, enrichment cannot be achieved for species 1 if the base-influent species is $n$, or for species $n$ if the base-influent species is 1.

## 2. *Trace-Component Pulses*

A special case deserving further comment is that of an influent pulse with all species other than $k$ at trace levels. With the trace species stated separately in parentheses:

$$k \quad \rightarrow \quad \begin{matrix} k \\ (1, \ldots, k-1, k+1, \ldots, n) \end{matrix} \quad \rightarrow \quad k \tag{4.66}$$

This case is of particular interest because it provides the closest link with the classical theories of chromatography.

The response combines the properties of systems with trace-component pulses in general (see Section III.B.3) with the unique distribution pattern produced by development with a single-component base influent, as shown in Fig. 4.18. In particular, the response shows the two chief characteristics of trace-component systems, namely, the predominance of the bulk species suppresses mutual interference of the trace species, and the composition variations are so small that self-sharpening and nonsharpening tendencies are virtually absent. For lack of interference, the response pattern is a mere superposition of single independent response pulses which can be calculated separately, as is taken for granted in the classical theories of chromatography. For lack of nonsharpening tendencies, the present treatment, granted its simplifying premises, leads to response pulses which retain their square-wave shapes indefinitely and do not attenuate; any disturbances neglected here will, however, result in bell-shaped, attenuating, nearly symmetrical response pulses (compare Section III.B.3 and Chapter 5, Section V.A.3).

As shown earlier for trace-component systems in general, concentration variations of a trace species advance at the respective trace-species velocity, which, for the trace species $j$ in the presence of $k$ as the bulk species, equals the separation factor $\alpha_{kj}$ (in terms of adjusted velocity; see Section I.G.2 and Chapter 3, Section IV.D.6). Hence, in the present case, the adjusted velocity of the response pulse of species $j$ (more precisely, of the flanks of this pulse) is

$$\mathscr{u}_{x_j} = \alpha_{kj} \tag{4.67}$$

Separation factors can thus be calculated from observed retention times or volumes of the response pulses. This is one of the various methods for chromatographic determination of equilibrium properties, to be surveyed briefly in Section III.E.

Because of the constancy of the pulse-flank velocities and the resulting linearity of the trajectories, resolution distances and times are readily calculated. The distance and adjusted time required for resolution of the arbitrary trace species $j$ and $l$ $(j < l)$ from one another are

$$z_{jl} = \frac{\alpha_{kl}}{\alpha_{jl}-1}\,\Delta\tau \qquad \text{and} \qquad \tau_{jl} = \frac{\alpha_{jl}}{\alpha_{jl}-1}\,\Delta\tau \tag{4.68}$$

[The derivation is analogous to that of Eqs. (4.53) and (4.54), with Eq. (4.67) instead of Eq. (4.38) used for the velocities.] These equations make no allowance for delays in resolution resulting from spreading of the pulse flanks because of disturbances; for practical purposes, therefore, the values merely constitute ideal lower limits.

Another feature of the response in trace-component systems is that enrichment or depletion of the species in their fully resolved flat-top pulses relative to the influent pulse, a phenomenon solely caused by interference, does not occur. Rather, each trace species preserves its influent pulse concentration in the flat top of its response pulse.

It is readily verified that the equations derived earlier for more general cases, after substitution of the appropriate root values, reduce to the forms or behavior stated here.

With the $h_i'$ values (4.57) and the condition of negligible root variations, (4.52), one obtains Eq. (4.67) from Eq. (4.49), and Eqs. (4.68) from Eqs. (4.53) and (4.54), and finds that the enrichment factors $R_j$ in Eqs. (4.62) and (4.63) become unity. It must be noted that the indices of pulse velocities and resolution distances and times in the earlier general equations refer to the pulse numbers, whereas in Eqs. (4.67) and (4.68) they refer to the trace species constituting the pulses; the trace species $j$ constitutes the $j$th pulse if $j < k$, and the $(j-1)$th pulse if $j > k$.

### D. Single-Component Influent Pulses. Vacancy Chromatography. Ghost Peaks

Another special case of practical interest is that of a single-component influent pulse of species $k$ interrupting a multicomponent base influent containing all species including $k$:

$$1, \ldots, n \quad \rightarrow \quad k \quad \rightarrow \quad 1, \ldots, n \tag{4.69}$$

Its characteristic properties will now be considered.

#### 1. *Properties of the Response Pattern*

The rules and criteria derived earlier for the general case of arbitrary influent pulses also apply in the case (4.69) and lead to a unique response pattern, i.e., one whose properties are independent of the relative concentrations in the influent. This pattern is shown schematically in Fig. 4.19. It consists of $n-1$ response pulses, each involving concentration maxima and minima of all species. The influent-pulse species, $k$, is the only one to have a concentration maximum in each response pulse. The other maxima and minima are arranged in the characteristic manner shown in the diagram. Each species other than $k$ is absent from the flat top of one response pulse, as indicated in Fig. 4.19 by the respective concentration profile's dipping to zero. (Of course, after attenuation has set in, the concentration minimum no longer reaches down to zero.)

The general properties of the pattern are readily derived as follows: With $k$ as the only species in the influent pulse, the $H$-function roots of the latter are all trivial:

$$\begin{aligned} h_i^0 &= \alpha_{1i} && \text{for all } i < k \\ h_i^0 &= \alpha_{1,i+1} && \text{for all } i \geqq k \end{aligned} \tag{4.70}$$

and with no species absent from the base influent, no root $h_i'$ equals a separation factor $\alpha_{1i}$ or $\alpha_{1,i+1}$, so that

$$\alpha_{1i} < h_i' < \alpha_{1,i+1} \tag{4.71}$$

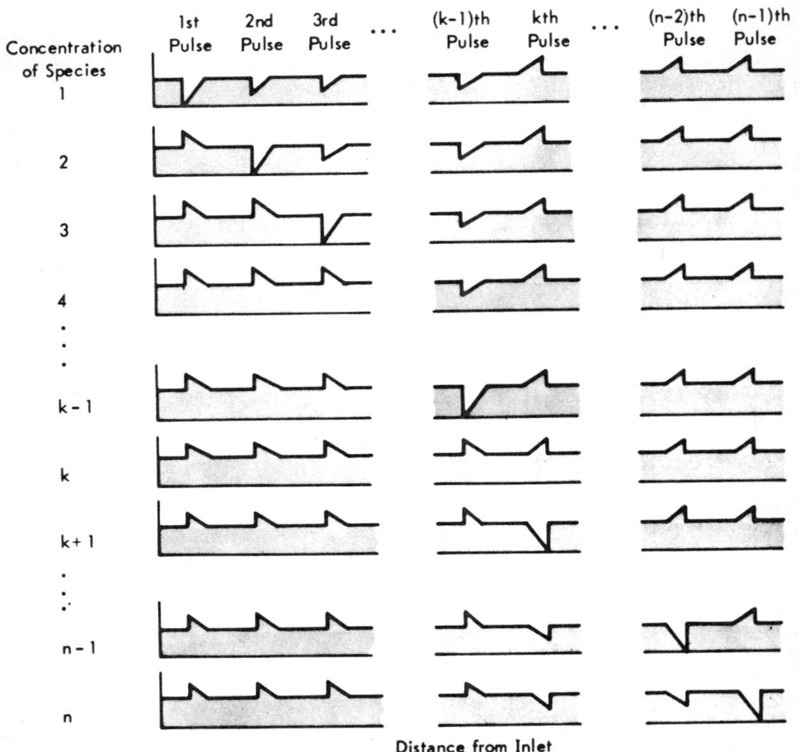


**Fig. 4.19.** Response pattern generated by single-component influent pulse of species $k$ in multicomponent base influent (schematic). Profiles for each pulse are shown at moment at which flat top disappears, i.e., before attenuation.

(see Chapter 3, Section IV.D.2). It follows that

$$\begin{aligned} h'_i &> h_i^0 \qquad \text{for all } i < k \\ h'_i &< h_i^0 \qquad \text{for all } i \geqq k \end{aligned} \tag{4.72}$$

With these inequalities, which are independent of the base-influent composition, condition (4.44) gives a pattern with the concentration maxima and minima and pulse shapes as indicated in Fig. 4.19. That species $j$ is absent from the flat top of the $j$th pulse if $j<k$, and from that of the $(j-1)$th pulse if $j>k$, follows from Eq. (4.45) after substitution of the root values (4.70).

A slightly different pattern develops if the influent-pulse species, $k$, is absent from the base influent:

$$1, \ldots, k-1, k+1, \ldots, n \quad \rightarrow \quad k \quad \rightarrow \quad 1, \ldots, k-1, k+1, \ldots, n \tag{4.73}$$

Species $k$ then is absent from the entire response pattern except for the $(k-1)$th or $k$th pulse (see rules for absence of species, Section III.A.4). Which of these two pulses carries species $k$ depends on the root values, and thus on the composition, of the base influent, so that the pattern is no longer unique. In all other respects, however, its properties are the same as in the case (4.69).

## 2. *Trace-Component Base Influents*

More interesting for practical applications is the response in a system containing all species other than $k$ at trace levels:

$$\frac{k}{(1, \ldots, k-1, k+1, \ldots, n)} \rightarrow k \rightarrow \frac{k}{(1, \ldots, k-1, k+1, \ldots, n)} \qquad (4.74)$$

The response in this case shows the characteristic features of trace-component systems. Since the composition variations are too small to cause significant self-sharpening or nonsharpening tendencies, attenuation of the response pulses results only from disturbances so far neglected in this treatment (see Section III.B.3). More importantly, mutual interference of the trace species is suppressed, so that the concentration variations of the latter advance independently and at the respective species velocities (e.g., see Section I.G.2 and Chapter 3, Sections III.B and IV.D.6). One may say that the influent pulse of species $k$ produces a "gap" in the otherwise continuous flow of the trace species, and that the gaps of the various species then advance independently in the same manner and at the same velocities as do the "peaks" in ordinary chromatography of trace species (i.e., the response pulses produced by trace-species injection into a single-component base influent; see Section III.C.2). In the response pattern, the sequence (counted in the direction of flow) of the gaps is that of decreasing affinity of the respective trace species, since the species velocities increase in this sequence. The pattern, shown schematically in Fig. 4.20, is an "upside down" replica of the normal chromatographic pattern in Fig. 4.18 with the concentration maxima and minima interchanged. Also, comparison with Fig. 4.19 reveals that the transition from finite to trace concentrations of the species other than $k$ has eliminated, in each response pulse, all concentration variations except those of species $k$ and of the species which is absent from the flat top.

As in the previously examined special cases, a proof for the properties of the pattern is readily derived from the general rules and criteria for arbitrary pulses after the appropriate root values have been substituted.

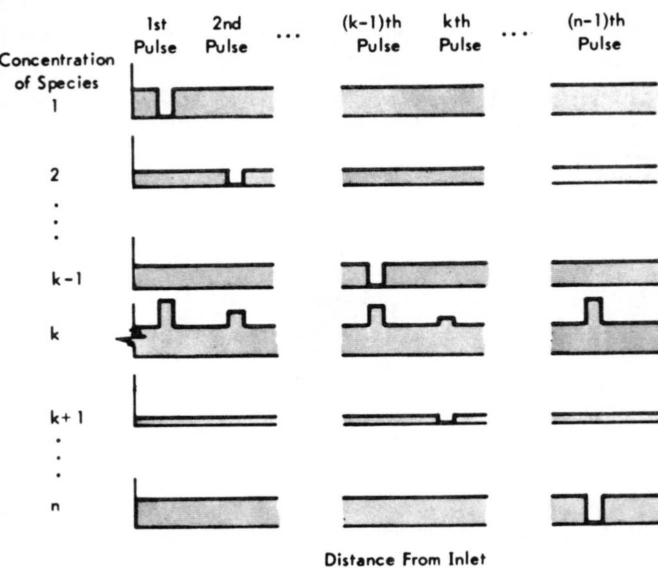


**Fig. 4.20.** Typical response pattern generated by single-component influent pulse of species $k$ in trace-component base influent with $k$ as bulk species (schematic).

In the present case, the $h_i^0$ values are given by Eqs. (4.70), and the $h_i'$ values differ only infinitesimally.

### 3. *Vacancy Chromatography*

In ordinary chromatography by elution development, where each species except that of the base influent has its own response pulse, material balance requires the area of each pulse in the effluent composition history to equal the injected amount of the respective species, and thus to be proportional to the concentration of that species in the injected mixture. This elementary principle is, of course, the basis for analytical applications. An analogous behavior is found in the type of operation just described, with a pulse of a species $k$ interrupting a base influent containing the other species at trace levels and $k$ as the bulk species [case (4.74)]. Again, each species other than $k$ has its own response pulse, although a negative one (see Fig. 4.20), and material balance requires the area (measured from the base-influent concentration) of each negative pulse in the effluent history to equal the missing amount of the respective species, i.e., the amount not received by the column because of the interruption of the base influent by the pulse of species $k$. The areas of the negative pulses in the effluent history therefore are proportional to the respective trace-species concentrations in the base influent. This mode of operation can thus be used as an alternative to conventional chromatography for determining the composition of a mixture. The technique has been called

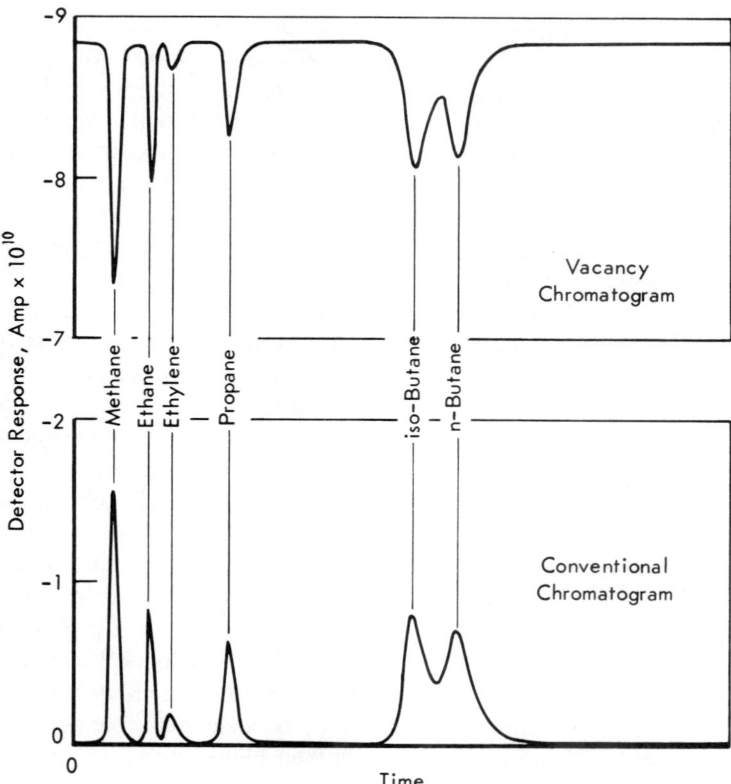


**Fig. 4.21.** Vacancy and conventional chromatograms of hydrocarbon gas mixture. Upper diagram: vacancy chromatogram generated by injection of clean air into air stream blended with small amount of hydrocarbon mixture; lower diagram: conventional chromatogram generated by injection of same mixture into clean air. (From results by B. O. Prescott and H. L. Wise [30].)

*vacancy chromatography* by Zhukhovitskii [28, 29], its inventor and chief proponent. Other names are *inverse*, *shadow*, *blank*, or *mirror chromatography*.

The method is most conveniently applied to gas chromatography. Here, pure carrier gas is used for the influent pulse, the detector in the effluent records the sum of the concentrations of all other species, and the negative pulses in this recording are evaluated. Figure 4.21 shows a comparison of such a "vacancy chromatogram" with that of conventional elution development (trace injection into carrier gas) of the same mixture, illustrating the close correspondence between the two techniques.

For the trace-component case (4.74) discussed so far, the present treatment and Zhukhovitskii's theory of vacancy chromatography [28, 29] are equivalent. For higher concentrations [i.e., case (4.69)], however, they are not, since Zhukhovitskii postulates composition-independent distribution ratios $y_i/x_i$, whereas the present treatment allows mutual interference to affect these ratios. The present treatment shows that the applicability of vacancy chromatography, in contrast to that of conventional chromatography, depends critically on the absence of interferences and therefore is essentially restricted to the trace- or at least low-concentration range. In conventional chromatography, interference at higher concentrations is of little concern because it merely affects the shapes and velocities of the response pulses, but does not obstruct resolution into pulses each containing only one species, so that the one-to-one correspondence between pulse areas and concentrations of the species in the injected mixture is preserved. In vacancy chromatography, in contrast, interference destroys this correspondence. As the pattern for the general case (4.69) in Fig. 4.19 shows, each response pulse then involves concentration variations of *all* species and can therefore no longer be correlated with any single species' concentration. Although operating with very short influent pulses helps to suppress interference effects in conventional chromatography by letting the concentrations of the injected species drop rapidly to very low levels, it does not do so in vacancy chromatography since, here, the concentrations in the pulses then rapidly approach their higher levels existing in the base influent. Zhukhovitskii's theory, which ignores interference effects, thus appears to be of doubtful value for other than very dilute base influents. In particular, his claim that the presence of other components in the negative pulse is advantageous because it suppresses tailing [28, 29] is based on contradictory premises; it is overlooked that interference, invoked as the cause of this beneficial effect, invalidates the principle on which the method is based.

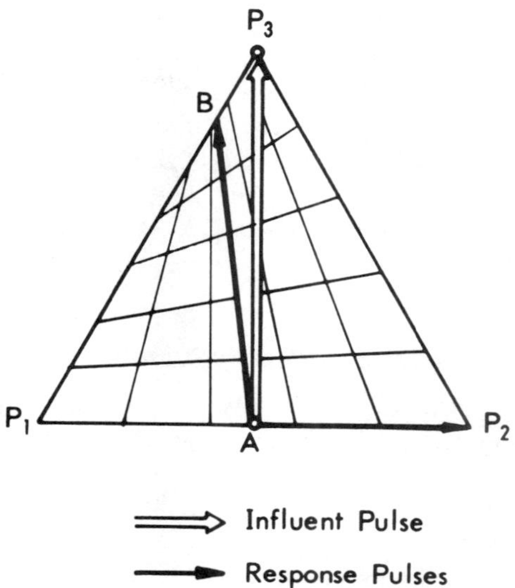


**Fig. 4.22.** Influent- and response-pulse vectors in three-component system with injection of species 3 into base influent of species 1 and 2.

The failure of vacancy chromatography in the event of interference is illustrated in Fig. 4.22 by the simple, but extreme, example of the case

$$1, 2 \rightarrow 3 \rightarrow 1, 2 \tag{4.75}$$

The composition diagram shows the influent-pulse vector $AP_3$ to be resolved into its components, the response-pulse vectors $AP_2$ (first pulse, $1 \mid 2$ cut) and AB (second pulse, $2 \mid 3$ cut). The influent-pulse species 3 is entirely absent from the first response pulse. A detector recording the sum of the concentrations of species 1 and 2 in the effluent, as would normally be used for vacancy chromatography of these two species, then misses the first pulse and shows only the second, giving an entirely misleading chromatogram. Even attenuation of the response pulses does not alter this situation.

### 4. *Ghost Peaks*

A phenomenon familiar from chromatographic practice and variously reported in recent years is the occurrence of "ghost peaks," i.e., negative pulses usually appearing near the air peak in otherwise normal gas chromatograms [30–34]. Such negative pulses must be expected if the sample is injected into an impure carrier gas. Sample injection then produces "gaps" in the flows of the carrier-gas impurities, so that a vacancy chromatography of the latter is superimposed on the regular elution development of the injected mixture. Although other causes cannot be entirely excluded, it is likely that most ghost peaks have this origin. Indeed, cases are known in which the ghost peaks were eliminated by careful purification of the carrier gas [30, 34] or by addition of appropriate amounts of the impurity to the sample [34].

### E. Concentration-Pulse and Tracer-Pulse Techniques

Concentration-pulse and tracer-pulse chromatography are two recent techniques for determining exchange or sorption isotherms. Although a comprehensive account cannot be given here, the fundamental difference between the two techniques, as seen in the light of the present treatment, deserves to be stressed.

In *concentration-pulse chromatography* [35–40], a small influent pulse is injected into a multicomponent base influent and is resolved into $n-1$ coherent response pulses, whose velocities are the velocity eigenvalues of the base-influent composition (see Section III.B.2; the use of a small influent composition step instead of a pulse has also been suggested [41]). In systems with only one sorbable component and in two-component systems with stoichiometric exchange, only a single response pulse appears, whose velocity is related to the isotherm slope $\mathrm{d}\bar{C}_i/\mathrm{d}C_i$ (see Chapter 3, Section III.B.1). Similarly, in multicomponent systems, the composition-velocity eigenvalues are related to the slopes $\partial\bar{C}_i/\partial C_i \mid_{\text{path}}$ of the multidimensional isotherm surface as measured in the directions of the $n-1$ composition paths. These slopes can thus be calculated from the observed response-pulse velocities or retention volumes. Partition ratios $\bar{C}_i/C_i$ of the various species are not directly obtained by this method. Rather, the isotherm surface must first be constructed from the slopes obtained for a large number of base-influent compositions. The complexity of the calculations increases rapidly with the number of components. Only for trace species, whose pulse velocities equal their species velocities (see Section III.B.2), can partition ratios be calculated directly with Eq. (3.10). This application to trace species dates back to 1956, when gas–liquid chromatography with small sample size was introduced as a method for determining limiting activity coefficients of solutes at infinite dilution [42, 43].

In *tracer-pulse chromatography* [38–41, 44–47), a distinguishable isotope is used to tag molecules of the species of interest. A small "tracer pulse" of tagged molecules is injected, with as little composition disturbance as possible, into a multicomponent base influent containing the same species untagged. To the extent that isotope effects are negligible (i.e., that tagging does not alter the affinity for the stationary phase), tagged and untagged molecules of the same species have the same species velocity. The injected tagged molecules advance at this velocity as a single pulse. The species velocity, regardless of the presence of other components and at any concentration, is directly related to the partition ratio $\bar{C}_i/C_i$ of the species by Eq. (3.10). The partition ratio of the tagged species at the base-influent composition can thus be calculated from the tracer-pulse velocity or retention volume. High concentrations of the tagged species and presence of even a large number of other components do not complicate the measurement and its evaluation.

According to these considerations, the two deceptively similar techniques differ in one fundamental aspect. Concentration-pulse chromatography, by observing composition changes, measures composition-velocity eigenvalues, whereas tracer-pulse chromatography, by observing tagged molecules, measures species velocities. The experimental verification of the differences between the two techniques [38] and the successful use of tracer-pulse chromatography in multicomponent systems [38, 44, 47] provide some of the strongest evidence supporting the conceptual basis of the present treatment, which rests largely on the distinction between concentration velocities and species velocities.

### F. Comparison with Other Theories

Pulse injection into a base influent of constant composition, as the most common chromatographic procedure, has been the subject of many theories, which differ in their basic premises. A brief attempt will be made to put the present treatment, as applied to influent pulses, into perspective by comparing it with the most typical other approaches.

#### 1. *Classical Theories of Elution Development*

The classical theories of chromatography (i.e., elution development) found in standard texts (e.g., [48–56]) are based on the premise of concentration-independent partition coefficients. While accounting, either directly or through the plate concept, for disturbances disregarded here, these "linear" theories ignore any interference effects, on which the present treatment has exclusively concentrated.

As far as the special case of very small sample size (i.e., pulse of low amplitude or short duration) is concerned, the discussion in Sections III.B and C has shown that interference is negligible. Here, the present treatment, if refined by inclusion of the various disturbances, would become equivalent to the classical theories and can serve no other purpose then to justify one of their premises.

On the other hand, for influent pulses of large amplitude and finite duration, there is only qualitative agreement between the present and the classical theories in that both predict complete eventual resolution into "peaks" of one species each (not counting the base-influent species). Here, the discussion in Section III.C.2 has shown that, not unexpectedly, interference during the period preceding resolution of the peaks can produce significant effects. In particular, interference increases the amplitudes of some peaks while decreasing those of others, as reflected by Eq. (4.61), an effect that cannot be explained, much less foreseen, on the basis of the classical theories. Although the predictions made with the present treatment still await experimental confirmation, they cast considerable doubt on the adequacy of the classical theories when applied to large pulses, e.g., to preparative separations involving high concentrations. The premise of invariant partition coefficients confines the validity even of the highly sophisticated modern developments [57–60] to the low-concentration range.

### 2. *Extensions to Nonlinear Isotherms*

Countless extensions to systems with concentration-dependent partition coefficients, and thus with nonlinear isotherms, have been undertaken. Early work includes that of DeVault [17] and Glueckauf [61]; typical more recent studies, in part with the use of computers, have included additional effects [62–66]. The theories of this group are concerned with the shapes of single-component peaks (on a background of carrier gas or solvent) and do not provide for interference.

While the restriction to constant partition coefficients has been removed, the applicability of these theories to chromatographic resolution remains somewhat doubtful. The positive or negative curvature of a nonlinear single-component sorption isotherm reflects what one may call "self-interference," since it reveals that solute already present in the sorbent either reduces or enhances sorption of further solute. It would be naive to expect that such interference takes place between molecules of the same species, but not between those of different species.

### 3. *Extensions to Interfering Solutes*

Attempts to devise theories for elution development of interfering species have so far been few. The earliest are "equilibrium theories" based on DeVault's [17] differential material-balance equations, which postulate local equilibrium and absence of disturbances. In this group, Glueckauf's treatment [19, 20] of elution development of two solutes with compound Langmuir sorption isotherms is mathematically equivalent to that of the three-component case with stoichiometric exchange and constant separation factors, discussed in Section III.C (see also Chapter 5, Section III.B for this equivalence). Glueckauf's formulas, however, appear more complex because he did not avail himself of the *H* function. Offord and Weiss [18] presented a controversial approach along somewhat similar lines (see also [67, 68] and footnote to p. 313). Sillén's work [7] on ion exchange with constant separation factors does not explicitly include the elution-development case, but is applicable to it and is mathematically identical to Glueckauf's, except for a different choice of variables. These three approaches, like all subsequent ones except numerical techniques, are based on premises which amount to postulating the absence of any finite regions of noncoherence in the distance-time field. In comparison, the present treatment needs no such postulate and is applicable also to cases involving finite regions of noncoherence. Such regions appear if the base-influent species $k$ is of intermediate affinity, $1 < k < n$.

Later work has failed to add significantly to these theories, except for a searching analysis of premises by Baylé and Klinkenberg [21] and numerical techniques to be discussed on p. 219.

### 4. *Theories for Pulses in Multicomponent Base Influents*

With broadening interest in chromatographic techniques other than elution development, various theories for, or applicable to, pulse injection into multicomponent base influents have been proposed in recent years.

One group of such theories is based on the premise, gleaned from the classical theories of elution development, that the species behave independently and have concentration-independent partition coefficients. Zhukhovitskii's theory of vacancy chromatography (see Section III.D.3) belongs to this group, and so does the intriguing work of Reilley *et al.* [69], who examined responses to a great variety of influent composition changes. Reilley's work includes a simultaneous and independent discovery of vacancy chromatography; in contrast to Zhukhovitskii's developments it also accounts for effects of disturbances, if in a somewhat unrealistic manner by postulating a Craig-distribution model for boundary broadening. By virtue of their premises, the applicability of these theories is essentially confined to the trace- or low-concentration range. They are equivalent to the treatment of pulses in trace-component systems (with the bulk species $k$ representing the carrier gas or solvent) given in Sections III.B.3 and D.2.

A second group of theories acknowledges the existence of interference effects, but confines itself to influent pulses of very low amplitudes. Thus, Stalkup *et al.* [35, 37] and Mangelsdorf [36] examined the response to such pulses, the former for the purpose of isotherm determination by "concentration-pulse chromatography" (see Section III.E), the latter to develop a sensitive difference method for analytical determinations. Neither treatment is confined to a particular type of isotherm, and neither gives explicit relations for the velocity eigenvalues and response-pulse compositions in the general case. Both were originally conceived as "equilibrium theories" (assuming local equilibrium and absence of disturbances), but more recent extensions include effects of finite mass-transfer rates, axial diffusion, and even interconversion of the species by chemical reactions [70, 71]. The theories in their original form, if applied to systems with constant separation factors, are equivalent to the treatment of the limiting case of infinitesimal pulses in Section III.B.2. The present treatment brings out clearly the two effects which make the theories inapplicable for finite pulses; these effects are the self-sharpening and nonsharpening tendencies of the pulse flanks and the variations of the crest velocities of finite response pulses, both arising from the variations of the velocity eigenvalues with composition.

For the same case of low-amplitude influent pulses in a multicomponent base influent containing the species at finite concentrations, a theory based on quite different premises has been proposed by Zhukhovitskii *et al.* [72–74] under the title "chromatography without carrier gas." As in this author's other theories [28, 29, 75–78], constant (i.e., composition-independent) partition coefficients are postulated. Despite the independence of these coefficients, however, the species are interdependent in their development behavior because the premise of constant total pressure in the gas phase has the effect that the partial pressure of a species depends not only on the stationary-phase concentration of that species, but on those of the other species as well. In other words, it is assumed that the species compete for room in the gas phase, but not in the stationary phase, a situation that can be envisaged only for very weakly sorbed gases. By virtue of the interference in the mobile phase, the equations of this theory bear some resemblance to those in the present treatment. In particular, a relation of the form of Eq. (3.57) exists for the mobile-phase concentrations, with partition coefficients and normalized response-pulse retention volumes substituted for separation factors and $H$-function roots, respectively. These analogies, however, are merely formal since the physical significance of the variables is different. Mobile-phase interference through coupling of partial pressures, as included by Zhukhovitskii, can indeed become significant in gas chromatography at high solute mole fractions. In the present treatment, this effect has so far been precluded by the premise of uniform bulk mobile-phase velocity. Relaxation of this premise will be discussed briefly in Chapter 5, Section VI.B. On the other hand, the present treatment clearly shows the grave consequences of interdependence of the partition coefficients, which Zhukhovitskii assumed to be independent constants. The applicability of the latter's theory thus is somewhat doubtful, particularly since the Russian group's own experimental results show clear evidence of such interdependence (e.g., see the reported sorption isotherms for propylene in the presence and absence of propane [79]). With respect to the limitation to low-amplitude pulses, the remarks stated earlier pertain to Zhukhovitskii's theory also.

No solutions for finite pulses in multicomponent base influents have so far been reported. However, the theories developed by Glueckauf [19, 20] and Sillén [7] as well as the numerical methods to be discussed below are, at least in principle, applicable to this case.

### 5. *Numerical Solution Methods*

With the advent of high-speed computers, it has become feasible to carry out a numerical "plate-by-plate" integration of the simultaneous differential material-balance equations for interfering solutes. Disturbances are readily included and, in a sense, even facilitate the computation because they prevent the formation of physically not realized "overhanging" concentration profiles in self-sharpening boundaries (compare Chapter 3, Section IV.E.1); regions of noncoherence then are no obstacle to the calculation. Such methods are applicable to all types of influent pulses as well as any other influent conditions. Results of such numerical calculations, still in a relatively primitive form, have recently been reported for selected elution-development cases with two solutes in a carrier gas [64].

It is difficult to see how quantitative calculations of complicated specific cases, say, with variable separation factors and several types of disturbances, could be handled in other ways than by numerical plate-by-plate integration. On the other hand, for deducing general rules the numerical methods are inherently less suited than explicit and conceptual approaches as taken here.

### 6. *Comparison with Present Treatment*

As the discussion above has shown, most of the other theories include one or the other effect disregarded here, but all are more limited in scope than the present treatment. With few exceptions (e.g., Zhukhovitskii's theory of chromatography without carrier gas), they reduce to various special cases of the present treatment if the simplifying premises of the latter are introduced. Because of its broader, unifying scope, the present treatment can serve to delineate the limits of applicability of the other theories, as was indicated above in a number of cases.

Also, the present treatment has been carried farther than any others in deriving predictive rules for arbitrary multicomponent systems. This has been possible in part through the use of $H$-function roots, which allow rules and criteria to be derived and stated in a simple manner, in part through its conceptual basis (species velocities, concentration velocities, coherence, composition paths), which still allows general qualitative rules to be deduced if the simplifying assumptions are relaxed and the $H$-function roots thus are inapplicable (see also Chapter 5).

## IV. Multiple Abrupt Influent Composition Changes

In many practical operations, the influent composition is changed several times in succession, usually abruptly, and in such a manner that the column receives a new influent before having come entirely to equilibrium with the previous one. This is the case, for example, if a column is used to remove various species from a stream and must be regenerated as soon as the first of these species begins to appear in the effluent. Other examples of practical importance are displacement development and selective displacement, to be discussed later in this section.

A virtually unlimited number of combinations of various successive influent composition changes can be conceived, and a comprehensive coverage will therefore not be attempted. Rather, the discussion will remain restricted to the common general properties of the responses and to a few selected special cases which are either of practical importance or particularly apt to illustrate certain characteristic features.

The following conventions as to notation will be adopted. Quantities referring to the various successive influents (such as concentrations, $H$-function roots) will be characterized by primes, double primes, etc.; for example, the compositions of four sequential influents will appear as

| | |
|---|---|
| $\mathbf{x}'$ | final influent |
| $\mathbf{x}''$ | later intermediate influent |
| $\mathbf{x}'''$ | earlier intermediate influent |
| $\mathbf{x}''''$ | presaturant |

To characterize special cases, the shorthand notation introduced in Section I.C.1 will be used, which lists the species in the various influents and indicates the sequence by arrows; for example,

$$1 \quad \rightarrow \quad 2, \ldots, n-1 \quad \rightarrow \quad n$$

will denote operation with species $n$ as the presaturant, a mixture of species $2, \ldots, n-1$ as the intermediate influent, and species 1 as the final influent.

## A. General Properties

The general qualitative features of the developing response pattern can be recognized in schematic distance-time diagrams. Each of the successive influent composition changes generates a set of coherent response boundaries; these, through crossovers, sort themselves out in such a manner that their eventual sequence (in the direction of flow) is that of increasing index numbers of the variable roots. This was discussed in detail in Section III.A, and illustrated in Fig. 4.11, for the response to a square-wave influent pulse, a special case of a system with two successive influent composition changes. In the schematic distance-time diagram, in which boundary spreading and velocity variations on crossovers are disregarded, nothing distinguishes the special case of an influent pulse from the general case of two arbitrary successive influent composition changes. If the influent composition is changed three, four, etc., times in succession, the respective schematic distance-time diagram then shows triplets, quadruplets, etc., of boundary trajectories.

The general properties of the final response pattern are readily predicted from the sets of $H$-function roots of the successive influents. The variation of each root in the final pattern may constitute a self-sharpening boundary, one or several nonsharpening boundaries, or a single or multiple pulse, depending on the type of variation of the root in the influent history. The general rules for development of arbitrary noncoherent boundaries, derived in Chapter 3, Section V.D and illustrated in Figs. 3.29 and 4.10 are applicable. For example, a single self-sharpening boundary eventually results if the value of the respective variable root is higher in the last influent than in the presaturant; this boundary forms either through merger of the self-sharpening boundaries generated by the various influent composition changes or, if the root variation in the influent history is not monotonic, through degeneration of a transient pulse. The responses for several types of root variations in the influent history are shown in Fig. 4.23. The complete final pattern is readily compiled once the form of variation has been established for each root.

The compositions of the plateau zones between the variations of the different roots in the final pattern are also readily established. In the final pattern, the entire variation of each root from its presaturant value

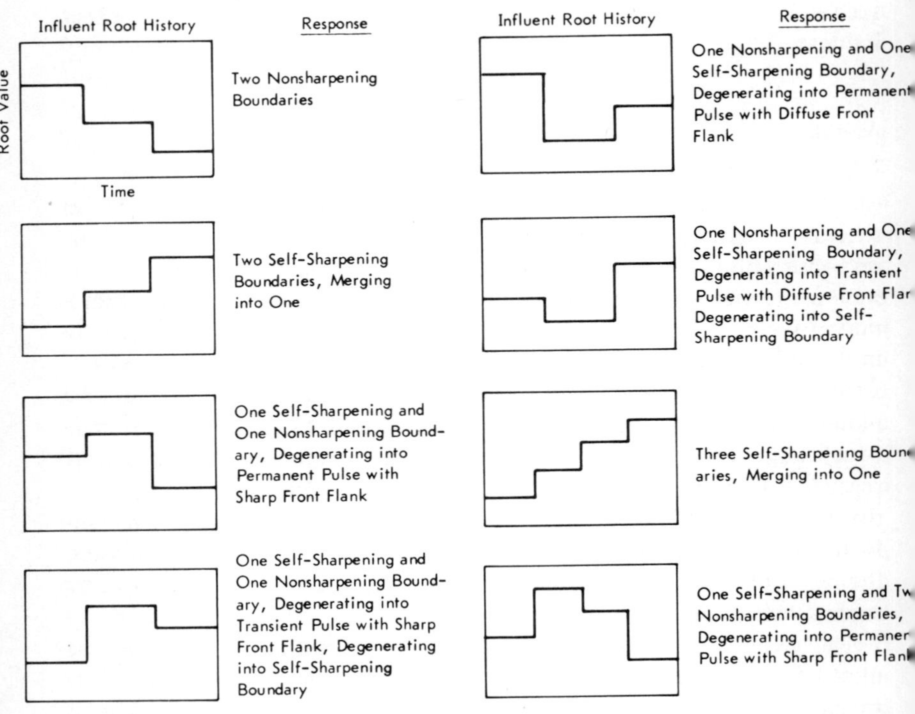


**Fig. 4.23.** Various types of influent histories of a root in successive abrupt influent composition changes, and resulting responses.

to its value in the last influent has become resolved from those of the other roots, in the same manner as in the response to a single influent composition change. The sets of roots in the plateau zones between variations of different roots, and thus the compositions of these zones, therefore are the same as they would be if the change from the initial to the final influent composition had been carried out in one instead of several steps. Accordingly, as in the pattern resulting from a single influent composition change or, for that matter, from development of any arbitrary initial noncoherent boundary, the root sets and plateau-zone compositions are given by Eqs. (3.92) to (3.94); merely the symbol $h_i''$ for the $H$-function roots of the presaturant must be replaced by that appropriate in the respective case (i.e., $h_i'''$ for two influent composition changes, $h_i''''$ for three changes, etc.).

As in responses to gradual influent composition changes, deviations from the normal resolution behavior can occur in exceptional cases involving compositions on watersheds. Specifically, a plateau zone may fail to form between variations of two vicinal roots, or the resolution of two such root variations may remain incomplete. The conditions for such behavior are as set forth in Chapter 3, Section V.F, for noncoherent boundaries in general.

These rules for the final pattern still leave much leeway for individual behavior of particular cases, as various examples in later parts of this section will show. The transient development behavior, moreover, is even less amenable to generalizations. To be sure, the development of each individual root variation from its initial to its final form is adequately described by the rules derived previously, but, except in a few simple cases, the sequence of the various boundary crossovers and mergers cannot be established without quantitative computation of the response. The quantitative numerical and graphical methods developed for the general case of arbitrary noncoherent boundaries are applicable here, too.

The general response behavior is illustrated in Fig. 4.24 by the composition-route and distance-time diagrams of three three-component cases, each with two successive influent composition changes. In each case, the compositions of the final, intermediate, and initial influent correspond to the points A, B, and C, respectively, in the route diagram, so that the influent history route (in the direction of progressing time) is CBA. The transient and final routes shown are profile routes, as have been used throughout in previous sections.

In the first case shown in Fig. 4.24, the two influent composition changes give rise to two self-sharpening response boundaries each, and the composition route immediately after the second influent composition change is ADBEC. As the boundaries DB and BE cross, the plateau zone B disappears and the route changes to ADFEC. The plateau zones D and E now find themselves between self-sharpening boundaries having the same variable root, and disappear when these boundaries merge. The final route is AFC, with two self-sharpening boundaries, and is the same as would have resulted from a single-step change from the presaturant C to the final influent A.

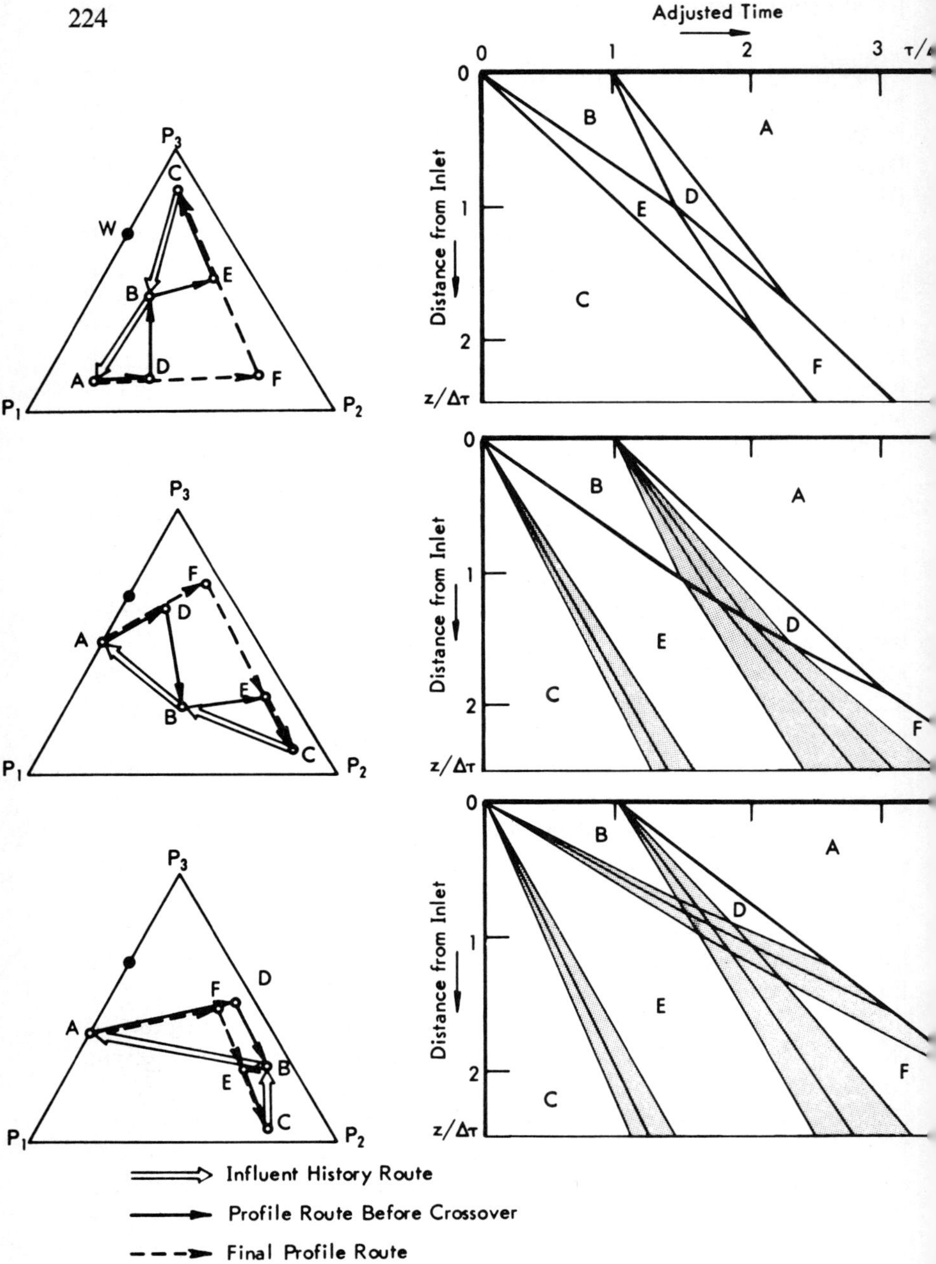

⟹ Influent History Route

→ Profile Route Before Crossover

- - → Final Profile Route

**Fig. 4.24.** Typical responses to two successive abrupt influent composition changes. Composition profile routes and distance-time diagrams for three three-component cases. (For $\alpha_{12} = 2$, $\alpha_{13} = 4$.)

Similarly, in the second case shown in Fig. 4.24, a final route develops via the transient routes ADBEC and ADFEC. Here, however, the plateau zone E, after crossover of the boundaries DB and BE, is between two nonsharpening boundaries having the same variable root and thus retains its length, so that the final route is AFEC.

In the third case shown in Fig. 4.24, the route again develops via the transients ADBEC and ADFEC. Here, the portion ADF corresponds to a flat-top pulse. The flat-top zone D disappears, and the pulse, transient because its self-sharpening flank involves the greater composition change, eventually degenerates into a single self-sharpening boundary, AF. The plateau zone E, after crossover of the boundaries DB and BE, is between two nonsharpening boundaries having the same variable root and thus retains its length. The final route in this case is AFEC.

These examples also illustrate how, at least in simple cases, the development behavior can be simply assessed by application of the general rules for responses to single influent composition changes, for interference of coherent boundaries, and for plateau zones between coherent boundaries, derived in various earlier sections. In the first two cases shown, the complete response can also be calculated algebraically with the method given in Appendix III; in the third case, numerical or graphical methods are required for constructing the trajectories of the diffuse boundaries on and after crossover.

## B. Displacement Development

A particular type of operation involving two successive influent composition changes is of practical importance for preparative separations. In this technique, called *displacement development* and first proposed by Tiselius [80], a column uniformly presaturated with a low-affinity species (or, in sorption chromatography, with the solvent or carrier gas) is used for separating a mixture of species of intermediate affinities by development with a high-affinity species [8, 11–13, 50, 81–84]. In shorthand:

$$1 \quad \rightarrow \quad 2, \ldots, n-1 \quad \rightarrow \quad n \tag{4.76}$$

Development resolves the mixture into separate plateau zones of the individual species. The technique is used, for example, in the commercial production of individual rare earths of high purity [85, 86].

### 1. *Final Pattern*

Displacement development gives a "unique" final pattern, i.e., one that is independent of the relative concentrations in the influents. In the shorthand notation introduced earlier (see Section I.C.1), this pattern is

$$S_2^1 S_3^2 S_4^3 \ldots S_{n-1}^{n-2} S_n^{n-1} \tag{4.77}$$

All boundaries are self-sharpening, and each plateau zone contains only a single species, the species having arranged themselves in the sequence of decreasing affinity (seen in the direction of flow).

How and why the final pattern (4.77) is attained is readily understood. The first influent composition change, 2, ..., $n-1 \rightarrow n$, develops a frontal-analysis pattern, in which the presaturant species, $n$, is displaced with a sharp and self-sharpening boundary (see Section I.F). The second influent

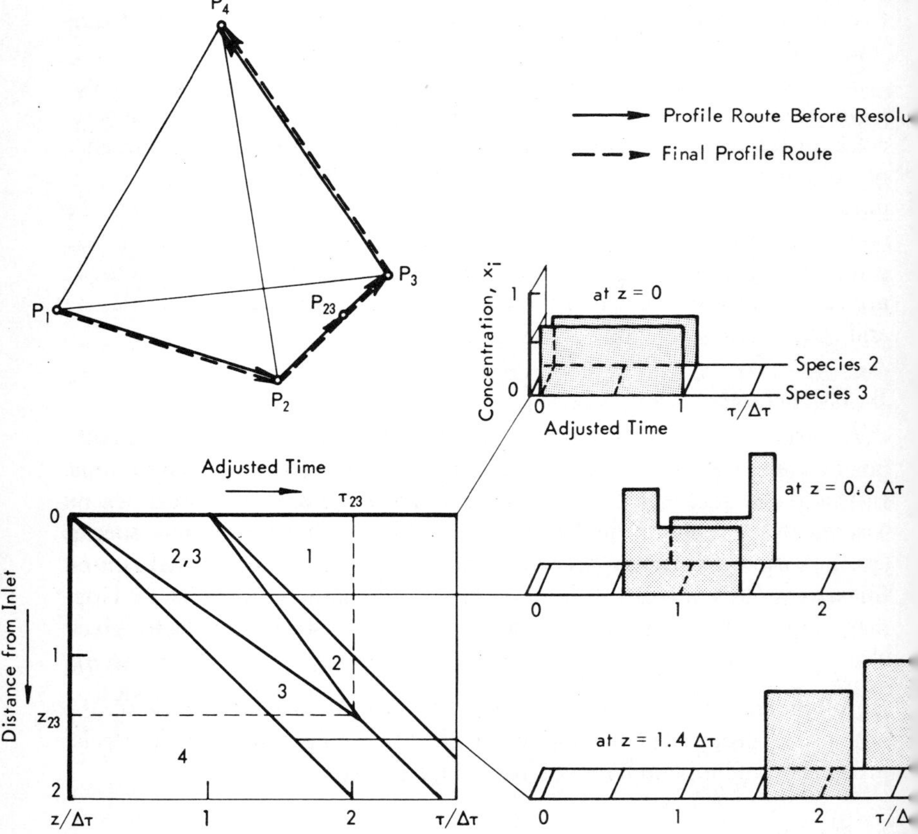


**Fig. 4.25.** Displacement development of two-component mixture. Composition profile routes, distance-time diagram, and concentration histories. Distance-time diagram shows species present in plateau zones. (For $\alpha_{23} = 2$; $x_2'' = 0.40$, $x_3'' = 0.60$.)

composition change, $1 \rightarrow 2, \ldots, n-1$, generates a pattern in which the development agent, species 1, by virtue of its higher affinity, displaces the species of the mixture with, again, a sharp and self-sharpening boundary. The species $2, \ldots, n-1$ of the mixture, while traveling down the column, thus remain confined between two sharp boundaries. Within this "band" between the two boundaries, the species sort themselves out in the order of their affinities because, of several species traveling with one another, those with the lower affinities have the higher species velocities (see Chapter 3, Section III.A). All boundaries in the final pattern are self-sharpening because each involves displacement of a species by another of higher affinity (see Chapter 3, Section IV.E).

As examples, Figs. 4.25 and 4.26 show the distance-time diagrams and composition histories at different column levels for displacement development of a two- and a four-component mixture. Details of the transient behavior and quantitative calculations will be discussed later in this section.

A characteristic and unusual feature of displacement development is that the band of the mixture travels with constant width. It does so because the total amount of species within the band remains constant, since no species, be it from the mixture, presaturant, or development agent, passes across either the front or rear boundary of the band in either direction. For an analogous reason, the plateau zones of the species $2, \ldots, n-1$, after their resolution, also travel with constant widths. Thus, with complete resolution, a so-called "constant pattern" is attained, that is, further development no longer produces any changes in the band, but merely shifts it as a whole in the direction of flow. In the distance-time diagram, this is reflected by the parallelity of the boundary trajectories in the final pattern. That plateau zones between coherent boundaries with different variable roots and different affinity cuts travel with constant widths is unusual, but was shown to be the case under the exceptional conditions which apply here (see the exception under 2a in Chapter 3, Section IV.F.4).

By virtue of its simplifying premises, the present treatment gives ideally sharp boundaries in the transient as well as the final response. In reality, disturbances neglected here let the boundaries acquire finite widths. However, since all boundaries are self-sharpening, a balance between the broadening effect of disturbances and the self-sharpening tendency is eventually established, so that the boundaries then travel with constant widths (see also Chapter 5, Section V.A.2). Disturbances thus do not impair the attainment of the constant pattern.

The final constant pattern, with the species in close-up square-wave pulses undiluted by the development agent and all traveling at the same

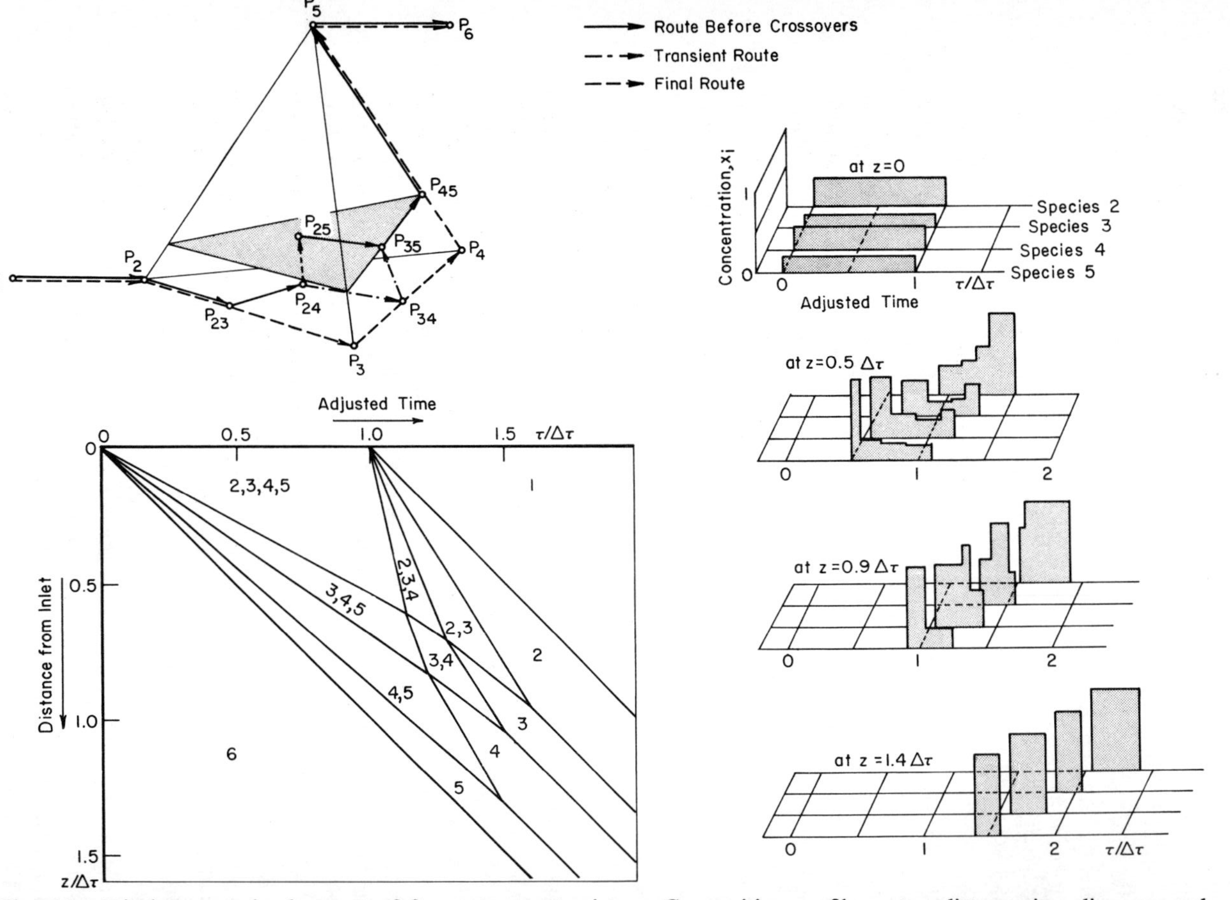


**Fig. 4.26.** Displacement development of four-component mixture. Composition profile routes, distance-time diagram, and

velocity, distinguishes displacement development sharply from the more familiar elution development, with the species in attenuating "peaks" traveling at different velocities (compare Section III.C). This characteristic difference and its implications for preparative separations were fully appreciated by Tiselius [80] when he first suggested displacement development as a useful technique.

The quantitative calculation of the final pattern is greatly facilitated by its time independence and by the fact that no more than two species are present at any boundary. Under the simplifying premises of the present treatment, the calculation is almost trivial in that the concentrations of the species in their respective square-wave pulses are $x_i = 1$, $y_i = 1$, and that material balance evidently requires the fractional band width (i.e., fraction of total width of band) occupied by the square-wave of any species $j$ to equal the fraction $x''_j$ of that species in the original mixture. The simplifying assumptions are readily relaxed in the present case, since theories covering the effects of disturbances on the constant-pattern shape of self-sharpening boundaries between zones of pure components have long been available ([1–4] and original work quoted in these reviews; [83, 87–89]).

### 2. *Transient Behavior*

The transient behavior of the response is illustrated in Figs. 4.25 and 4.26 by distance-time diagrams, profile routes, and concentration histories.

The simplest possible displacement-development case,

$$1 \rightarrow 2, 3 \rightarrow 4 \tag{4.78}$$

is shown in Fig. 4.25. The first influent composition change, at the time $\tau = 0$, generates two self-sharpening boundaries, resulting in the composition-profile route $P_{23}P_3P_4$, where $P_{23}$ is the composition point of the intermediate influent.* The second influent composition change, at the time $\tau = \Delta\tau$, also generates two self-sharpening boundaries, and the route then is $P_1P_2P_{23}P_3P_4$. The plateau zone $P_{23}$, now between two self-sharpening boundaries with the same variable root and affinity cut, shrinks, and disappears as the two boundaries merge. This completes the resolution, and the final route $P_1P_2P_3P_4$ is attained.

As a more complex example, the case

$$1 \rightarrow 2, 3, 4, 5 \rightarrow 6 \tag{4.79}$$

* The following notation is used: point $P_j$ is the composition point $x_j = 1$; point $P_{jk}$ is that of a plateau zone containing species $j, \ldots, k$.

is illustrated in Fig. 4.26. In the route diagram, the points $P_1$ and $P_6$ in reality are outside the three dimensions of the tetrahedron.

For the transient response in $n$-component systems in general, the following rules, recognizable in the distance-time diagram of Fig. 4.26 and derived subsequently, can be stated:

1. The band of species 2, ..., $n-1$ travels with constant width between two self-sharpening boundaries advancing at unit adjusted velocity.
2. All boundaries in the entire response are self-sharpening. Those generated by the first influent composition change have variable roots $h_2, \ldots, h_{n-1}$, and those generated by the second, $h_1 \ldots, h_{n-2}$.
3. As the boundaries with variable $h_2, \ldots h_{n-2}$ from the two influent

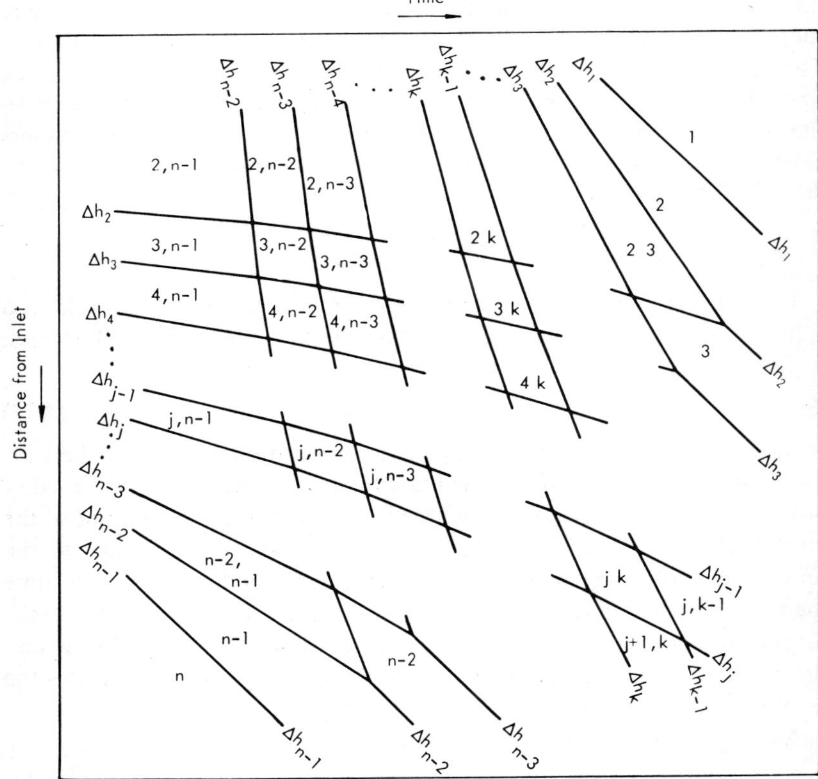


**Fig. 4.27.** Topology of boundary trajectories and transient plateau zones in distance-time plane of $n$-component displacement development (schematic). Pairs of numbers in plateau zones are index numbers; arbitrary plateau zone $jk$ contains species $j, \ldots, k$.

composition changes cross one another, those from the first accelerate while those from the second decelerate. Thus, on crossover, the trajectories of the two crossing boundaries are deflected toward one another.

4. Each transient plateau zone consists of species adjacent to one another in the affinity sequence, e.g., $j, j+1, j+2, \ldots, k-1, k$ (with $j<k$). No two zones contain the same set of species.
5. Plateau zones adjacent to one another in the distance-time plane contain related sets of species. From the zone with species $j, \ldots, k$, when crossing a boundary generated by the first influent composition change in the direction of flow ("south-southeast" in the distance-time diagram), one enters the zone with species $j+1, \ldots, k$; and when crossing a boundary generated by the second influent composition change in the direction of increasing time ("east-southeast" in the distance-time diagram), one enters the zone with species $j, \ldots, k-1$ (see Fig. 4.27).

The initial response patterns generated by the two influent composition changes are unique, and so is the topology of the transient plateau zones in the distance-time plane. The transient profile and history routes, however, are not, since the exact sequence of crossovers depends on the relative concentrations in the original mixture and on the separation factors.

Derivations of the rules above are given in the following paragraphs.

That the mixed band travels with constant width was shown earlier by a material-balance argument. Its front and rear boundaries travel with unit adjusted velocity because, in both cases, the sole species on one side is absent from the other side; the boundaries thus involve concentration changes from $x_i = y_i = 1$ on one side to $x_i = y_i = 0$ on the other, and according to Eq. (3.36) the adjusted velocity of such a step, with $\Delta x_i = \Delta y_i = 1$, is unity.

To derive the other properties of the transient pattern, we first establish the relations between the values of the $H$-function roots in the three influents. The final influent $\mathbf{x}'$, contains only species 1, and the presaturant, $\mathbf{x}'''$, only species $n$; their $H$-function roots, all trivial, thus are

$$h_i' = \alpha_{1,i+1} \qquad \text{for all } i \tag{4.80}$$

$$h_i''' = \alpha_{1i} \qquad \text{for all } i \tag{4.81}$$

[see condition (3.55)]. From the intermediate influent, $\mathbf{x}''$, only species 1 and $n$ are absent, so that

$$\begin{aligned} &h_1'' = \alpha_{11}, \qquad h_{n-1}'' = \alpha_{1n} \\ &\alpha_{1i} < h_i'' < \alpha_{1,i+1} \qquad \text{for all } i = 2, \ldots, n-2 \end{aligned} \tag{4.82}$$

Thus,

$$\begin{aligned} h_1' &> h_1'' &&= h_1''' \\ h_i' &> h_i'' &&> h_i''' \qquad \text{for all } i = 2, \ldots, n-2 \\ h_{n-1}' &= h_{n-1}'' &&> h_{n-1}''' \end{aligned} \tag{4.83}$$

Conditions (4.83) show for both influent composition changes that each nonconstant root has a higher value in the later than in the earlier influent; all boundaries generated by these influent composition changes therefore are self-sharpening [see condition (3.74)]. Furthermore, the first and the second influent composition change are seen to leave the roots $h_1$ and $h_{n-1}$, respectively, unaltered and thus generate only boundaries with variable $h_2, \ldots, h_{n-1}$ and $h_1, \ldots, h_{n-2}$, respectively.

Since all boundaries generated at the inlet are self-sharpening, and since crossovers do not alter the sharpening characteristics (see Chapter 3, Section VI.A.1), all boundaries in the entire response are self-sharpening. That, upon crossover, the (slower) boundaries from the earlier influent composition change accelerate while the (faster) ones from the later change decelerate has been shown for self-sharpening boundaries in general (see case 1 in Fig. 3.39).

The distribution of the species over the various transient plateau zones can be established as illustrated in Fig. 4.27, which shows portions of a distance-time diagram. In the distance-time plane, the boundaries generated by each of the two influent changes are in the sequence of increasing index numbers of their variable roots (in the direction of flow, i.e., downward in the diagram). The plateau zone having the composition of the intermediate influent, $\mathbf{x}''$, is bounded by the boundaries with variable $h_2$ and $h_{n-2}$ from the first and second influent composition change, respectively, as the slowest and fastest boundaries of their respective sets. According to Eqs. (4.82) the set of roots in this zone is

$$\mathbf{h}'' = \{\alpha_{11}, h_2'', \ldots, h_{n-2}'', \alpha_{1n}\} \tag{4.84}$$

Across each boundary generated by the first influent composition change, in the direction of flow, the respective variable root $h_i$ changes its value from $h_i''$ to its presaturant value $h_i'''$, which, according to Eq. (4.81), equals $\alpha_{1i}$. Similarly, across each boundary generated by the second influent composition change, in the direction of increasing time, the respective variable root changes its value from $h_i''$ to its final-influent value $h_i'$, which, according to Eqs. (4.80), equals $\alpha_{1,i+1}$. This leads to the pattern shown in Fig. 4.27, in which the plateau zones are characterized by double indices $jk$ in such a manner that their root sets are

$$\mathbf{h}_{jk} = \{\alpha_{11}, \ldots, \alpha_{1,j-1}, h''_j, \ldots, h''_{k-1}, \alpha_{1,k+1}, \ldots, \alpha_{1n}\} \tag{4.85}$$

In accordance with condition (3.55), any species $i$ whose separation factors $\alpha_{1i}$ appear as root values are absent; the plateau zone $jk$, with the root set $\mathbf{h}_{jk}$ given in Eq. (4.85), thus contains only the species $j, \ldots, k$. Adjacent to the zone $jk$ is the zone $j+1, k$ in the "south-southeast" and the zone $j, k-1$ in the "east-southeast" on the distance-time plane. These two zones contain only the species $j+1, \ldots, k$ and $j, \ldots, k-1$, respectively, as stated earlier without proof.

### 3. *Compositions of Transient Plateau Zones*

The compositions of the various transient plateau zones in displacement development are readily calculated from the $H$-function roots $h''_i$ of the intermediate influent. As shown earlier, the set of $H$-function roots of the arbitrary plateau zone $jk$, containing species $j, \ldots, k$, is given by Eq. (4.85). For the mobile- and stationary-phase concentrations of the arbitrary species $l$ (with $j \leqq l \leqq k$) in that zone, one obtains from the general relations (3.57) and (3.58) after substitution of the root values:

$$x_{l,jk} = \prod_{i=j}^{k-1} (h''_i - \alpha_{1l}) \Big/ \prod_{\substack{i=j \\ i \neq l}}^{k} (\alpha_{1i} - \alpha_{1l}) \tag{4.86}$$

$$(j \leqq l \leqq k)$$

$$y_{l,jk} = \prod_{i=j}^{k-l} \left(\frac{1}{h''_i} - \alpha_{l1}\right) \Big/ \prod_{\substack{i \neq l \\ i=j}}^{k} (\alpha_{i1} - \alpha_{l1}) \tag{4.87}$$

(For $l < j$ or $l > k$ one obtains $x_{l,jk} = y_{l,jk} = 0$.)

### 4. *Resolution Distances and Times*

The resolution distances and times for displacement development with any number of species can be calculated algebraically. Explicit equations for the general case can be written, but are unwieldy. Instead, a simple iteration procedure will be given. An explicit approximation yielding upper limits for resolution distances and times will be derived also.

Of all plateau zones containing both of the arbitrary species $j$ and $k$, that forming and disappearing latest and farthest away from the inlet is the zone $jk$, as can be seen in Fig. 4.27. The distance $z_{jk}$ and adjusted time $\tau_{jk}$ required for resolution of the arbitrary species $j$ and $k$ from one another therefore are the coordinate values of the distance-time point at which this zone disappears.

A "pedigree" as shown in Fig. 4.28 can be set up to illustrate the

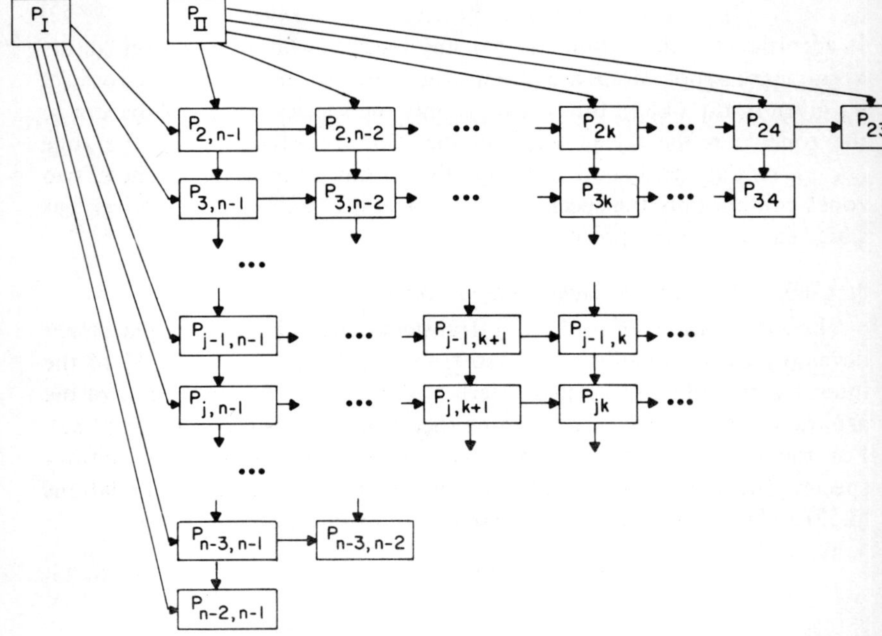


**Fig. 4.28.** Topology of points of disappearance of transient plateau zones in distance-time plane. $P_{ij}$ is point at which zone *ij* disappears; $P_I$ and $P_{II}$ are points of first and second influent composition change, respectively; arrows (pointing in direction of flow) signify boundary trajectories.

topology of the points of disappearance of the various transient plateau zones and the connecting boundary trajectories. The points of resolution of species 2 from others, and of species $n-1$ from others (top row and first column, respectively), are points at which one boundary is intersected for the first time. Their coordinate values are readily given in closed form:

$$z_{2k} = \frac{h_2''}{\alpha_{1,n-1}-\alpha_{12}} \prod_{i=3}^{k-1}\left(\frac{h_i''}{\alpha_{1i}}\right) \prod_{i=k}^{n-2}\left(\frac{h_i''-\alpha_{12}}{\alpha_{1i}-\alpha_{12}}\right)\Delta\tau \tag{4.88}$$

$$\tau_{2k} = \left[1+\frac{\alpha_{12}}{\alpha_{1,n-1}-\alpha_{12}} \prod_{i=k}^{n-2}\left(\frac{h_i''-\alpha_{12}}{\alpha_{1i}-\alpha_{12}}\right)\right]\Delta\tau \tag{4.89}$$

$$z_{j,n-1} = \frac{h_j''}{\alpha_{1,n-1}-\alpha_{1j}} \prod_{i=2}^{j-1}\left(\frac{\alpha_{1,n-1}-h_i''}{\alpha_{1,n-1}-\alpha_{1i}}\right) \prod_{i=j+1}^{n-2}\left(\frac{h_i''}{\alpha_{1i}}\right)\Delta\tau \tag{4.90}$$

$$\tau_{j,n-1} = \frac{\alpha_{1,n-1}}{\alpha_{1,n-1}-\alpha_{1j}} \prod_{i=2}^{j-1} \left( \frac{\alpha_{1,n-1}-h_i''}{\alpha_{1,n-1}-\alpha_{1i}} \right) \Delta\tau \tag{4.91}$$

where $\Delta\tau$ is the adjusted time required for introduction of the mixture into the column. Once these $z$ and $\tau$ values have been established, the others can be calculated row by row from left to right with the following recursion formulas:

$$z_{jk} = \frac{1}{\alpha_{j1}-\alpha_{k1}} \left[ \left( \alpha_{j1} - \frac{1}{h_k''} \right) z_{j,k+1} + \left( \frac{1}{h_{j-1}''} - \alpha_{k1} \right) z_{j-1,k} - \left( \frac{1}{h_{j-1}''} - \frac{1}{h_k''} \right) z_{j-1,k+1} \right] \tag{4.92}$$

$$\tau_{jk} = \frac{1}{\alpha_{1k}-\alpha_{1k}} [(h_k''-\alpha_{1j})\tau_{j,k+1} + (\alpha_{1k}-h_{j-1}'')\tau_{j-1,k} - (h_k''-h_{j-1}'')\tau_{j-1,k+1}] \tag{4.93}$$

A computer program for calculating resolution distances and times and plotting the distance-time diagram has been made available [90, 91] and its application to a practical fifteen-component case has been described [91].

*Derivation.* Equations (4.88) to (4.91) can be derived with the "triangle formula" below, which gives the distance-time coordinates of the intersection of two boundaries whose velocities and points of origin are known. For two boundaries originating from $(z_A, \tau_A)$ and $(z_B, \tau_B)$, traveling at the adjusted velocities $u_A$ and $u_B$, and intersecting one another at $(z, \tau)$ (see Fig. 4.29), one has

$$\frac{z-z_A}{\tau-\tau_A} = u_A \quad \text{and} \quad \frac{z-z_B}{\tau-\tau_B} = u_B \tag{4.94}$$

and, after solving for $z$ and $\tau$,

$$\tau = \frac{\tau_B u_B - \tau_A u_A - z_B + z_A}{u_B - u_A}$$
$$z = z_A + (\tau-\tau_A)u_A \quad \text{or} \quad z = z_B + (\tau-\tau_B)u_B \tag{4.95}$$

When applied to the calculation of a particular pair of coordinate values $z_{jk}$ and $\tau_{jk}$ $(2 \leqq j < k \leqq n-1)$, the triangle formula (4.95) involves the boundaries separating the zone $jk$ from its neighbors $j+1, k$ and $j, k-1$, as a comparison of Fig. 4.29 with 4.27 shows. With the root values for these zones substituted according to Eq. (4.85), the general relation (3.71) for step velocities gives

$$u_A = \prod_{i=j}^{k-1} (h_i''\alpha_{i+1,1}) \quad \text{and} \quad u_B = \prod_{i=j}^{k-1} (h_i''\alpha_{i1}) \tag{4.96}$$

To calculate $z_{2,n-1}$ and $\tau_{2,n-1}$, one has to substitute in Eq. (4.95):

$$z = z_{2,n-1}, \quad z_A = 0, \quad z_B = 0,$$
$$\tau = \tau_{2,n-1}, \quad \tau_A = 0, \quad \tau_B = \Delta\tau \tag{4.97}$$

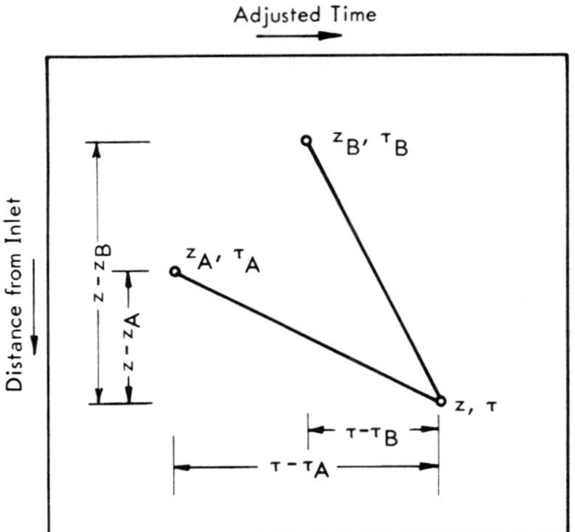


**Fig. 4.29.** Trajectories of two intersecting sharp boundaries with arbitrary points of origin.

since the two intersecting boundaries arise directly from the two influent composition changes; the values of $u_A$ and $u_B$, also needed in Eq. (4.95), are obtained from Eqs. (4.96) with $j = 2$ and $k = n-1$.

To calculate all further $z_{j,n-1}$ and $\tau_{j,n-1}$, the triangle formula (4.95) is applied with the substitutions

$$
\begin{aligned}
z &= z_{j,n-1}, \qquad z_A = 0, \qquad z_B = z_{j-1,n-1}, \\
\tau &= \tau_{j,n-1}, \qquad \tau_A = 0, \qquad \tau_B = \tau_{j-1,n-1}
\end{aligned}
\tag{4.98}
$$

and the values of $u_A$ and $u_B$ are obtained from Eqs. (4.96) with $k = n-1$.

For the calculation of the $z_{2k}$ and $\tau_{2k}$, the appropriate substitutions in the triangle formula are

$$
\begin{aligned}
z &= z_{2k}, \qquad z_A = z_{2,k+1}, \qquad z_B = 0, \\
\tau &= \tau_{2k}, \qquad \tau_A = z_{2,k+1}, \qquad \tau_B = \Delta\tau
\end{aligned}
\tag{4.99}
$$

and Eqs. (4.96) apply with $j = 2$.

With the substitutions stated above, the triangle formula gives coordinate values which can be written in general form as shown in Eqs. (4.88) to (4.91).

In principle, all further $z_{jk}$ and $\tau_{jk}$ can also be calculated with the triangle formula as recursion formula. However, the separate iterative calculation of the $z$ and $\tau$ values by means of the independent Eqs. (4.92) and (4.93) is more convenient. Equation (4.92), which permits $z_{jk}$ to be calculated if the $z$ values of the other three corner points of the zone $jk$ and the velocities of its boundaries are known, can be derived as follows. With the time intervals $(\Delta\tau)_A$, etc., defined by

$$
\begin{aligned}
(\Delta\tau)_A &\equiv \tau_{jk} - \tau_{j,k+1}, \qquad (\Delta\tau)_B \equiv \tau_{jk} - \tau_{j-1,k}, \\
(\Delta\tau)_C &\equiv \tau_{j,k+1} - \tau_{j-1,k+1}, \qquad (\Delta\tau)_D \equiv \tau_{j-1,k} - \tau_{j-1,k+1}
\end{aligned}
\tag{4.100}
$$

and with boundary velocities obtained from Eq. (3.71) with the appropriate root values, the following equations can be written for the four sides of the zone $jk$:

$$\begin{aligned}
\frac{z_{jk}-z_{j,k+1}}{(\Delta\tau)_A} &= u_A = \prod_{i=j}^{k-1} (h_i''\alpha_{i+1,1}) \\
\frac{z_{jk}-z_{j-1,k}}{(\Delta\tau)_B} &= u_B = \prod_{i=j}^{k-1} (h_i''\alpha_{i1}) \\
\frac{z_{j,k+1}-z_{j-1,k+1}}{(\Delta\tau)_C} &= u_C = \prod_{i=j}^{k} (h_i''\alpha_{i1}) \\
\frac{z_{j-1,k}-z_{j-1,k+1}}{(\Delta\tau)_D} &= u_D = \prod_{i=j-1}^{k-1} (h_i''\alpha_{i+1,1})
\end{aligned} \tag{4.101}$$

(see Fig. 4.30). Furthermore, by virtue of the definitions (4.100) of the time intervals:

$$(\Delta\tau)_A+(\Delta\tau)_C-(\Delta\tau)_B-(\Delta\tau)_D = 0 \tag{4.102}$$

Equations (4.101) and (4.102) constitute five linear algebraic equations in the five un-

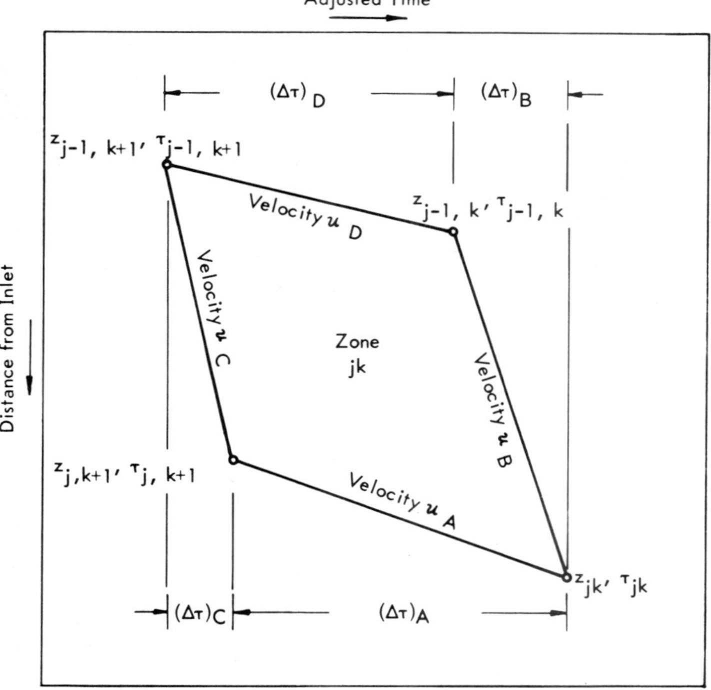


**Fig. 4.30.** Plateau zone $jk$ in distance-time plane of $n$-component displacement development.

knowns $z_{jk}$, $(\Delta\tau)_A$, $(\Delta\tau)_B$, $(\Delta\tau)_C$, and $(\Delta\tau)_D$. Solving for $z_{jk}$ one obtains Eq. (4.92). The derivation of Eq. (4.93) is analogous, with $z$ and $\tau$ interchanged and with the velocities, roots, and separation factors replaced by their reciprocals.

For the "binary" case $1 \to 2, 3 \to 4$, the general equations (4.88) to (4.91) reduce to

$$z_{23} = \frac{h_2''}{\alpha_{13}-\alpha_{12}}\,\Delta\tau = \left(\frac{1}{\alpha_{23}-1}+x_2''\right)\Delta\tau$$
$$\tau_{23} = \frac{\alpha_{13}}{\alpha_{13}-\alpha_{12}}\,\Delta\tau = \frac{\alpha_{23}}{\alpha_{23}-1}\,\Delta\tau \qquad (4.103)$$

[$h_2''$ has been expressed in terms of $\mathbf{x}''$ by means of Eq. (A.6) in Appendix I.] For the "ternary" case $1 \to 2, 3, 4 \to 5$, too, explicit solutions exclusively in terms of the influent concentrations and separation factors can be given, since the $H$-function roots of the three-component mixture can be obtained as the roots of a quadratic equation (see Appendix I).

Equations (4.103) for the binary case, and the corresponding ones for the ternary, are at variance with earlier derivations by various other authors [7, 92–96]. The cause of the discrepancy is that the others have calculated a case with different initial and boundary conditions, namely, development of a mixed band of uniform composition existing in the column at zero time. This variant is briefly discussed in Section IV.B.5.

For systems with many components, the iterative calculation of the resolution distances and times is lengthy. However, the following inequalities, giving upper limits for the resolution distance and time of an arbitrary pair of species, can be used for a rough estimate of the column requirements:

$$z_{jk} < \frac{h_j''}{\alpha_{1k}-\alpha_{1j}} \prod_{i=j+1}^{k-1} \left(\frac{h_i''}{\alpha_{1i}}\right) \Delta\tau \qquad (4.104)$$

$$\tau_{jk} < \frac{\alpha_{jk}}{\alpha_{jk}-1}\,\Delta\tau \qquad (4.105)$$

(For the extreme pair $j = 2$, $k = n-1$, equalities replace the inequalities.)

The inequalities (4.104) and (4.105) can be derived as follows. The expressions on the right-hand sides are obtained if, for calculating $z_{jk}$ and $\tau_{jk}$, the triangle formula is used with the substitutions (4.96) and (4.97). This procedure would give the correct values if both boundaries were traveling all the way from the inlet at the velocities they have acquired when intersecting one another. The actual velocity difference between the two boundaries, however, is greater initially and decreases with every crossover because of the way the trajectories are deflected toward one another (e.g., see Fig. 4.26). Since the calculation thus is based on the smallest velocity difference occurring before resolution, instead of on an appropriate average difference, it gives upper limits for the resolution distance and time.

It is worth noting that any resolution distance, but no resolution time, can be smaller than the band width $\Delta\tau$.* This is readily established for the resolution distances with the inequalities (4.104) and (3.50), and for the resolution times with Eq. (4.89) applied to $\tau_{2,n-1}$ as the necessarily shortest. The reason is that separation by a frontal-analysis mechanism begins, for every entering portion of the mixture, right at the column inlet and can give, with a column shorter than the band width, an effluent of pure species $n-1$ even while the mixture is still being introduced at the inlet. Under favorable conditions, a small amount of development agent and a column shorter than the band may suffice to achieve complete resolution. On the other hand, resolution can obviously not become complete until after the introduction of the mixture has ceased, that is, until a time longer than $\Delta\tau$ has elapsed. The distance-time diagram shows these effects at a glance (e.g., see Fig. 4.26).

5. *Displacement Development of Uniform Initial Band*

In certain applications of displacement development, a mixed band of virtually uniform composition exists in the top part of the column at the time development is initiated. This is the case, for example, in the usual procedure for separating rare earths; here, the rare earths are first sorbed unselectively by the top layer of the bed, then to be developed with a complexing agent which penetrates the band and is virtually the sole cause of the affinity differences [85, 86, 92–95].

The distance-time diagram for such an operation is of the type shown in Fig. 4.31a, zero time being taken as the moment development begins. Although it does not fall into the category of responses of uniformly presaturated columns to influent composition changes, this variant deserved a few comments, particularly since it constitutes the case treated in the previous theories.

It is apparent that the main features of displacement development stem from the fact that a band of constant width, confined between a high-affinity displacing agent and a low-affinity presaturant, is being developed. The change in the initial and boundary conditions does not alter this, and operation with a uniform initial band in the column thus is very similar to the standard mode of operation. The main difference is that, in operation

* Since the front and rear boundaries of the band travel at unit adjusted velocity, the band width $\Delta\tau$ and length $\Delta z$ are equal.

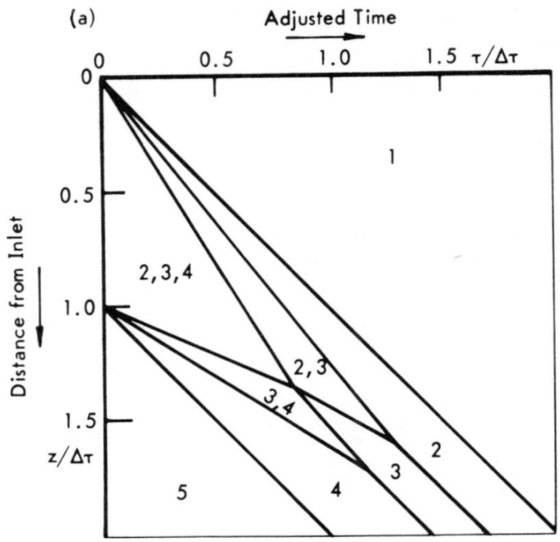


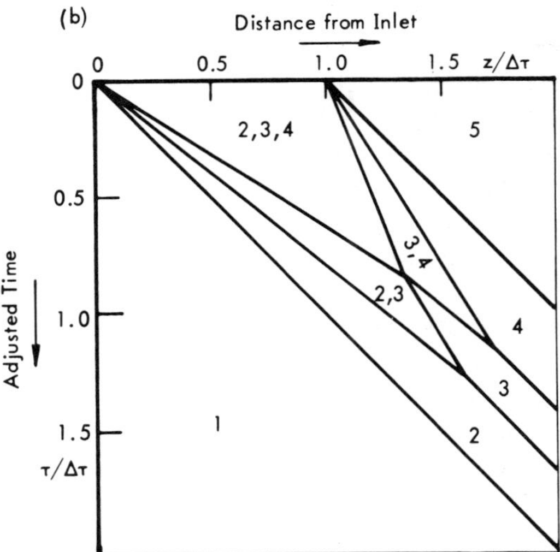


**Fig. 4.31.** Displacement development of uniform initial band of three-component mixture. (a) Distance-time diagram; (b) distance-time diagram after interchange of coordinates. Numbers in plateau zones designate species present. (For $\alpha_{23} = 2$, $\alpha_{24} = 4$; mixture composition $y_2^0 = 0.32$, $y_3^0 = 0.24$, $y_4^0 = 0.44$.)

with a uniform initial band, the resolution distances are longer (obviously always longer than the band length $\Delta z$), while the resolution times are shorter, at least if the time needed to sorb the mixture is not included.

The mathematical approach developed in this section, by virtue of certain symmetry properties, is readily adapted to the different initial and boundary conditions. As illustrated in Fig. 4.31a and b, a distance-time diagram of the familiar appearance, i.e., with two points of origin of the various boundary trajectories on the abscissa on top, is obtained from that for the development of a uniform initial band if the ordinate and abscissa are interchanged. The resolution distances and times can be calculated as the coordinate values of trajectory intersections in the new diagram with the same equations as for the standard displacement-development case. However, because of the interchange of $z$ and $\tau$, the trajectory slopes in the new diagram correspond to reciprocal velocities rather than velocities. Since the velocities appear as products of ratios of $h_i''$ and $\alpha_{1i}$, these quantities must be replaced by their reciprocals. Also, since the relations between $h_i$, $\alpha_{1i}$, and $x_i$ are the same as those between $1/h_i$, $\alpha_{i1}$, and $y_i$ (see Chapter 3, Section IV.D.2), the exchange of roots and separation factors for their reciprocals entails that any $x_i$ must be replaced by $y_i$. Furthermore, as a comparison of Figure 4.31b with the standard diagrams shows, the sequence of the species in the zones (as read left to right in the diagram) has been reversed by the interchange of coordinates; therefore, species as well as $H$-function roots must be numbered in reverse order. Specifically, as is confirmed by calculations not reproduced here, all relations derived for the standard conditions apply to the case of development of a uniform initial band with the following replacements:

$$\begin{array}{llllll} \text{replace} & z & \text{by} & \tau, & u & \text{by} \quad 1/u, \\ & \tau & \text{by} & z, & \alpha_{1i} & \text{by} \quad \alpha_{n+1-i,1}, \\ & z_{ij} & \text{by} & \tau_{n+1-j,n+1-i}, & h_i'' & \text{by} \quad 1/h_{n-i}^0, \\ & \tau_{ij} & \text{by} & z_{n+1-j,n+1-i}, & x_i'' & \text{by} \quad y_{n+1-i}^0 \end{array} \tag{4.106}$$

where $y_i^0$ and $h_i^0$ refer to the uniform initial band. For example, the distance and adjusted time for resolution of a binary mixture are obtained from Eqs. (4.103) with the appropriate substitutions:

$$\begin{aligned} \tau_{23} &= \frac{1/h_2^0}{\alpha_{21}-\alpha_{31}}\,\Delta z = \left(\frac{1}{\alpha_{23}-1}+y_3^0\right)\Delta z \\ z_{23} &= \frac{\alpha_{21}}{\alpha_{21}-\alpha_{31}}\,\Delta z = \frac{\alpha_{23}}{\alpha_{23}-1}\,\Delta z \end{aligned} \tag{4.107}$$

The results obtained in this manner are in agreement with the previously mentioned derivations by other authors (see also Section IV.F.2).

### 6. *Invariance to Properties of Presaturant and Development Species*

The transient and final response behavior of displacement development of a given mixture is independent of the properties of the species constituting the presaturant and development agent (provided these species have lower and higher affinity, respectively, than those of the mixture, as is implied when the operation is called displacement development). This becomes apparent if one considers that these species, $n$ and 1, only serve to confine the mixture between self-sharpening boundaries in its band; at no time does the band contain species 1 or $n$, nor does its velocity depend on the properties of these species.

Being independent of the properties of the presaturant and development species, the response can be expressed in terms of quantities unrelated to these species. The equations given in this section for resolution distances and times and for concentrations in transient plateau zones contain separation factors $\alpha_{12}, \ldots, \alpha_{1,n-1}$ and roots $h_2'', \ldots, h_{n-2}''$ defined by a function involving such separation factors, which depend on the relative affinity of species 1. All these equations, however, are readily rearranged to contain roots and separation factors exclusively as ratios $h_i''/\alpha_{1j}$, which can be replaced as follows:

$$\frac{h_i''}{\alpha_{1j}} = \frac{h_i''/\alpha_{12}}{\alpha_{1j}/\alpha_{12}} = \frac{g_{i-1}}{\alpha_{2j}} \tag{4.108}$$

where

$$g_{i-1} \equiv h_i''/\alpha_{12} \tag{4.109}$$

Thus, the relations remain valid with the replacements:

$$\text{replace} \quad \alpha_{1i} \quad \text{by} \quad \alpha_{2i} \qquad \text{and} \qquad h_i'' \quad \text{by} \quad g_{i-1} \tag{4.110}$$

One can obtain the $g_i$ directly from the composition $\mathbf{x}''$ of the mixture without first calculating the $h_i''$. As is readily verified with Eqs. (4.109), (3.54), and (2.9), the values $g_1, \ldots, g_{n-3}$ are the roots of the equation

$$\sum_{i=2}^{n-1} \frac{x_i''}{g - \alpha_{2i}} = 0 \tag{4.111}$$

that is, they are the $H$-function roots calculated for the mixture if the existence of species 1 and $n$ is ignored. Since neither the $g_i$ nor the $\alpha_{22}, \ldots, \alpha_{2,n-1}$ depend on properties of species 1 or $n$, the substitutions

(4.110) eliminate all quantities relating to the presaturant or development species.

The quantitative relations for displacement development could have been derived from the outset in terms of the $g_i$ and $\alpha_{22}, \ldots, \alpha_{2,n-1}$ [90, 91]. This has not been done here, to avoid obscuring the derivations by a notation differing from that in other sections.

### C. Selective Displacement

Selective displacement is a mode of operation often used in analytical separations, particularly in ion-exchange chromatography. From a mixture sorbed at the top of the column, the species of lowest affinity is displaced preferentially by a development agent which, having merely intermediate affinity, leaves the species of high affinity in the column, to be displaced later one by one or in groups by development agents of higher affinities ([3, 77, 82, 84, 97] and original work quoted in these reviews).

#### 1. *Binary Separations*

The simplest example of selective displacement is

$$1 \rightarrow 3 \rightarrow 2, 4 \rightarrow 5 \tag{4.112}$$

The mixture to be resolved is that of species 2 and 4. The first development agent, species 3, displaces species 4. Finally, species 1 is used as the second development agent to displace species 2. Only this simple case will be discussed in some detail. The reader will have no difficulty in extending the treatment to more complex situations.

As in displacement development, the species of the lowest and highest affinity constitute the presaturant and final influent, respectively, and are absent from the intermediate influents. As shown in the previous section, the consequences are as follows. The species of the intermediate influents are confined to a band of constant width, traveling at unit adjusted velocity, between the presaturant and final influent. Within this band, the species sort themselves out in the sequence of their affinities. A final, constant pattern identical to that of the corresponding displacement development case is eventually attained. For the case (4.112), the pattern is

$$S_2^1 S_3^2 S_4^3 S_5^4$$

with close-up square-wave pulses of species 2, 3, and 4, as would have resulted from the corresponding displacement development case $1 \rightarrow 2, 3, 4 \rightarrow 5$.

Whereas the purpose of displacement development usually is to achieve complete resolution of all species in the mixed band, necessitating attainment of the final constant pattern, the purpose of selective displacement merely is to resolve the species of the mixture, 2 and 4 in the example (4.112), a separation achieved well before the final pattern is attained. The transient behavior therefore is of particular interest in selective displacement and differs, of course, significantly from that in displacement development.

The qualitative features of the transient behavior are readily understood. While the mixture of species 2 and 4 enters the column, partial separation by a frontal-analysis mechanism takes place, that is, species 4 accumulates in a plateau zone by itself at the front of the mixed band and displaces the presaturant with a self-sharpening boundary, while the rear portion of the mixed band is in equilibrium with the mixed influent (see Section I.F). What the first development agent, species 3, essentially achieves is to concentrate the entire amount of species 4 in a square-wave pulse at the front of the mixed band. In doing so, this agent partially bypasses species 2, which has the higher affinity, so that resolution of species 2 and 4 is eventually attained. The second development agent and final influent, species 1, is not needed for the separation, but serves to suppress "tailing" of species 2 by displacing, with a self-sharpening boundary, any portions of species 2 that would otherwise fall more and more behind.

Further details of the transient behavior can be gathered from an analysis of the transient composition routes. The transient behavior even in the simple case (4.112) turns out not to be "unique," in that development takes either of two courses, depending on the separation factors and relative concentrations of species 2 and 4. Distance-time diagrams, profile routes, and concentration histories at different column levels are shown in Figs. 4.32 and 4.33 for the two potential courses of development.

The following qualitative deduction of the transient behavior in the two cases is based on the general rules for coherent boundaries and pulses and for interference of such boundaries, as derived in Chapter 3, Sections IV and VI.

The composition point $P_{24}$ of the mixture of species 2 and 4 may be located on either side of the watershed in the composition diagram $P_2P_3P_4$ ($P_i$ is the composition point $x_i = 1$). In the case shown in Fig. 4.32, $P_{24}$ is on the side toward $P_2$. Here, the composition route generated by introduction of the mixture is $P_{24}P_4P_5$, becoming $P_3P_2P_{24}P_4P_5$ when development with species 3 is initiated. The front $P_3P_2$ of the development agent is nonsharpening, but nevertheless accumulates species 2 in a plateau zone

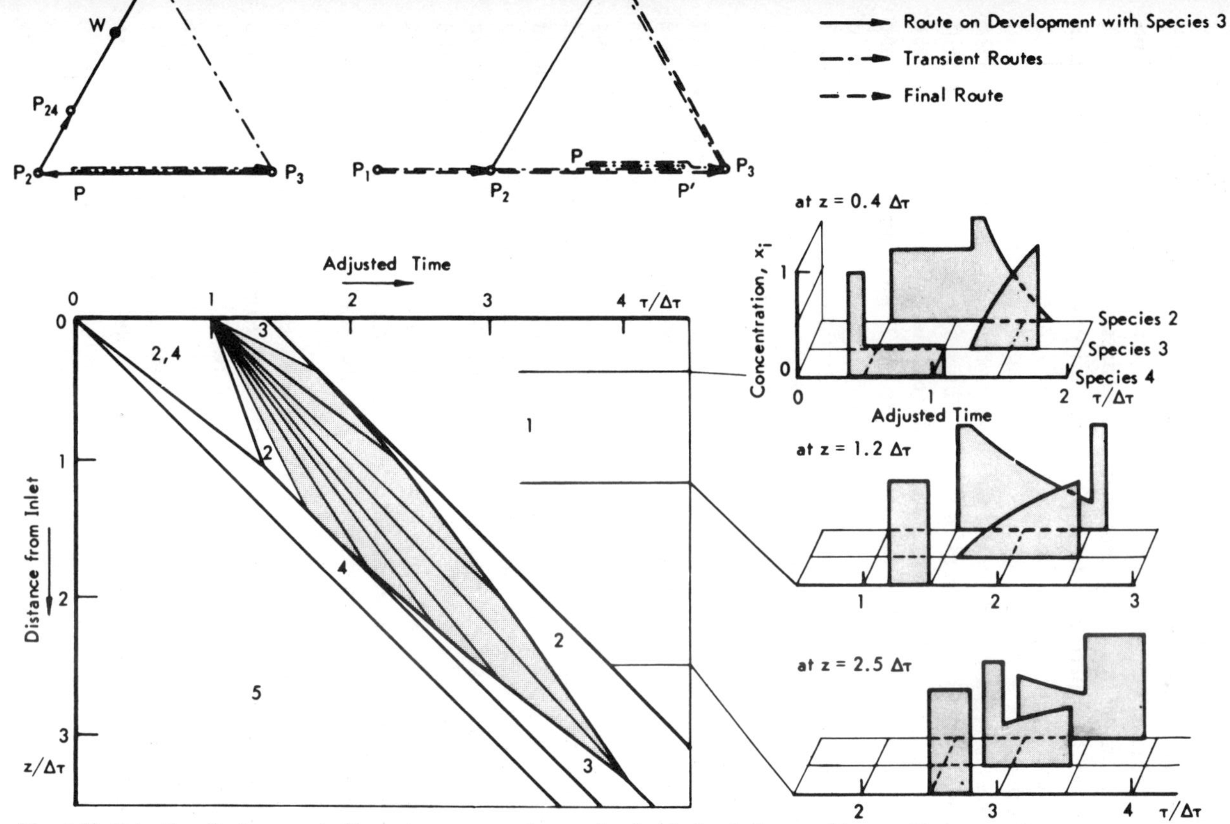


**Fig. 4.32.** Selective displacement of two-component mixture, $1 \to 3 \to 2, 4 \to 5$. Composition profile routes, distance-time diagram, and concentration histories. Distance-time diagram shows species present in plateau zones; sharp boundaries and pulse flanks shown as heavy lines; diffuse boundaries, shaded. (For $\alpha_{23} = 2$, $\alpha_{24} = 4$; $x_2''' = 0.7$, $x_4''' = 0.3$, $\Delta\tau' = 0.4\Delta\tau$.) On development with species 3, plateau zone containing only species 2 is formed upstream of zone of mixture.

by itself. The plateau zone $P_{24}$ is now between two self-sharpening boundaries having the same affinity cut, and disappears upon further development, so that the route becomes $P_3P_2P_4P_5$. This route is also a transient one because it crosses a watershed. With the disappearance of the shrinking plateau zone $P_2$, adjacent to the self-sharpening boundary whose route crosses the watershed, species 3 begins to bypass the pulse of species 2, and resolution of species 2 and 4 is achieved. The route is now $P_3P\ P_3P_4P_5$, where P (not a plateau zone) is the attenuating maximum of the pulse of species 2. This would be the final route, except for introduction of the second development agent, species 1. This agent advances with a self-sharpening front $P_1P_3$, which eventually reaches and begins to displace the nonsharpening "tail" of the pulse of species 2, generating a plateau zone of species 2 alone. The route now is $P_1P_2P'P\ P_3P_4P_5$, where P′ (not a plateau zone) is the attenuating concentration maximum of species 3 between $P_2$ and P. The double pulse $P_2P'P\ P_3$ along this route is transient, and its eventual degeneration into a single self-sharpening boundary produces the final route $P_1P_2P_3P_4P_5$.

In the other case, shown in Fig. 4.33, the composition point $P_{24}$ of the mixture is on the other side of the watershed. Here, again, the route generated by introduction of the mixture is $P_{24}P_4P_5$, but changes to $P_3P_{23}P_{24}P_4P_5$ when development with species 3 is initiated. In contrast to the other case, the advancing front of the development agent does not generate a plateau zone of species 2. Rather, some of the development agent from the outset joins species 2 in displacing species 4. The plateau zone $P_{24}$ now is shrinking, and its disappearance completes the resolution of species 2 and 4 and lets the route change to $P_3P_{23}P_3P_4P_5$. Here, the zone $P_{23}$ constitutes the flat top of a pulse $P_3P_{23}P_3$ of species 2. The flat top disappears as the front and rear flanks of this pulse merge. From then on, the route and further development are qualitatively the same as in the case shown in Fig. 4.32.

As a comparison of Figs. 4.32 and 4.33 shows, the main difference between the two cases is that the plateau zone preceding the front of the first development agent contains species 2 alone in the first case (Fig. 4.32) and species 2 and 3 in the second (Fig. 4.33). Which alternative will be realized under given conditions is readily predicted as follows. The location of the watershed point of the composition triangle $P_2P_3P_4$ is given by

$$x_2 = \frac{\alpha_{23}-1}{\alpha_{24}-1}, \qquad x_3 = 0 \tag{4.113}$$

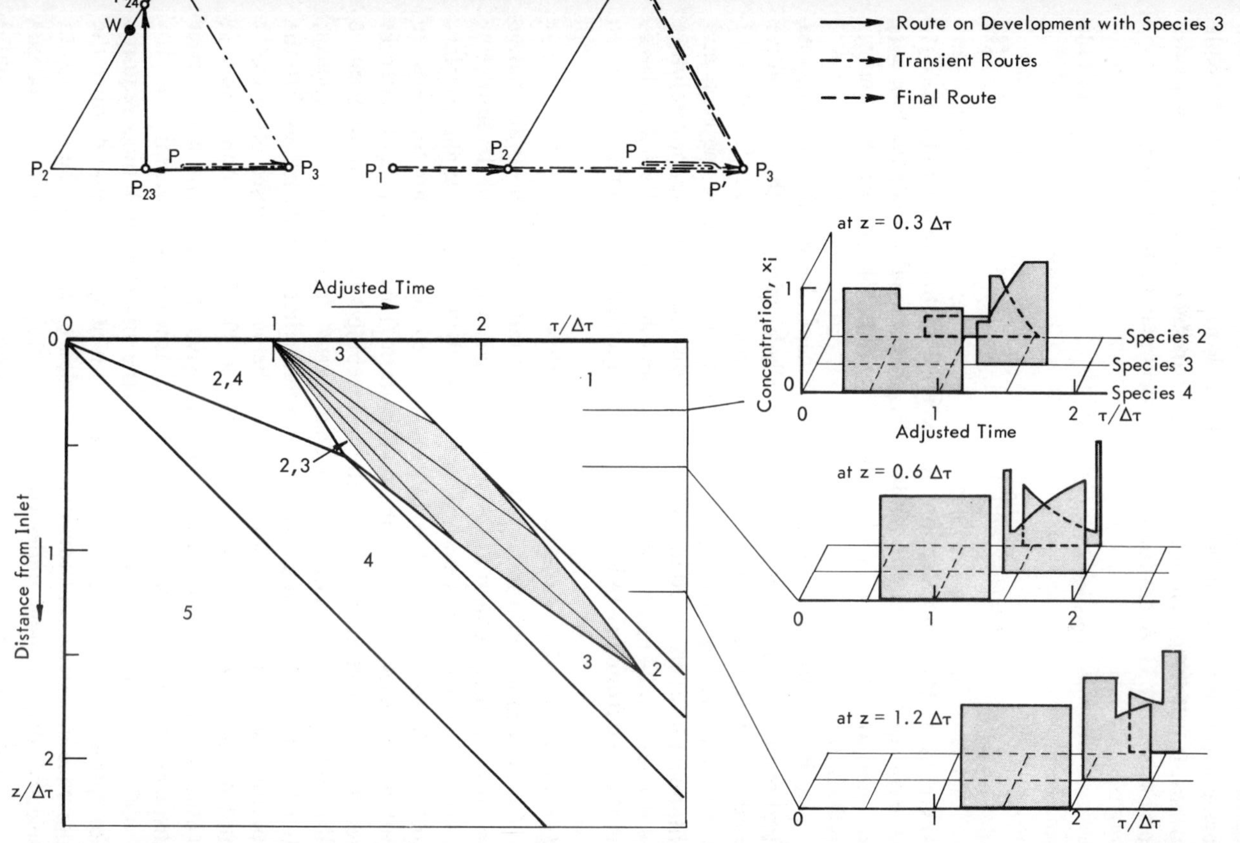


**Fig. 4.33.** Selective displacement of two-component mixture. Diagrams and conditions as in Fig. 4.32, except for composition of mixture ($x_2''' = 0.2$, $x_4''' = 0.8$). On development with species 3, plateau zone containing both species 2 and 3 is formed upstream of zone of mixture.

[compare Eq. (3.47)]. On the side toward $P_2$, $x_2$ is higher than at the watershed point, and on the other side, $x_2$ is lower. With $P_{24}$ on the side toward $P_2$, development is as in Fig. 4.32, and with $P_{24}$ on the other side, it is as in Fig. 4.33. Accordingly, the transient plateau zone contains

$$\text{species 2 alone} \qquad \text{if} \qquad x_2''' > \frac{\alpha_{23}-1}{\alpha_{24}-1} \tag{4.114}$$

$$\text{species 2 and 3} \qquad \text{if} \qquad x_2''' < \frac{\alpha_{23}-1}{\alpha_{24}-1} \tag{4.115}$$

where $x_2'''$ is the concentration of species 2 in the mixed influent, $P_{24}$. In terms of the $H$-function roots at $P_{24}$, the trivial roots equalling $\alpha_{11}$, $\alpha_{13}$, and $\alpha_{15}$ are $h_1'''$, $h_2'''$, and $h_4'''$ in the former case, and $h_1'''$, $h_3'''$, and $h_4'''$ in the latter.

For both potential courses of development, the entire transient and final response can be calculated algebraically by means of the quantitative methods illustrated in Appendix III. The resulting coordinate values of the various distance-time points of boundary crossovers, mergers, and disappearances are listed in Table 4.4; the distance and time for resolution of species 2 and 4 in a "ternary" displacement development $1 \rightarrow 2, 3, 4 \rightarrow 5$, as obtained from Eqs. (4.88) and (4.89), are included for comparison.

### 2. *Comparison with Carrier Displacement Development*

For separation of a mixture of species 2 and 4, an obvious alternative to selective displacement as discussed above is displacement development, $1 \rightarrow 2, 4 \rightarrow 5$. Here, however, disturbances will cause a slight overlap, undesirable for analytical purposes, between the close-up square-wave pulses of species 2 and 4 in the final pattern. This overlap is obviated in selective displacement through the intervening species 3. An alternative also obviating the overlap is so-called *carrier displacement development*, $1 \rightarrow 2, 3, 4 \rightarrow 5$, in which a species 3 of intermediate affinity is added to the mixture before the operation, to appear between the pulses of species 2 and 4 when these are resolved [11, 12, 98, 99]. The equations listed in Table 4.4 allow the efficiencies of selective displacement and carrier displacement development to be compared, as will now be shown.

Of interest for this comparison are the distance and time required for species 3 to intervene between the resolved species 2 and 4, since the presence of species 3 at this place is needed to eliminate the overlap. In selective displacement under condition (4.115) as well as in carrier displacement development, species 3 intervenes as soon as species 2 and 4

... Times for Resolution of Species and Disappearance of Transient Plateau Zones in Separation of Species 2 and 4 by Selective Displacement and Carrier Displacement Development. (Indices *ij* Refer to Resolution of Species *i* and *j*; Indices *Pi* (or *Pij*), to Disappearance of Plateau Zones of Species *i* (or *i* and *j*); $\Delta\tau$ and $\Delta\tau'$ Are Adjusted Times Required for Introduction of Mixture and First Development Agent, Respectively.)

| Selective displacement | | Carrier displacement development |
|---|---|---|
| If $x_2''' > (\alpha_{23}-1)/(\alpha_{24}-1)$ | If $x_2''' < (\alpha_{23}-1)/(\alpha_{24}-1)$ | |
| $z_{24} = \dfrac{\alpha_{24}x_2''' + x_4'''}{\alpha_{24}-1}\Delta\tau$ <br> $\tau_{24} = \dfrac{\alpha_{24}}{\alpha_{24}-1}\Delta\tau$ | | $z_{24} = \dfrac{\alpha_{24}x_2'' + \alpha_{23}x_3'' + x_4''}{\alpha_{24}-1}\Delta\tau$ <br> $\tau_{24} = \dfrac{\alpha_{24}}{\alpha_{24}-1}\Delta\tau$ |
| $z_{P2} = \dfrac{\alpha_{23}x_2'''}{\alpha_{23}-1}\Delta\tau$ <br> $\tau_{P2} = \left(1 + \dfrac{x_2'''}{\alpha_{23}-1}\right)\Delta\tau$ | $z_{P23} = \dfrac{[(\alpha_{24}-1)x_2''' + 1]^2(\alpha_{23}-1)}{\alpha_{23}(\alpha_{24}-1)^2 x_2'''}\Delta\tau$ <br> $\tau_{P23} = \left(1 + \dfrac{\alpha_{23}-1}{(\alpha_{24}-1)^2 x_2'''}\right)\Delta\tau$ | — |
| $z_{P3} = \dfrac{1}{\alpha_{23}-1}\Delta\tau'$ <br> $\tau_{P3} = \Delta\tau + \dfrac{\alpha_{23}}{\alpha_{23}-1}\Delta\tau'$ | | — |
| $z_{23} = \dfrac{[(\alpha_{23}x_2'''\,\Delta\tau)^{1/2} + \Delta\tau'^{1/2}]^2}{\alpha_{23}-1}$ <br> $\tau_{23} = \Delta\tau + \dfrac{[(x_2'''\,\Delta\tau)^{1/2} + (\alpha_{23}\,\Delta\tau')^{1/2}]^2}{\alpha_{23}-1}$ | | $z_{23} = \dfrac{h_2''(h_3''-\alpha_{12})}{(\alpha_{14}-\alpha_{12})(\alpha_{13}-\alpha_{12})}\Delta\tau$ <br> $\tau_{23} = \left(1 + \dfrac{\alpha_{12}(h_3''-\alpha_{12})}{(\alpha_{14}-\alpha_{12})(\alpha_{13}-\alpha_{12})}\right)\Delta\tau$ |

have been resolved (see Figs. 4.26 and 4.33), so that the distance and adjusted time of interest are $z_{23}$ and $\tau_{23}$. In selective displacement under condition (4.114), however, species 3 intervenes only at a later stage, namely, when the transient plateau zone $P_2$ has disappeared (see Fig. 4.32); here, accordingly, the distance and adjusted time of interest are $z_{P2}$ and $\tau_{P2}$.

For proper comparison, the expressions in Table 4.4 must be transformed to show amounts $Q_i$ instead of concentrations and, for displacement development, to account for the fact that the time for introduction of the mixture is longer by $\Delta\tau'$ because of the prior addition of species 3. For separations under condition (4.115), selective displacement then turns out to give the shorter resolution distance and time than carrier displacement development, since

$$\frac{\alpha_{24}Q_2+Q_4}{(\alpha_{24}-1)(Q_2+Q_4)}\,\Delta\tau \quad < \quad \frac{\alpha_{24}Q_2+\alpha_{23}Q_3+Q_4}{(\alpha_{24}-1)(Q_2+Q_3+Q_4)}\,(\Delta\tau+\Delta\tau') \tag{4.116}$$

$$\frac{\alpha_{24}}{\alpha_{24}-1}\,\Delta\tau \quad < \quad \frac{\alpha_{24}}{\alpha_{24}-1}\,(\Delta\tau+\Delta\tau')$$

[It is assumed here that the total concentration is the same in all influents, so that $(Q_2+Q_4)/\Delta\tau=(Q_2+Q_3+Q_4)/(\Delta\tau+\Delta\tau')$.] On the other hand, for separations under condition (4.114), either technique may be more efficient, depending on the relative concentrations in the mixture and the separation factors.

No general judgment of values should be formed on the basis of the simple comparison above, since many factors disregarded here are of potential importance in practical applications.

## D. Development without Resolution

The resolution in displacement development and similar operations is commonly explained with the plausible contention that, in a band with several species, the "faster" species of low affinities should advance into the forward portion while the "slower" species of high affinities should fall back into the rear. This argument is an oversimplification, valid only if the species are confined to the band, i.e., do not move across its front or rear boundary in either direction. This particular condition is met in displacement development and, for example, in the case of selective displacement discussed in the previous section. If this condition is not met, development may fail to produce resolution, as will now be illustrated with two examples.

First, consider a system with several influent composition changes such that a species $k$ is absent from the presaturant and final influent while being present in at least one intermediate influent, e.g.,

$$1, \ldots, k-1, k+1, \ldots, n \quad \to \quad 1, \ldots, n \quad \to \quad 1, \ldots, k-1, k+1, \ldots, n \tag{4.117}$$

If the trivial root equalling $\alpha_{1k}$ is $h_{k-1}$ in the final influent, and $h_k$ in the presaturant, then the variations of both $h_{k-1}$ and $h_k$ in the final pattern will constitute self-sharpening boundaries, between which species $k$ is confined. The portion of the composition route corresponding to this part of the final profile is as shown in Fig. 3.19c. The zone containing species $k$ travels with constant length between self-sharpening boundaries, yet retains its uniform composition indefinitely, i.e., no resolution whatsoever of the species in the zone takes place. The explanation in terms of physical processes is simple; instead of accumulating in different parts of the zone, the faster species $1, \ldots, k-1$ steadily pass through both boundaries and the entire zone in the direction of flow, while the slower species $k+1, \ldots, n$ do so in the opposite direction (relative to the traveling boundaries).

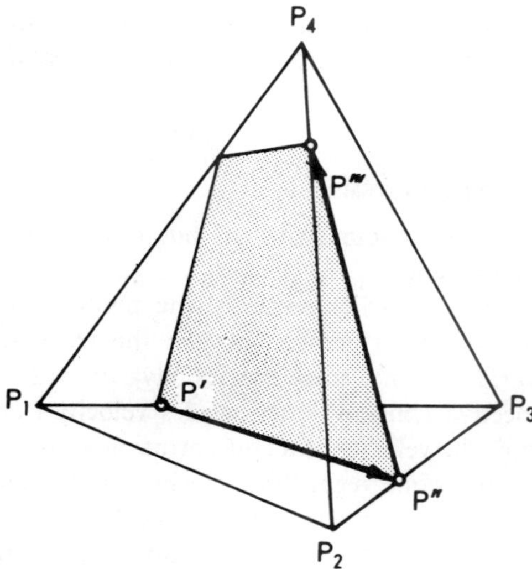


**Fig. 4.34.** Development without resolution. Composition profile route for four-component case 1, 3 → 2, 3 → 2, 4 with composition points of all zones on same common plane (shaded). Plateau zone P″ travels without resolution between P′ (influent composition) and P‴ (presaturant composition).

The second example, more similar to displacement development, is shown in Fig. 4.34. The system is of the type

$$1, 3 \rightarrow 2, 3 \rightarrow 2, 4 \tag{4.118}$$

and differs from "binary" displacement development $1 \rightarrow 2, 3 \rightarrow 4$ only in that the development agent and presaturant contain small amounts of species 3 and 2, respectively. If the compositions are matched in such a way that the composition point of the intermediate influent falls on the intersection of composition paths from those of the final influent and presaturant, as shown in Fig. 4.34, then displacement does not produce any resolution. (In terms of $H$-function roots, the condition is that $h_2' = h_2'' = h_2'''$, $h_2$ being the existing root in each influent.) Again, the explanation in physical terms is simple; the tendency of the "slow" species 2 to accumulate in the rear portion of the zone is balanced by intrusion of species 3 from the displacement agent through the rear boundary, and the tendency of the "fast" species 3 to accumulate in the front portion is similarly balanced by intrusion of species 2 from the presaturant through the front boundary. Thus, the zone of species 2 and 3 lengthens, but maintains its uniform composition indefinitely.

These examples are another illustration of the fact that concentration velocities and species velocities manifest themselves in different ways, and of how these concepts can help to understand physical causes and seemingly paradoxical situations.

## E. Trace-Component Systems

The special case of systems with all but one species at trace levels warrants a brief comment.

As shown in various earlier sections, the predominance of the bulk species suppresses all interference between the trace species, and any concentration variations of a trace species advance without loss of sharpness at the respective constant trace-species velocity (see Sections I.G.2 and III.B.3). Thus, the velocity of a concentration variation $\Delta x_j$ of a given trace species $j$ is the same regardless of whether it extends from $x_j = 0$ to $\Delta x_j$ or from any concentration $x_j$ to $x_j + \Delta x_j$, i.e., the presence or absence of a background of species $j$ has no effect on the velocity of a concentration variation of this species. One may say that the predominance of the bulk species also suppresses interference of any trace species with itself.

Because of the lack of mutual as well as self-interference of the trace species, the responses generated by the various successive influent composition changes do not interfere with one another as their boundaries cross, and the overall response is an additive superposition of all the individual concentration variations propagated from the various influent composition changes.

## F. Comparison with Other Theories

The following brief survey cites typical theories for comparison, without attempting to provide complete coverage.

### 1. *General Theories*

Relatively few other theories have concerned themselves with responses to multiple influent composition changes. Two noteworthy exceptions are the theories by Reilley *et al.* [69] and Glueckauf [19, 20], both discussed previously in connection with influent pulses (compare Section III.F. 3–4). Reilley's work covers a great variety of influent composition changes, but presupposes absence of interference and ideal additive superposition of all concentration variations, so that its applicability is essentially confined to trace-component systems (see Section IV.E). Glueckauf's pioneer work includes applications to various systems with two successive influent composition changes, such as sorptive displacement development. His quantitative treatment is for nonstoichiometric sorption systems with compound Langmuir isotherms, the mathematical equivalent of stoichiometric exchange systems with constant separation factors. The solutions given for special two- and three-component cases, although seemingly more complex because they lack the economy provided by the use of $H$-function roots, are equivalent to the corresponding special cases contained in the more general solution given here. The main shortcoming, which Glueckauf's theory shares with all others, lies in a basic premise equivalent to postulating the absence of any finite regions of noncoherence in the distance-time plane. As the present treatment shows, this postulate confines the applicability to a relatively small number of cases.

### 2. *Displacement Development*

Not surprisingly, displacement development as a highly successful technique has attracted considerable theoretical attention. Both nonstoichiometric sorptive and stoichiometric exchange displacement development have been treated. These systems are not completely equivalent.

Rather, in the terms used here, the mathematical equivalent of sorptive displacement development is

$$1, n \quad \rightarrow \quad 2, \ldots, n \quad \rightarrow \quad n \tag{4.119}$$

(see also Chapter 5, Section III.A).

A quantitative description of the final pattern in sorptive displacement development was already given by Tiselius [80], who first proposed the technique. The assumptions are those of the classical equilibrium theories.

The first description of the transient behavior was provided in Glueckauf's previously mentioned work [19, 20, 87, 88]. Independently, Sillén [7] advanced a treatment of ion-exchange displacement development of a uniform initial band, assuming constant separation factors and local equilibrium and using an approach similar to the operation with distance-time diagrams in the present treatment. Apparently unaware of Sillén's work, various later authors essentially duplicated his solutions for binary and ternary separations [92–96, 100, 101]. The latest of these publications [101] already reflects the influence of the present treatment in that the method of calculating resolution requirements from intersections of boundary trajectories was adopted, and is also one of the very few to cover standard displacement development [i.e., case (4.76)] as well as development of a uniform initial band. None of these approaches covers, or is conveniently extended to, separation of more than three species, some are confined to low concentrations (i.e., $C \ll \bar{C}$), and all share with Glueckauf's the unproven postulate of absence of finite regions of non-coherence. With these qualifications, the solutions are equivalent to those derived here.

Only few attempts to account for disturbances in displacement development have as yet been made [83, 87, 88, 96], and most of these have been restricted to the final pattern. A more general treatment is so far lacking.

One of the early theoretical studies of displacement development [99] has also dealt with deviations caused by selectivity reversals. These have so far been excluded from the present treatment by virtue of the premise of constant separation factors (see Chapter 5, Section IV.A.3, for relaxation of this premise).

## V. Trace Species in Systems with Variable Bulk Composition

In various previous sections, attention has been given to the exceptional behavior of trace-component systems in the narrow sense, as defined in

Chapter 2, Section I, i.e., systems with only *one* bulk species, all others being at trace levels. In such systems, the bulk composition, made up of a single species, is constant. However, in a number of practical applications, particularly in various radiochemical separations, the behavior of trace species in the presence of more than one bulk species is of interest. Here the bulk composition may vary significantly since the relative concentrations of the various bulk species may do so. The response behavior in such systems will now be examined.

As the previous discussion of trace-component systems has shown, the behavior of trace species has its peculiarities, but constitutes a limiting case of the general theory rather than obeying different basic equations. The general treatment set forth in the preceding sections thus comprises cases with several trace and bulk species along with systems of all other types. However, systems with trace species can be approached in a simpler manner. As this section will show, species which are not present at higher than trace concentrations in any influent (including the presaturant) virtually do not interfere with the bulk species and with one another. Therefore, one can calculate the bulk-species response disregarding any such trace species, and can subsequently fill in the trace-species responses one by one. The reduction in mathematical complexity is considerable. For example, in a seven-component system with three bulk and four trace species, the calculation of the bulk-species response is merely a three-component problem, which, in many cases, can be solved algebraically. A derivation, illustrated by practical cases, is given in the following.

### A. *H*-Function Roots in Systems with Trace Species

A convenient way to establish rules for the response behavior in systems with trace species is provided by the *H*-function roots. In this context, the following properties of the roots in such systems are important.

In the general case of a composition with one or several bulk species $k$ and one or several absent or trace species $j$, there is one root $h = \alpha_{1j}$ (trivial root) for each absent species $j$ and one root $h \simeq \alpha_{1j}$ for each trace species $j$ (see Chapter 3, Section IV.D.2). Thus, a particular root which equals, or deviates only infinitesimally from, $\alpha_{1j}$ can be attributed to each absent or trace species $j$. These "trace-species roots," with index numbers $j-1$ or $j$, will be designated $h(j)$. The other roots, not attributable to any particular species, will be called "bulk-system roots" and designated $h_l$.

The distinction between bulk-system and trace-species roots is practical because the latter can often be ignored. Thus, the bulk composition (i.e., the concentrations of the bulk species $k$) is completely determined by the bulk-system roots $h_l$ alone, and vice versa; similarly, the composition of any particular trace species $j$ is completely determined by the $h_l$ and its trace-species root $h(j)$. [This can be shown with Eq. (3.57), in which all factors involving other roots cancel.] Furthermore, all species, concentration, composition, step, and root velocities are determined by the bulk-system roots alone and are independent of the trace-species roots, as is apparent from the respective equations (see Chapter 3, Sections IV.D. 6–7 and V.A). These invariances with respect to the trace-species roots reflect the fact that the trace species do not interfere with the bulk species and with one another.

To calculate the $H$-function roots for a given composition, the following procedure, based on the properties stated above, can be used. First, one calculates the bulk-system roots $h_l$ ignoring any trace species, i.e., omitting from the $H$-function all terms involving trace concentrations (as well as those of absent species). Next, for any trace species $j$ separately, one calculates the respective root $h(j)$ from $x_j$ and the $h_l$ with the following relation, obtained by solving Eq. (3.57), in the notation used here, for $h(j)$:

$$h(j) = \alpha_{1j} + x_j \prod_k (\alpha_{1k} - \alpha_{1j}) \Big/ \prod_l (h_l - \alpha_{1j}) \tag{4.120}$$

For any absent species $j$, a root is equated to $\alpha_{1j}$. Finally, the index numbers of all roots are established with the general condition $\alpha_{1i} \leqq h_i \leqq \alpha_{1,i+1}$. The index number of the root $h(j)$ of any absent or trace species $j$ may be $j-1$ or $j$, depending on which intervals are occupied by the $h_l$ and whether the second term in Eq. (4.120) is negative or positive.

For deducing responses to influent composition changes, it will furthermore be important that all compositions in a response arise exclusively from combinations of $H$-function roots, one of each index number, of the presaturant and influent compositions. This property of the roots is not restricted to systems with trace species and was discussed in detail in Chapter 3, Section V.A.

### B. Systems with Constant Bulk Influent Composition

We can now proceed to examine responses to influent composition changes. First, we consider the relatively simple case of systems in which the concentrations of the bulk species in the influent are kept constant at their presaturant levels, and only the trace concentrations are varied. Here, the bulk-system roots $h_l$ in the influent, being solely determined by the constant bulk influent composition, remain constant. Since no other than their influent values can appear in the response, the $h_l$ then are constant in the entire response also. Being completely determined by the $h_l$, the bulk composition then is also constant in the entire response. The trace species thus move on a uniform background of the bulk influent composition.

For a constant bulk composition, the species and concentration velocities of all trace species, being independent of the trace-species roots and concentrations, are also constant. From the general equations for the adjusted velocities one obtains, for any particular trace species $j$,

$$u_j = u_{x_j} = \alpha_{1j} \prod_l h_l \prod_k \alpha_{k1} \tag{4.121}$$

[derived from Eqs. (3.67) and (3.68) with conditions (3.55)] or, in terms of concentrations,

$$u_j = u_{x_j} = \sum_k \alpha_{kj} x_k \tag{4.122}$$

[derived from Eqs. (3.35) and (3.7)]. These equations comprise the previously derived special case of trace velocities in systems with a single bulk species $k$ [compare Eqs. (3.69) and (3.70)].

Because of the constancy of the trace-species velocities, the response of the trace species to any type of influent composition change is the same as in the previously discussed trace-component systems with a single bulk species (e.g., see Sections I.G.2, III.B.3, and IV.E). Specifically, the influent concentration variations of the trace species are propagated at constant rates, without mutual interference, and without self-sharpening or nonsharpening tendencies. The propagation rates, of course, depend on the bulk composition, as shown by Eq. (4.122).

## C. Systems with Varying Bulk Influent Composition

Systems with varying bulk influent composition can be divided into two groups, according to the behavior of the index numbers of their trace-species roots. As pointed out earlier, any trace-species root $h(j)$ may assume the index number $j-1$ or $j$ (see Section V.A). With a variation in bulk composition, the index number may change from one to the other. It will be convenient to distinguish between systems with and without such index-number changes in the influent, as they differ in other properties as well.

In the composition space, the significance of the index number of a trace-species root $h(j)$ is that composition points with $h(j)=h_{j-1}$ and $h(j)=h_j$ are on opposite sides of the watershed of $x_j=0$. Composition routes with an index-number change of $h(j)$ thus cross over this watershed, whereas routes without such a change remain on one side (see also Fig. 4.36).

### 1. *Systems with Constant Index Numbers of Trace-Species Roots*

The simpler case of constant index numbers of the roots in the influent will be examined first. Here, species at trace levels in the influent are not accumulated to finite concentrations anywhere in the response pattern, and the bulk-species response is entirely independent of the trace species. This can be shown as follows. In the influent, each trace-species root $h(j)$ has its fixed index number and deviates but infinitesimally from the respective value $\alpha_{1j}$. Since all compositions in the response arise exclusively from combinations of the influent root values, one of each index number, any such composition then involves roots $h(j) \simeq \alpha_{1j}$, one for each influent trace species $j$, and therefore has all these species at no higher than trace concentrations. (It will later be seen that this is not necessarily so for cases with index-number changes.) The independence of the bulk-species response now becomes evident, since the trace-species roots or concentrations neither enter into the relation between the bulk composition and the bulk-system roots nor affect the velocities of the latter.

Its invariance with respect to the trace species permits the bulk-species response to be calculated separately, i.e., disregarding any species not exceeding trace levels in the influent. Furthermore, once the bulk-species response has been established, the response of each trace species can be calculated separately, i.e., disregarding any other trace species; this is possible because neither the relation between the concentration of a given trace species $j$ and the roots $h_l$ and $h(j)$ nor the root velocity of this root $h(j)$ are affected by the roots or concentrations of any other trace species. The further discussion can therefore be restricted to the response behavior of a single trace species.

In the general case, the response of a trace species $j$ can be calculated by means of computing the trajectories of the values of its root $h(j)$ in the distance-time plane. The root velocity, and thus the trajectory slope, of this root at any point in this plane is

$$u_{h(j)} = \alpha_{1j} \prod_l h_l \prod_k \alpha_{k1} \tag{4.123}$$

[see Eq. (3.89)]. The concentration of species $j$ at any distance-time point can then be calculated in the usual manner from the local values of $h(j)$ and the $h_l$.

A qualitative examination of the development behavior gives the following picture: The root velocity of $h(j)$, like those of the $h_l$, is subject to the general rule that roots of higher index numbers have higher velocities [see condition (3.90)]. Therefore, the variation of $h(j)$ participates in the general "sorting out" of root variations upon development, and is eventually resolved from the others when coherence is attained (see Chapter 3, Section V.B). In the final coherent pattern, the variation of $h(j)$ is between those of the roots with the next lower and higher index numbers. The trace species $j$ in the final pattern thus has a boundary or pulse of its own [that with variable $h(j)$], across which the concentrations of all other species are constant. In addition, the concentration of this trace species varies across each boundary or pulse of the bulk-species response, but remains constant across the boundaries and pulses contributed by any other trace species.

The variation of a trace-species root $h(j)$ does not possess self-sharpening or nonsharpening tendencies of its own, because the root velocity is virtually independent of the value of $h(j)$. Once resolved from other root variations, that of $h(j)$ thus travels without sharpening or spreading (except for spreading caused by disturbances neglected here). However, since the velocity of $h(j)$ depends on the bulk-system roots, sharpening or spreading can be induced by the latter prior to resolution. For example, when crossing a diffuse coherent, self-sharpening or nonsharpening

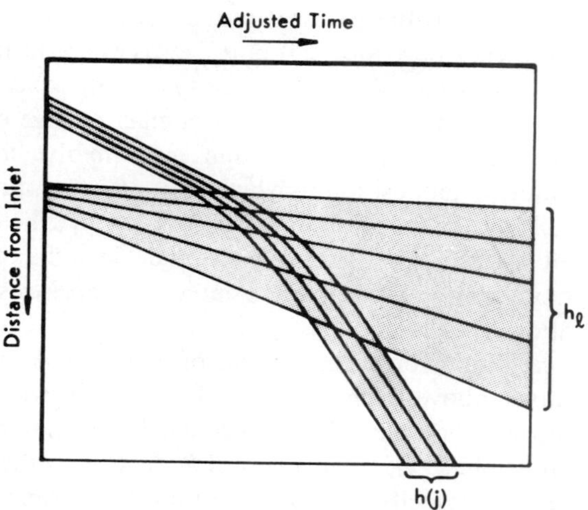


**Fig. 4.35.** Spreading of variation of trace-species root $h(j)$ on crossover with nonsharpening bulk-system boundary with variable $h_l$.

boundary, the variation of $h(j)$ sharpens or spreads as this boundary does, since the velocities of $h(j)$ and of the variable root of the boundary then vary in the same direction [see Eq. (4.123) and Fig. 4.35]. In the general case of noncoherent behavior, the velocity of $h(j)$ is proportional to the product of the bulk-system roots [see Eq. (4.123)], and the variation of $h(j)$ therefore sharpens where this product decreases, and spreads where this product increases, in the direction of flow. In any event, the variation of $h(j)$ becomes distorted by interference and does not retain its influent form, as it does in systems with constant bulk composition.

2. *Systems with Varying Index Numbers of Trace-Species Roots*

The chief complication arising from changes of index numbers of trace-species roots in the influent history is that trace species (i.e., species not exceeding trace levels in the influent) may be accumulated to finite concentrations in certain portions of the response. Material balance does not preclude such accumulations but confines them to infinitesimal portions of the response. As will be shown, however, the existence of such portions has but a negligible effect on the response behavior of the other species, so that the overall response can be computed in essentially the same manner as for systems without index-number changes.

That and how a trace species $j$ can reach finite concentrations in the response can be shown as follows. It is true that all compositions in the response arise exclusively from combinations of influent root values, one of each index number. In systems without index-number changes, this guarantees the presence of a root $h(j) \simeq \alpha_{1j}$ everywhere in the response, and thus rules out any accumulation of species $j$ to finite concentrations, as previously shown. In systems with an index-number change of $h(j)$ in the influent, however, a response composition may involve a root value $h_{j-1}$ from an influent composition in which $h(j) = h_j$, and a root value $h_j$ from another influent composition in which $h(j) = h_{j-1}$; the composition then has no root $h(j) \simeq \alpha_{1j}$ in its set of roots and, since the other roots in general differ more than infinitesimally from $\alpha_{1j}$, comprises species $j$ at a finite concentration.

A necessary condition for accumulation of a trace species $j$ can now be stated. As shown above, such an accumulation arises where two roots $h_{j-1}$ and $h_j$, neither of them a trace-species root $h(j)$, encounter one another in the distance-time plane. Of these, the root $h_{j-1}$, having the lower index number, has the lower velocity and must therefore have originated from an earlier part of the influent history than had $h_j$. This is possible only if the trace-species root $h(j)$ has the index number $j$ in the earlier part, and

$j-1$ in the later part, of the influent history. Accumulation of a trace species to a finite concentration anywhere in the response thus can occur only if the index number of its root somewhere in the influent history decreases with time.

The accumulation of a trace species $j$ may be transient or permanent. It is permanent if the index number of $h(j)$ is $j$ in the presaturant, and is $j-1$ in the final influent, since the final pattern in this case comprises a plateau zone containing species $j$ at a finite concentration; this zone is the

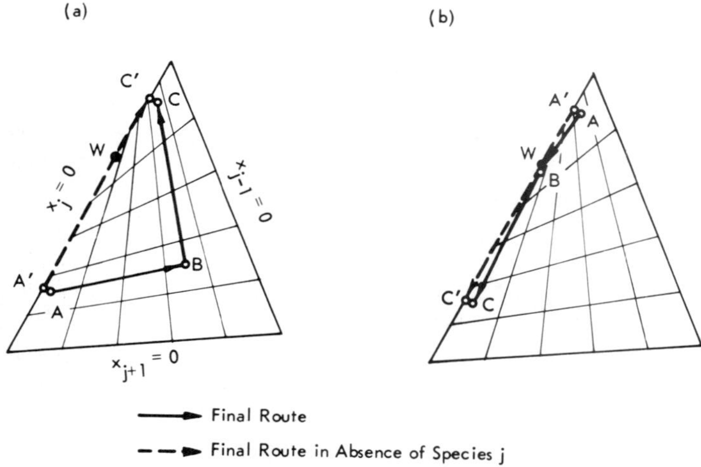


**Fig. 4.36.** Response in systems with varying bulk composition and varying index number of trace-species root $h(j)$. Segment of final composition profile route comprising boundaries with variable $h_{j-1}$ and $h_j$, on common plane. Accumulation of species $j$ to finite concentrations occurs in (a), but not in (b). Final route segment in system with species $j$ entirely absent is shown for comparison.

$j$th (i.e., between the boundaries with variable $h_{j-1}$ and $h_j$) and, according to the root sets (3.92), takes its $h_{j-1}$ value from the presaturant and its $h_j$ value from the final influent, neither root being a trace-species root $h(j)$ in the present case.

The relevant segment of the final composition route, comprising the boundaries with variable $h_{j-1}$ and $h_j$, for the case of permanent accumulation of species $j$ is shown in Fig. 4.36a.* Because of the index-number

* For the sake of clarity, Figs. 4.36 to 4.39 are drawn for low but finite rather than for infinitesimal concentrations of the trace species in the influent.

change of $h(j)$, the composition points A and C upstream and downstream of this route segment are on opposite sides of the watershed. The plateau zone B, containing species $j$ at a finite concentration and restricted to infinitesimal length by material balance, is between two self-sharpening boundaries with variable $h_{j-1}$ and $h_j$ and with equal velocities, given by

$$u_{\Delta h_{j-1}} = u_{\Delta h_j} = \prod_{i=1}^{j-1} h_i'' \prod_{i=j}^{n-1} h_i' \prod_{i \neq j} \alpha_{i1} \tag{4.124}$$

as obtained from Eq. (4.17) with $h'_{j-1}$ and $h''_j$ being trace-species roots $h(j) \simeq \alpha_{1j}$. This common velocity is the same as that of the single, coherent, self-sharpening boundary A′C′ with bracketing affinity cut $j-1 \mid j+1$ and root variation extending over $h_{j-1}$ and $h_j$, which would appear in the place of AB and BC if species $j$ were entirely absent. Thus, the only effect of the presence of species $j$ on the final pattern of the other species is that an infinitesimal plateau zone B, with the species at concentrations different than elsewhere, and with sharp flanks AB and BC, takes the place of the single sharp boundary A′C′ with bracketing affinity cut.

For comparison, the respective route segment for a case in which the index number of $h(j)$ changes from $j-1$ in the presaturant to $j$ in the final influent is shown in Fig. 4.36b. Here, the positions of the plateau zones A and C relative to the watershed are reversed. No accumulation of the trace species $j$ develops. The compositions and composition velocities along the route ABC differ but infinitesimally from those along the route A′C′, which the system would take if species $j$ were entirely absent. The only effect of the presence of species $j$ on the final pattern is to split the single, coherent, nonsharpening boundary A′C′ with bracketing affinity cut $j-1 \mid j+1$ in two such boundaries AB and BC with nonbracketing cuts; the intervening plateau zone B remains of infinitesimal width, its growth rate being negligible because its composition point is virtually on the watershed, where the difference between the composition velocities for the paths of the two different cuts is nil (e.g., see Chapter 3, Section V.E.1). Thus, there is virtually no interference of species $j$ with the other species.

The following brief examination of the transient behavior in cases with eventual accumulation of a trace species will show that here, too, such a species merely inserts an infinitesimal zone into the otherwise unchanged response of the other species.

As was shown above, accumulation of a trace species $j$ to finite concentrations occurs only if the initial composition route crosses over the watershed of $x_j = 0$ in the same direction as in Fig. 4.36a. The development behavior of such an initial route is illustrated in Fig. 4.37*; for representation in a two-dimensional diagram, the projection (along composition paths) onto a common plane of paths with variable $h_{j-1}$ and $h_j$ is shown. Development of route projections on such a plane reflect the shifts of $h_{j-1}$ and $h_j$ values relative to one another, regardless of the behavior of the other roots, and is subject to the same qualitative rules as for three-component systems (see Chapter 3, Section V.E). Accumulation of species $j$ to finite concentrations occurs when the route projection moves a finite distance away from the border $x_j = 0$. The geometry of the

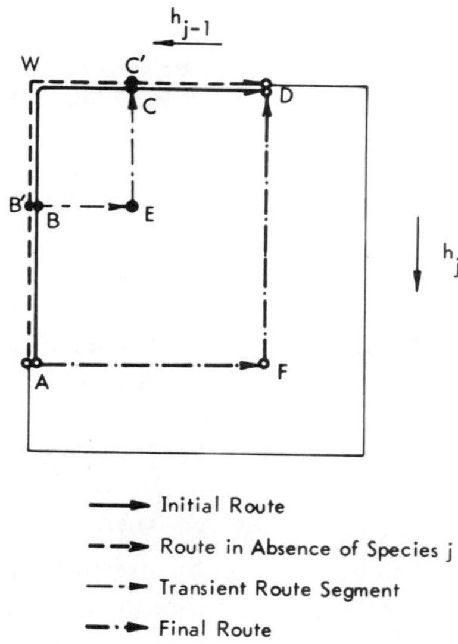


**Fig. 4.37.** Development of composition profile route in case of accumulation of influent trace species $j$. Projections of initial, transient, and final routes on $h$-space common plane with variable $h_{j-1}$ and $h_j$. Route projection for system with species $j$ entirely absent is shown for comparison.

path grid is such that the first route-projection segment to do so must be one spanning the watershed (e.g., segment BC in Fig. 4.37). However, in the immediate vicinity of the watershed, the resolution of $h_{j-1}$ from $h_j$ values and, thus, development of the route projection are extremely slow. At a faster pace, the root variations of $h_{j-1}$ and $h_j$ sharpen, and once a route-projection segment BC spanning the watershed has become sharp, the compositions in the immediate vicinity of the watershed are no longer in existence; the segment then immediately develops into BEC, since resolution of abrupt root variations is instantaneous. The two root steps BE and EC are sharp and self-sharpening and are readily shown both to have the same velocity as the step B′C′,

* See footnote to p. 261.

which would appear in their place if species $j$ were entirely absent. Further development eventually brings the route projection to its final form AFD. In all intermediate stages, the composition with finite concentration of species $j$ is between sharp and self-sharpening root steps both having the same velocity as the corresponding single step would have if species $j$ were absent, as is readily shown with equations analogous to Eq. (4.124).

According to these considerations, the transient behavior can be summarized as follows. A finite concentration of species $j$ first appears when the initial noncoherent boundary has become sharp in a portion whose route extends across the watershed of $x_j = 0$. From then on, an infinitesimal zone with finite concentration of species $j$ and sharp flanks takes the place which, in the absence of species $j$, would be occupied by a sharp boundary with route crossing the watershed. This behavior is consistent with the material-balance requirement that a species not exceeding trace levels in the influent can occur at finite concentrations in only infinitesimal portions of the response.

The considerations in this section have shown that (1) species not exceeding trace levels in the influent do not interfere with the response behavior of others, except where they are accumulated to finite concentrations, and (2) such finite concentrations of influent trace species are restricted to infinitesimal regions of the response, which is in no other way affected by their presence. Therefore, even if influent trace species are accumulated, one can calculate the overall response by use of the previously described simple procedure of calculating first the bulk-species response (ignoring any influent trace species) and then filling in the trace-species responses one by one (see Section V.B); one must merely correct the so obtained response of each species by inserting the infinitesimal zones, if any, in which other trace species are accumulated to finite concentrations.

### D. Two Examples: Frontal Analysis and Displacement Development

The preceding discussion will be illustrated by two examples of practical interest, namely, frontal analysis and displacement development of mixtures containing a trace species. Both techniques have been used for determination, enrichment, and isolation of trace components [77, 88, 92, 102–104].

#### 1. *Frontal Analysis*

First, consider frontal analysis of a three-component mixture with the species of intermediate affinity at trace level:

$$1, (2), 3 \quad \rightarrow \quad 4 \qquad (4.125)$$

The composition routes for two cases with different compositions of the mixture, but with the same separation factors are shown in Fig. 4.38.* Accumulation of the trace species to a finite concentration results in one case, but not in the other, depending on the position of the composition point of the mixture relative to the watershed of $x_2 = 0$. It is interesting to note that the relative concentrations of species 1 and 3, but not of 2, in the mixture determine whether a zone with species 2 at finite concentration is formed.

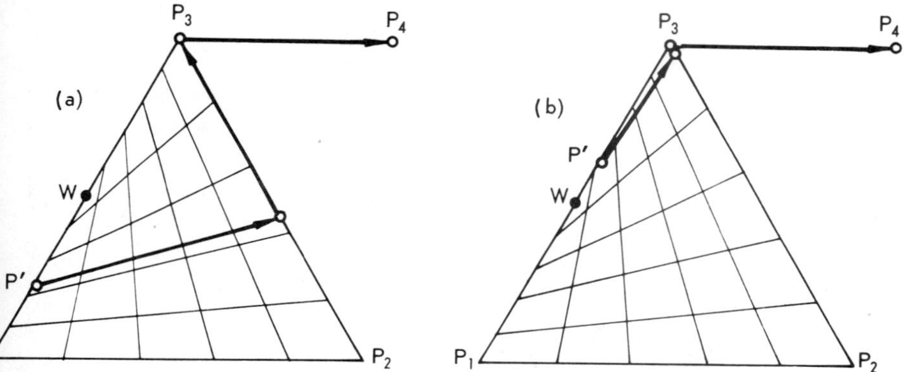


**Fig. 4.38.** Composition profile routes for frontal analysis of three-component mixture with species 2 at trace level. Accumulation of species 2 to finite concentration occurs in (a), but not in (b).

For the general case of an $n$-component system with an arbitrary species $j$ at trace level,

$$\underset{(j)}{1, \ldots, j-1, j+1, \ldots, n-1} \quad \rightarrow \quad n \tag{4.126}$$

the following criterion can be used to predict whether or not species $j$ will be accumulated to a finite concentration. As shown earlier in this section, such accumulation occurs if the index number of the root $h(j) \simeq \alpha_{1j}$ changes from $j$ in the presaturant (here species $n$) to $j-1$ in the final influent (here the mixture). The requirement that $h(j)$ has the index number $j$ in the presaturant is met in any frontal-analysis case [compare Eqs. (4.32)]. The index number of $h(j)$ for the mixture can be determined

* See footnote to p. 261.

with a criterion analogous to conditions (3.64); for a composition with $x_j \ll 1$ instead of $x_k = 0$, these conditions are

$$h_{j-1} \simeq \alpha_{1j} \quad \text{if} \quad \sum_{i \neq j} \frac{x_i}{1-\alpha_{ji}} > 0, \qquad x_j \ll 1 \tag{4.127}$$

$$h_j \simeq \alpha_{1j} \quad \text{if} \quad \sum_{i \neq j} \frac{x_i}{1-\alpha_{ji}} < 0, \qquad x_j \ll 1 \tag{4.128}$$

Thus, the root $h(j) \simeq \alpha_{1j}$ is $h_{j-1}$ under condition (4.127), and $h_j$ under condition (4.128). Accordingly, in frontal analysis, species $j$ is accumulated to a finite concentration if the mixture meets the former, and is not if the mixture meets the latter, condition. Also, no accumulation takes place in the singular case in which the sum in these conditions is zero (composition on watershed).

Any trace species except 1 and $n-1$ may or may not be accumulated to a finite concentration, depending on the composition of the mixture. Species $n-1$ always is accumulated, because a plateau zone with $n-1$ as the only species invariably leads the pattern; species 1 never is, because it is present only in the plateau zone having the composition of the mixture (see Section I.F).

### 2. *Displacement Development*

Displacement development of a mixture containing an arbitrary trace species $j$

$$1 \quad \rightarrow \quad \begin{matrix} 2, \ldots, j-1, j+1, \ldots, n-1 \\ (j) \end{matrix} \quad \rightarrow \quad n \tag{4.129}$$

obeys essentially the same rules as ordinary displacement development with all species at finite concentrations. It is readily verified that the derivation of the final, constant pattern with the species of the mixture in the sequence of their affinities and each in a separate plateau zone (see Section IV.B.1) is valid for systems with trace species also. Thus, regardless of the composition of the mixture being developed, the trace species is eventually accumulated in a zone by itself. (In practice, of course, the slight diffuseness of the self-sharpening boundaries, caused by disturbances neglected here, will let a trace species appear as a bell-shaped peak, rather than as a narrow and high square-wave pulse, between the plateau zones of the neighboring species [50, 88, 102]; see Fig. 5.22.)

In keeping with the fact that accumulation of the trace species $j$ to finite concentrations invariably takes place, the index number of the root

$h(j) = \alpha_{1j}$ changes from $j$ in the presaturant to $j-1$ in the final influent [see Eqs. (4.81) and (4.80)], as required for such accumulation. For the mixture constituting the intermediate influent, $h(j)$ may have either index number. If the composition of the mixture obeys condition (4.127), the index number is $j-1$, and the index-number change from $j$ to $j-1$ thus accompanies the first of the two influent composition changes. In this case, there is accumulation of species $j$ to finite concentrations in the transient response to the first, but not in that to the second, influent composition change. On the other hand, if the composition of the mixture obeys condition (4.128), $h(j)$ in the mixture has the index number $j$; the index-number change then accompanies the second influent composition change, and species $j$ appears at finite concentrations in the transient response to that change.

For illustration, the distance-time diagram of a five-component displacement development with one trace species,

$$1 \quad \rightarrow \quad \begin{matrix} 2, 4, 5, 6 \\ (3) \end{matrix} \quad \rightarrow \quad 7 \tag{4.130}$$

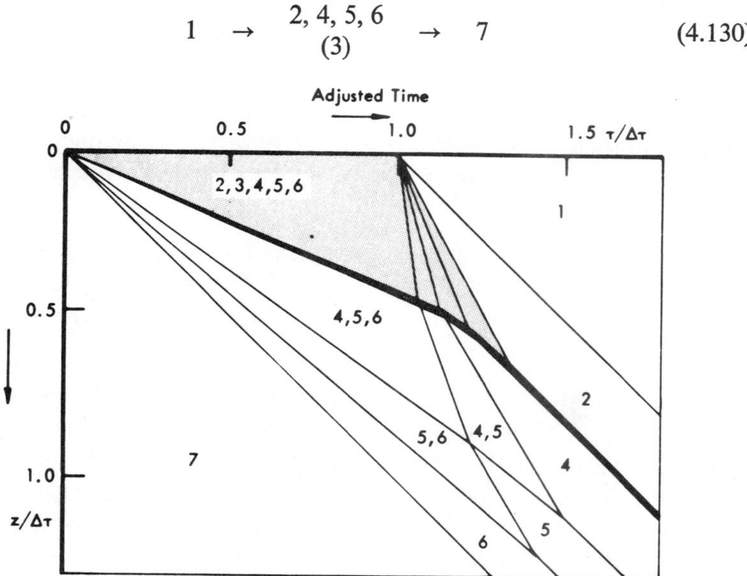


**Fig. 4.39.** Distance-time diagram for displacement development of typical five-component mixture with one species at trace level.* Region in which trace species 3 is present is shaded; region in which this species accumulates to finite concentrations is darkly shaded. (For $\alpha_{23} = 2$, $\alpha_{24} = 4$, $\alpha_{25} = 8$, $\alpha_{26} = 16$; $x_2'' = 0.30$, $x_3'' = 0.02$, $x_4'' = 0.28$, $x_5'' = 0.20$, $x_6'' = 0.20$.)

* See footnote to p. 261.

is shown in Fig. 4.39. The case shown is one in which the composition of the mixture obeys condition (4.127), so that the trace species, 3, is accumulated to finite concentrations in the transient response to the first influent composition change.

## E. SUMMARY

The most important conclusions arrived at in this section can be summarized as follows:

1. Species not exceeding trace levels in the influent history (presaturant included) do not interfere with the bulk species or with one another, except where they are accumulated to finite concentrations. Such accumulations are, however, restricted to infinitesimal regions of the response, which is not affected in any other way by their presence.
2. Because of this lack of interference by the influent trace species, one can calculate the bulk-species response ignoring any such trace species, and can then fill in the trace-species responses one by one, correcting, where necessary, for any infinitesimal regions in which a trace species is accumulated.
3. In systems with constant bulk influent composition, the bulk composition is also constant in the response. The response behavior of the trace species then is the same as in the previously discussed systems with a single bulk species.
4. In systems with variable bulk influent composition, an influent trace species $j$ may be accumulated to finite concentrations in certain infinitesimal regions of the response. Any such region in the transient and the final response appears where the system in the complete absence of species $j$ would have a sharp and self-sharpening boundary with affinity cut bracketing species $j$ and composition route extending across the watershed of $x_j = 0$.
5. Aside from such potential regions of accumulation, each influent trace species in general may have, in the final coherent pattern, its own response boundary or pulse across which all other concentrations are constant. In addition, the concentration of each influent trace species varies across all boundaries and pulses produced by variations of the bulk composition.

## VI. Review and Perspective

This section has dealt in detail with the response of a uniformly presatured column to various types of influent composition changes, selected mainly for comparison with experience in typical chromatographic operations. Conventional chromatography (elution development) with small sample size appears as a limiting case with insignificant interference. For frontal analysis and displacement development, the results are in complete agreement with known behavior. For other cases, such as elution development with large sample size, vacancy chromatography, and concentration-pulse chromatography, the treatment correctly reproduces the known qualitative features of the patterns and leads to various as yet untested predictions concerning the effects of interference.

Numerous other influent composition changes can be imagined and be treated with essentially the same methods. Furthermore, with simple substitutions the treatment can be made to cover the response of a column with nonuniform initial composition to development with an influent of constant composition. The symmetry properties on which this simple extension is based were discussed, and an example given, in connection with displacement development [see Section IV.B.5, substitutions (4.106)]. The extension to systems with both nonuniform initial and varying influent composition is readily achieved with the methods indicated in Chapter 3, Section V.

Little would be gained through examination of further special cases under the same premises. The relaxation of the premises, however, still poses challenges, which will be taken up next.

## REFERENCES

1. T. Vermeulen, in *Advances in Chemical Engineering* (T. B. Drew and J. W. Hoopes, Jr., eds.), Vol. 2, Academic Press, New York, 1958, p. 147.
2. F. Helfferich, *Angew. Chem. Intern. Ed. Engl.*, **1**, 440 (1962); *Chem. Ing.-Tech.*, **34**, 269 (1962).
3. F. Helfferich, *Ion Exchange*, McGraw-Hill, New York, 1962, Chap. 9.
4. N. K. Hiester, T. Vermeulen, and G. Klein, in *Chemical Engineers' Handbook* (R. H. Perry, C. H. Chilton, and S. D. Kirkpatrick, eds.), 4th ed., McGraw-Hill, New York, 1963, Sec. 16.
5. D. Tondeur and G. Klein, *Ind. Eng. Chem. Fundamentals*, **6**, 351 (1967).
6. A. Tiselius, *Arkiv Kemi Mineral. Geol.*, **B 14**, No. 22 (1940).
7. L. G. Sillén, *Arkiv Kemi*, **2**, 477 (1950).
8. S. Claesson, *Arkiv Kemi Mineral Geol.*, **A 23**, No. 1 (1946).
9. S. Claesson, *Discussions Faraday Soc.*, **7**, 34 (1949).

10. R. C. Brimley and F. C. Barrett, *Practical Chromatography*, Reinhold, New York, 1953, pp. 20, 22, 54, 56.
11. H. G. Cassidy, *Fundamentals of Chromatography* (Techniques of Organic Chemistry, Vol. X, A. Weissberger, ed.), Interscience, New York, 1957, Chap. IV.
12. E. Lederer and M. Lederer, *Chromatography*, 2nd ed., Elsevier, New York, 1957, pp. 4, 8.
13. A. I. M. Keulemans, *Gas Chromatography*, 2nd ed., Reinhold, New York, 1959, pp. 205, 208.
14. A. B. Littlewood, *Gas Chromatography*, Academic Press, New York, 1962, p. 234.
15. H. Purnell, *Gas Chromatography*, J. Wiley, New York, 1962, pp. 78, 80.
16. V. V. Rachinskii, *The General Theory of Sorption Dynamics and Chromatography*, English translation, Consultants Bureau, New York, 1965, Chap. 4.
17. D. DeVault, *J. Am. Chem. Soc.*, **65**, 532 (1943).
18. A. C. Offord and J. Weiss, *Nature*, **155**, 725 (1945); *Discussions Faraday Soc.*, **7**, 26 (1949).
19. E. Glückauf, *Proc. Roy. Soc.* (*London*), **A 186**, 35 (1946).
20. E. Glueckauf, *Discussions Faraday Soc.*, **7**, 12 (1949).
21. G. G. Baylé and A. Klinkenberg, *Rec. Trav. Chim.*, **73**, 1037 (1954).
22. D. O. Cooney and E. N. Lightfoot, *Ind. Eng. Chem. Process Design Develop.*, **5**, 25 (1966).
23. G. Klein, D. Tondeur, and T. Vermeulen, *Ind. Eng. Chem. Fundamentals*, **6**, 339 (1967).
24. N. K. Hiester and T. Vermeulen, *Chem. Eng. Progr.*, **48**, 505 (1952).
25. D. W. Simpson and W. C. Bauman, *Ind. Eng. Chem.*, **46**, 1958 (1954).
26. D. R. Asher and D. W. Simpson, *J. Phys. Chem.*, **60**, 518 (1956).
27. W. C. Bauman, R. M. Wheaton, and D. W. Simpson, in *Ion Exchange Technology* (F. C. Nachod and J. Schubert, eds.), Academic Press, New York, 1956, Chap. 7.
28. A. A. Zhukhovitskii and N. M. Turkeltaub, *Dokl. Akad. Nauk SSSR*, **143**, 646 (1962).
29. A. A. Zhukhovitskii, N. M. Turkeltaub, M. Gaier, M. N. Lagashkina, L. A. Malyasova, and G. P. Shlepuzhnikova, *Zavodsk. Lab.*, **29**, 8 (1963).
30. B. O. Prescott and H. L. Wise, in *Advances in Gas Chromatography 1965* (A. Zlatkis and L. S. Ettre, eds.), Preston Techn. Abstr., Evanston, 1966, p. 89.
31. S. Ahuja, G. D. Chase, and J. G. Nikelly, *Anal. Chem.*, **37**, 840 (1965).
32. F. Steinbach, *J. Chromatog.*, **15**, 432 (1964).
33. R. T. Parkinson and R. E. Wilson, *J. Chromatog.*, **24**, 412 (1966).
34. G. Castello and G. d'Amato, *J. Chromatog.*, **32**, 625 (1968).
35. F. I. Stalkup and H. A. Deans, *A. I. Ch. E. J.*, **9**, 106 (1963).
36. P. C. Mangelsdorf, Jr., *Anal. Chem.*, **38**, 1540 (1966).
37. F. I. Stalkup and R. Kobayashi, *A. I. Ch. E. J.*, **9**, 121 (1963).
38. K. T. Koonce, H. A. Deans, and R. Kobayashi, *A. I. Ch. E. J.*, **11**, 259 (1965).
39. R. Kobayashi, P. S. Chappelear, and H. A. Deans, *Ind. Eng. Chem.*, **59**, No. 10, 63 (1967).
40. J. R. Conder, in *Progress in Gas Chromatography* (Advances in Analytical Chemistry and Instrumentation, Vol. 6, J. H. Purnell, ed.), Wiley-Interscience, New York, 1968, p. 209.
41. F. Helfferich and D. L. Peterson, *Science*, **142**, 661 (1963).
42. P. E. Porter, C. H. Deal, and F. H. Stross, *J. Am. Chem. Soc.*, **78**, 2999 (1956).

43. J. R. Anderson, *J. Am. Chem. Soc.*, **78**, 5692 (1956).
44. H. B. Gilmer and R. Kobayashi, *A. I. Ch. E. J.*, **11**, 702 (1965).
45. D. L. Peterson, F. Helfferich, and R. J. Carr, *A. I. Ch. E. J.*, **12**, 903 (1966).
46. J. C. Giddings and K. L. Mallik, *Ind. Eng. Chem.*, **59**, No. 4, 18 (1967).
47. S. Masukawa and R. Kobayashi, *J. Chem. Eng. Data*, **13**, 197 (1968).
48. A. J. P. Martin and R. L. M. Synge, *Biochem. J.*, **35**, 1358 (1941).
49. S. W. Mayer and E. R. Tompkins, *J. Am. Chem. Soc.*, **69**, 2866 (1947).
50. E. Glueckauf, in *Ion Exchange and its Applications*, Society of Chemical Industry, London, 1955, p. 34.
51. J. J. van Deemter, F. J. Zuiderweg, and A. Klinkenberg, *Chem. Eng. Sci.*, **5**, 271 (1956).
52. Ref. 13, Chap. 4.
53. S. Dal Nogare and R. S. Juvet, Jr., *Gas–Liquid Chromatography*, J. Wiley, New York, 1962, Chap. 5.
54. Ref. 14, Chap. 2.
55. Ref. 15, Chap. 8.
56. J. C. Giddings, in *Chromatography* (E. Heftmann, ed.), 2nd ed., Reinhold, New York, 1967, Chap. 3.
57. J. C. Giddings, *Dynamics of Chromatography*, Part I, Marcel Dekker, New York, 1965.
58. J. Kragten, *J. Chromatog.*, **37**, 373 (1968).
59. O. Grubner, in *Advances in Chromatography* (J. C. Giddings and R. A. Keller, eds.), Vol. 6, Marcel Dekker, New York, 1968, p. 173.
60. P. Schneider and J. M. Smith, *A. I. Ch. E. J.*, **14**, 762, 886 (1968).
61. E. Glueckauf, *J. Chem. Soc.*, **1947**, 1302.
62. J. E. Funk and G. Houghton, *Nature*, **188**, 389 (1960); *J. Chromatog.*, **6**, 281 (1961).
63. G. Houghton, *J. Phys. Chem.*, **67**, 84 (1963).
64. R. S. Henly, A. Rose, and R. F. Sweeny, *Anal. Chem.*, **36**, 744 (1964).
65. F. T. Dunckhorst and G. Houghton, *Ind. Eng. Chem. Fundamentals*, **5**, 93 (1966).
66. P. C. Haarhoff and H. J. van der Linde, *Anal. Chem.*, **38**, 573 (1966).
67. E. Glückauf, *Nature*, **156**, 205 (1945); *Discussions Faraday Soc.*, **7**, 50, 53 (1949).
68. A. C. Offord and J. Weiss, *Nature*, **156**, 570 (1945); *Discussions Faraday Soc.*, **7**, 51 (1949).
69. C. N. Reilley, G. P. Hildebrand, and J. W. Ashley, Jr., *Anal. Chem.*, **34**, 1198 (1962).
70. C. G. Collins and H. A. Deans, *A. I. Ch. E. J.*, **14**, 25 (1968).
71. C. A. Barrere and H. A. Deans, *A. I. Ch. E. J.*, **14**, 280 (1968).
72. A. A. Zhukhovitskii, M. L. Sazonov, A. F. Shlyakhov, and A. I. Karymova, *Zavodsk. Lab.*, **31**, 1048 (1965).
73. A. A. Zhukhovitskii, M. L. Sazonov, A. F. Shlyakhov, and V. P. Shvartsman, *Zh. Fiz. Khim.*, **41**, 2640 (1967).
74. A. A. Zhukhovitskii, N. M. Turkeltaub, L. A. Malyasova, A. F. Shlyakhov, V. V. Naumova, and T. I. Pogrebnaya, *Zavodsk. Lab.*, **29**, 1162 (1963).
75. A. A. Zhukhovitskii and N. M. Turkeltaub, *Dokl. Akad. Nauk SSSR*, **144**, 829 (1962).
76. A. A. Zhukhovitskii, M. S. Selenkina, N. M. Turkeltaub, V. P. Shvartsman, A. F. Shlyakhov, and I. A. Smirnova, *Zavods. Lab.*, **30**, 1308 (1964).

77. A. A. Zhukhovitskii, N. M. Turkeltaub, A. I. Karymova, and R. I. Koreshkova, *Zavodsk. Lab.*, **32**, 133 (1966).
78. A. A. Zhukhovitskii, M. L. Sazonov, L. M. Lapkin, and M. S. Selenkina, *Zavodsk. Lab.*, **33**, 131 (1967).
79. S. Z. Roginskii, O. V. Altshuler, O. M. Vinogradova, M. I. Yanovskii, and O. P. Krivoruchko, *Izv. Akad. Nauk SSR, Ser. Khim.*, **1965**, 214.
80. A. Tiselius, *Arkiv Kemi Mineral. Geol.*, **A 16**, No. 18 (1943).
81. Ref. 3, Secs. 9-6, 9-7, 9-9.
82. O. Samuelson, *Ion Exchange Separations in Analytical Chemistry*, Almqvist and Wiksell, Stockholm, 1963, pp. 115, 193, 219.
83. Ref. 16, Chap. 6.
84. J. Inczédy, *Analytical Applications of Ion Exchangers*, Pergamon Press, Oxford, 1966, pp. 91, 156.
85. F. H. Spedding and J. E. Powell, in *Ion Exchange Technology* (F. C. Nachod and J. Schubert, eds.), Academic Press, New York, 1956, Chap. 15.
86. W. L. Silvernail and N. J. Goetzinger, *Rare Earth Elements and Compounds*, in Kirk-Othmer, *Encyclopedia of Chemical Technology*, 2nd ed., Vol. 17, A. Standen, ed., Interscience, New York, 1968, p. 143.
87. E. Glueckauf and J. I. Coates, *J. Chem. Soc.*, **1947**, 1315.
88. E. Glueckauf, K. H. Barker, and G. P. Kitt, *Discussions Faraday Soc.*, **7**, 199 (1949).
89. L. G. Sillén, *Arkiv Kemi*, **2**, 499 (1950).
90. D. B. James and F. Helfferich, *Proceedings of the Seventh Rare Earth Research Conference, San Diego, October 1968*, Vol. I, International Harvester Co., Solar Division, San Diego, Calif., 1968, p. 397.
91. F. Helfferich and D. B. James, *J. Chromatog.*, in press.
92. F. H. Spedding, J. E. Powell, and H. J. Svec, *J. Am. Chem. Soc.*, **77**, 6125 (1955).
93. J. E. Powell and F. H. Spedding, *Chem. Eng. Progr. Symp. Ser.*, **55**, No. 24, 101 (1959).
94. J. E. Powell, in *Progress in the Science and Technology of the Rare Earths* (L. Eyring, ed.), Vol. 1, Pergamon Press, Oxford, 1964, p. 62.
95. J. E. Powell, H. R. Burkholder, and D. B. James, *J. Chromatog.*, **32**, 559 (1968).
96. J. Shankar and T. R. Bhat, *J. Chromatog.*, **9**, 461 (1962).
97. F. Helfferich, in *Advances in Chromatography* (J. C. Giddings and R. A. Keller, eds.), Vol. 1, Marcel Dekker, New York, 1965, Chap. 1, p. 24.
98. A. Tiselius and L. Hagdahl, *Acta Chem. Scand.*, **4**, 394 (1950).
99. L. Hagdahl, R. J. P. Williams, and A. Tiselius, *Arkiv Kemi*, **4**, 193 (1952).
100. V. V. Veselov, *Zh. Priklad. Khim.*, **39**, 2679 (1966).
101. D. B. James, J. E. Powell, and H. R. Burkholder, *J. Chromatog.*, **35**, 423 (1968).
102. E. Glueckauf, in *Separations of Isotopes* (H. London, ed.), George Newnes, London, 1961, Sec. 5.6; *J. Chim. Phys.*, **60**, 73 (1963).
103. V. S. Mirzayanov, A. A. Zhukhovitskii, V. G. Berezkin, and N. M. Turkeltaub, *Zavodsk. Lab.*, **29**, 1166 (1963).
104. V. G. Berezkin and V. S. Mirzayanov, *Izv. Akad. Nauk SSSR, Ser. Khim*, **1967**, 1202.

# 5

# RELAXATION OF SIMPLIFYING PREMISES

The treatment to this point has been subject to the severely restrictive premises listed in Chapter 3, Section I, introduced to bring out the basic effects of interference unobscured by the many complications bound to occur in more realistic systems. The present section will deal with the consequences of relaxing these premises. In some instances this is straightforward and has no significant effect on the treatment, while in many others the quantitative treatment is greatly complicated but the overall qualitative picture remains largely unchanged. In still other but fortunately exceptional instances, principal difficulties arise, and we shall enter areas that are as yet largely uncharted.

The variety of combinations of possible effects is too great and the present knowledge too fragmentary for a systematic coverage to be attempted at this stage, and a discussion in depth does not seem warranted as long as more suitable and economical descriptions are still likely to be found. Thus, types of effects and methods of deducing predictive rules will be emphasized rather than rules themselves. A truly comprehensive coverage must await further theoretical progress, which, we hope, this survey will help to stimulate.

With few exceptions, deviations from the original premises make it impossible or at least impracticable to derive algebraic solutions. This is particularly true once the effective range of the $h$ transformation is exceeded. While quantitative calculations for specific cases can be left to numerical techniques, the deduction of general rules must then rely largely on conceptual arguments. Fortunately, the concepts of velocities and coherence, introduced in Chapter 3, Sections III and IV, provide adequate means for handling almost any type of situation with only minimum recourse to mathematics. In fact, the earliest version of the present treatment [1] was a qualitative conceptual construction on this basis and of more general scope than the preceding sections.

In the following, effects and complications will be discussed in roughly the order of increasing complexity. Physically unrelated effects, even if they are not additive, will be examined separately, and their interactions will largely be ignored. In many instances, the reader will be referred to theories of chromatography without interference, as they are applicable here, too.

If the theoretical predictions to this point may be considered somewhat daring despite their solid foundation in mathematics, many of those from here on must be called speculative, since it is too early to judge whether any essential feature may as yet have escaped attention.

## I. Variable Flow Rate, Total Concentrations, and Geometrical Factors. Generalized Distance and Time Variables. Gradient Elution

The premises easiest to remove are those of constant linear flow rate, constant and uniform total concentrations in the mobile and stationary phases, and uniform cross-sectional area and fractional interstitial void volume of the column. They were introduced mainly for convenience, to allow deductions and rules to be stated in simple and readily visualized terms of distances, times, and velocities. At a small sacrifice in ease of perception and evaluation, one can replace distance and time as the independent variables by others that make the treatment invariant to the behavior of the mentioned parameters, as has been done at least in part in many other theories of chromatography and fixed-bed behavior (see Chapter 2, Section IV.B). It must be assumed, however, that the variations of the parameters, whatever their physical causes, do not entail deviations from (1) stoichiometric exchange of species between the phases, (2) composition-independent separation factors, (3) constant phase volumes at any given location, (4) uniform volumetric flow rate (e.g., in $cm^3/sec$) throughout the column at any given time, and (5) ideal plug flow.

The physical effects caused by variations of the parameters and the transformation of variables to accommodate them are neither novel nor particular to systems with interference. In view of their practical importance, however, a brief discussion appears warranted.

To achieve the desired invariance, distance and time as the independent variables are replaced as follows. One uses the amount $\bar{m}$ of stationary phase upstream of any given location as a measure of the latter's distance from the inlet, and the amount of $m^0$ of influent which the column has received from $t = 0$ on as a measure of elapsed time, both amounts being expressed in terms of the number of moles (or equivalents, or whatever scale is used to define concentrations) of sorbable species contained. These variables are a natural choice: Whenever there is transfer between two phases moving relative to one another, what matters is the *amounts* that have come in contact, and dynamics and geometry are immaterial except for their effect on disturbances not as yet considered here.

The relations of the generalized variables to distance and time are

$$\bar{m}(z) = \int_0^z \bar{C}(z)S(z)\,\mathrm{d}z \qquad \text{and} \qquad m^0(t) = \int_0^t c^0(t)\dot{V}(t)\,\mathrm{d}t \tag{5.1}$$

where $S$ is the cross-sectional area, $c^0$ is the total concentration of the influent, and

$$\dot{V} = \varepsilon S u_0 \tag{5.2}$$

is the volumetric flow rate, $\varepsilon$ being the fractional interstitial void volume. Furthermore, for constant $c^0$:

$$c^0 = C/\varepsilon \tag{5.3}$$

The scale of the generalized time variable must be "adjusted" so that, like $\tau$, it starts to count not from $t = 0$, but from the moment at which the respective location $z$ or $\bar{m}$ is reached by the "solvent front" (or a nonadsorbable marker in the mobile phase) having entered the column at $t = 0$ (see Chapter 2, Section IV.B). As will later become apparent, the adjustment is a necessity in the general case, not merely a convenient option as under the initial premises. The so adjusted time variable $m$ is related to distance and time by

$$m(z, t) = \int_{t_z}^t c^0(\vartheta)\dot{V}(t)\,\mathrm{d}t \qquad (\vartheta \equiv t - t_z) \tag{5.4}$$

where $t_z$ is the time at which the solvent front reaches $z$ and, in the general case, is given implicitly by

$$\int_0^{t_z} \dot{V}(t)\,\mathrm{d}t = \int_0^z \varepsilon(z)S(z)\,\mathrm{d}z \tag{5.5}$$

For variable flow rate but constant $\bar{C}$, $c^0$, and $\varepsilon$, Eq. (5.4) reduces to

$$m(z, t) = m^0(t) - (C/\bar{C})\bar{m}(z) \tag{5.6}$$

For variable influent concentration but constant $\bar{C}$, $\dot{V}$, and $\varepsilon$, the time $t_z$ can be stated explicitly:

$$t_z(z) = (\varepsilon/\dot{V}\bar{C})\bar{m}(z) \tag{5.7}$$

Except for having the dimension of an amount instead of a distance, the quantity $m$ has the same characteristic properties as the adjusted time $\tau$ defined by Eq. (2.18).

In terms of the new variables $\bar{m}$ and $m$, the quantities corresponding to the adjusted velocities $\partial z/\partial \tau$ are of the form $\partial \bar{m}/\partial m$. For these generalized adjusted velocities of species, concentrations, and steps one finds

$$\begin{aligned} \partial \bar{m}/\partial m|_i &= x_i/y_i \\ (\partial \bar{m}/\partial m)_{x_i} &= (\partial x_i/\partial y_i)_{\bar{m}} \\ (\partial \bar{m}/\partial m)_{y_i} &= (\partial x_i/\partial y_i)_m \\ \partial \bar{m}/\partial m|_{\Delta x_i} &= \Delta x_i/\Delta y_i \end{aligned} \tag{5.8}$$

that is, the same expressions as for the adjusted velocities $\partial z/\partial \tau$ in systems with constant values of the parameters [see Eqs. (3.33) to (3.36); for derivation, see below]. Thus, the replacement of $z$ and $\tau$ by $\bar{m}$ and $m$, defined without the premises of constant linear flow rate, total concentrations, and geometrical factors, has made the treatment independent of variations of these parameters, and the response $\mathbf{x}(\bar{m}, m)$ to a given input $\mathbf{x}(\bar{m}, 0)$, $\mathbf{x}(0, m)$ is invariant to $\dot{V}(t)$, $\bar{C}(z)$, $c^0(t)$, $S(z)$, and $\varepsilon(z)$. Otherwise the treatment remains unchanged, except that expressions such as lengthening and shortening, faster and slower, etc., in previous sections can no longer be taken literally, but must be understood in terms of distance and time measured in amounts of stationary and mobile phase.

Equations (5.8) can be obtained in essentially the same manner as the corresponding equations in systems with constant parameter values. To derive the relations for the concentration velocities, Eqs. (3.15) and (3.16) must be modified to account for variations in the cross-sectional area and void fraction:

$$SJ_i = \dot{V}c_i \tag{5.9}$$

$$S\left(\frac{\partial(\bar{C}_i + \varepsilon c_i)}{\partial t}\right)_z = -\operatorname{div}(SJ_i) = -\dot{V}\left(\frac{\partial c_i}{\partial z}\right)_t \tag{5.10}$$

[In view of the variable $\varepsilon$, mobile-phase concentrations $c_i$ per unit volume of the mobile phase rather than of the column are used. The conversion is $\varepsilon c_i = C_i$; see Eqs. (2.1)].

With the premise of sorption and desorption by stoichiometric exchange, one has

$$\sum_i (\partial \bar{C}_i/\partial t)_z = 0 \tag{5.11}$$

and summation of Eq. (5.10) over all species then gives, with the rule (3.17) and after rearrangement,

$$(\partial z/\partial t)_c = \dot{V}/\varepsilon S = u_0 \tag{5.12}$$

where

$$c \equiv \sum_i c_i = C/\varepsilon \tag{5.13}$$

With Eq. (5.12) and introduction of $x_i$, $y_i$, $\bar{m}$, and $m$, Eq. (5.10) can be reduced to

$$(\partial y_i/\partial m)_{\bar{m}} + (\partial x_i/\partial \bar{m})_m = 0 \tag{5.14}$$

from which the concentration velocities are obtained with the rule (3.17).

That the response $\mathbf{x}(\bar{m}, m)$ of a column should be independent of the flow rate, total concentrations, and geometrical factors is contingent on the other premises. Once disturbances through finite sorption-desorption rates, axial diffusion, etc., are allowed, a direct dependence on actual time and thus on the linear velocities reappears, and the transformation of the independent variables loses much of its appeal.

In contrast to the response $\mathbf{x}(\bar{m}, m)$, which is not affected, the response $\mathbf{c}(z, t)$ reflects the physical effects of variations of the flow rate, total influent concentration, and geometrical factors. Some of these effects are illustrated in Figs. 5.1 and 5.2 by distance-time diagrams for operation with a column having a smaller diameter in its lower part, and with increasing flow rate in one case and increasing influent concentration in the other. Lines of constant $\bar{m}$ and $m$, for evenly spaced $\bar{m}$ and $m$ values, are shown. The effect of the variations of the parameters is reflected in the behavior of a representative trajectory so chosen that it would be linear if the parameters were constant. Since transformation of $z$ and $t$ into $\bar{m}$ and $m$ would restore its linearity, the trajectory is readily constructed with the aid of the lines of constant $\bar{m}$ and $m$.

It is instructive to compare the effects of increasing flow rate (Fig. 5.1) and increasing influent concentration (Fig. 5.2). In both cases, the rate at which sorbable species are fed to the column is increased and the operation is correspondingly accelerated, as reflected by the increasing steepness of the trajectory. The difference between the two modes of operation is that an increase in the mobile-phase flow is propagated instantaneously through the entire column, whereas an increase in the total concentration of the mobile phase, according to Eq. (5.12), is propagated at the rate of mobile-phase flow (i.e., along a line of constant $m$)

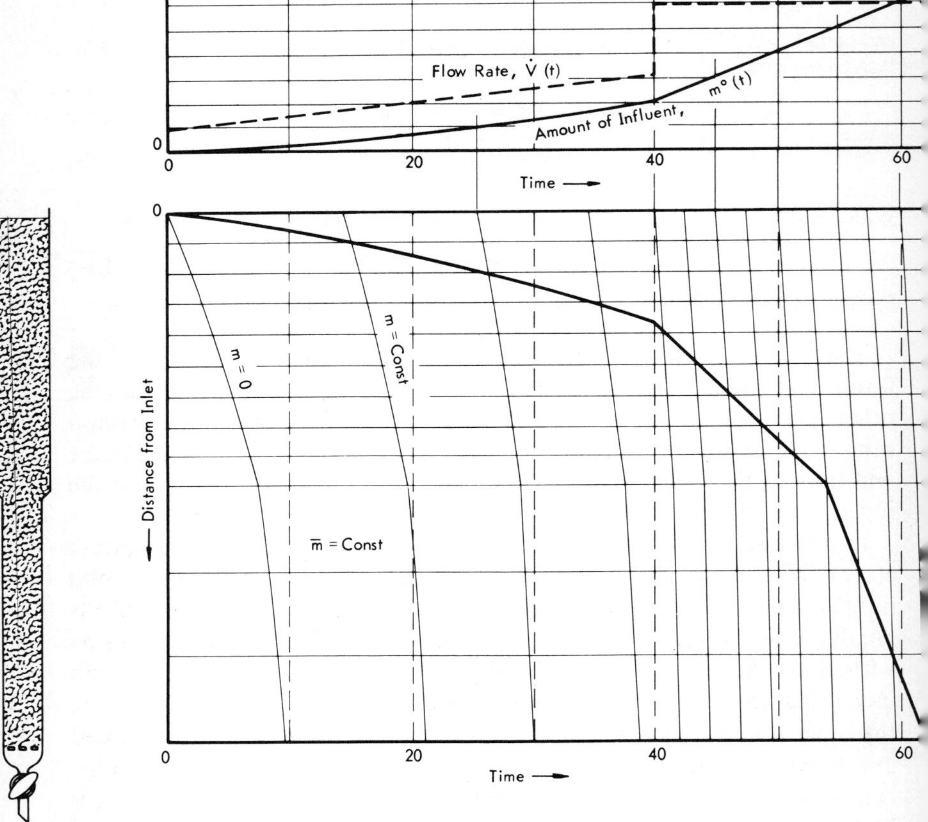


**Fig. 5.1.** Operation with variable flow rate and nonuniform column cross section. Influent program and distance-time diagram with trajectory of entity having constant generalized velocity $\partial\bar{m}/\partial m$. Loci of constant $\bar{m}$ and $m$ are shown as thin lines.

and thus incurs a delay before acting on a trajectory in the column. For example, in Fig. 5.2 the trajectory responds to the sharp increase in $c^0$ at $t = 40$ with a delay of about one time unit, whereas in Fig. 5.1 the trajectory responds immediately to the equivalent increase in $\dot{V}$.

In Eq. (5.4), the difference in propagation rate between variations of the flow rate and total concentration of the mobile phase is reflected in the arguments of $\dot{V}$ and $c^0$: The former counts without delay from $t = 0$, whereas the latter counts with a delay of $t_z$

relative to the inlet. The difference in propagation rate is what makes the unadjusted time variable $m^0$, stated in Eq. (5.1), inadequate for the general case, because this variable fails to discriminate between the relative contributions of $\dot{V}$ and $c^0$ to the amount of sorbable species introduced. In Figs. 5.1 and 5.2, for example, the same influent history $m^0(t)$ under identical initial conditions will generate different responses $\mathbf{x}(\bar{m}, m^0)$, as the differences in the behavior of the trajectory demonstrate.

Aside from this difference in the promptness of evoked reaction, variations of the flow rate and the total concentration of the influent

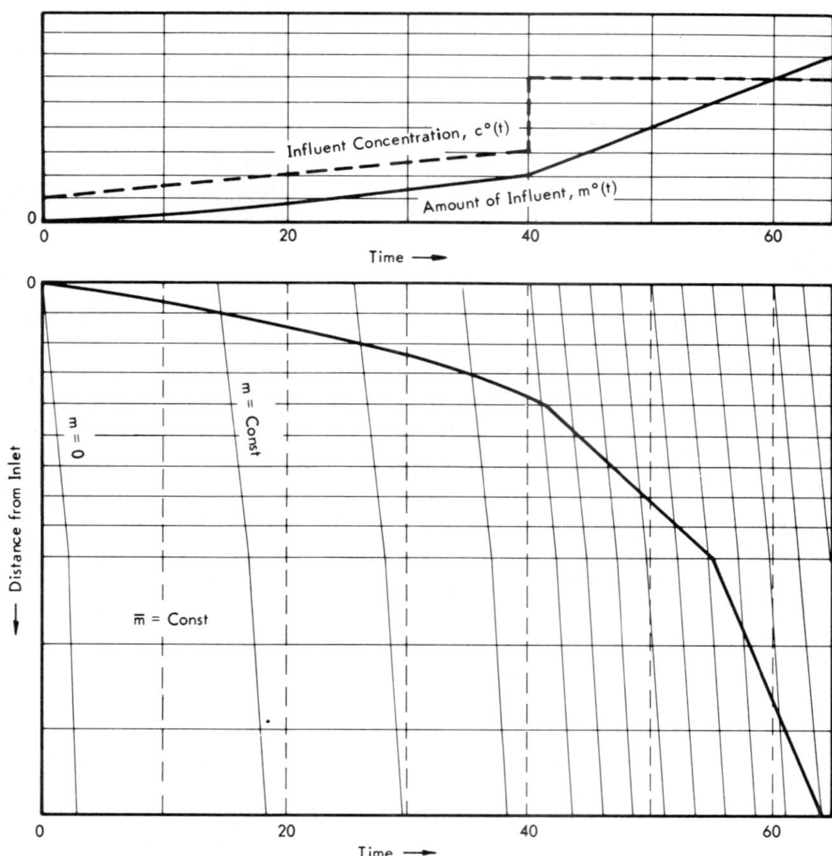


**Fig. 5.2.** Operation with variable total concentration of influent and nonuniform column cross section. Influent program and distance-time diagram with trajectory of entity having constant generalized velocity $\partial\bar{m}/\partial m$. Column geometry, $m^0(t)$, and generalized velocity of entity are same as in Fig. 5.1.

achieve equivalent effects on the response $\mathbf{x}(z, t)$. In particular, a variation of the total concentration $c$ (at constant $\mathbf{x}$), like a flow variation, does not produce noncoherence nor interacts with variations of the species concentrations $x_i$ in any way other than to affect their velocities $\partial z/\partial t$.

The absence of interactions between the species concentrations $x_i$ and the total concentration $c$ greatly facilitates the calculation of the response in cases with varying total concentration of the influent. On the response $\mathbf{x}(\bar{m}, m)$, which is invariant to changes in the total concentration, a response $c(\bar{m}, m)$ is superimposed. Since any variation of $c$ is propagated at the rate of mobile-phase flow, and thus along a line of constant $m$, this response is

$$c(\bar{m}, m) = c^0(m) \tag{5.15}$$

With this simple relation to the influent history, the response $\mathbf{x}(\bar{m}, m)$, calculated as in cases with constant parameters, is readily translated into absolute concentrations $c_i$:

$$\mathbf{c}(\bar{m}, m) = \mathbf{x}(\bar{m}, m)c^0(m) \tag{5.16}$$

The response of practical interest, $\mathbf{c}(x, t)$, can now be obtained with Eqs. (5.1) and (5.4) to (5.7).

The expedient of increasing, during elution, the influent concentration $c^0$ of an eluent to suppress excessive "tailing" of peaks is used in liquid chromatography under the name *gradient elution* [2–4]. The gradual acceleration of the elution process sharpens peaks with diffuse rear flanks—the most common form of peak asymmetry—by compressing their long tails into narrower effluent fractions. This principle and the essential features of the quantitative treatment are also applicable to systems with interference, since the response $c(\bar{m}, m)$ of the total concentration is responsible for the effect and, as shown above, is independent of $\mathbf{x}(\bar{m}, m)$ and thus of interference of the species with one another.

## II. Variation of Separation Factors with Total Concentration, Solvent, or Co-ion

Changes of the total concentration of the mobile phase, the solvent, the co-ion in ion-exchange operations, or other nonsorbable constituents of the mobile phase may entail variations of the separation factors. One example encountered earlier is the usual procedure for displacement development of rare earths, where, after nonselective sorption, introduction of a chelating agent alters the separation factors (see Chapter 4,

Section IV.B.5). The effect also arises in many practical operations when exhausted columns are switched to regeneration, or in stepwise or gradient elution with eluents of different concentrations, solvent compositions, complexing agents, or pH.

To see the effect of altered separation factors unencumbered by other phenomena, we examine the response of a column to a single, abrupt influent composition change involving, say, a change $\Delta c$ of total concentration with attendant change of the separation factors. The influent composition change generates various boundaries, which are subject to the relations

$$u_{\Delta c} \equiv \partial z/\partial t|_{\Delta c} = u_0 \tag{5.17}$$

$$u_{\Delta C_i} = \frac{u_0}{1+\Delta\bar{C}_i/\Delta C_i} \qquad \text{for all } i \tag{3.19}$$

[see Eq. (5.12) and Chapter 3, Section III.B]. The variation $\Delta c$ of the total concentration, and with it that of the separation factors, thus is propagated at the rate of mobile-phase flow $u_0$. The boundary involving these variations is the fastest, since any variations of $H$-function roots have root velocities lower than $u_0$ [see Eqs. (3.89) with conditions (2.27)]. Across this boundary, none of the stationary-phase concentrations vary, since Eq. (3.19) requires

$$\Delta\bar{C}_i = 0 \qquad \text{if} \quad u_{\Delta C_i} = u_0 \tag{5.18}$$

The stationary-phase composition, still in equilibrium with the presaturant on the downstream side of the boundary, thus is the same on the upstream side. The mobile-phase composition, however, being in equilibrium with that of the stationary phase on both sides, varies across the boundary because the separation factors vary. The remainder of the pattern, upstream of the boundary with varying total concentration, is subject to the rules derived in Chapter 4, Section I, for single abrupt influent composition changes: In general, there will be $n-1$ coherent boundaries, each involving concentration variations of all sorbable species. Thus, the variation of the separation factors with the influent composition change adds to the normal response pattern an $n$th boundary, which advances at the rate of mobile-phase flow, is faster than all others, carries the variation of the total concentration and the separation factors, and involves variations of the mobile-phase but not of the stationary-phase concentrations.

The quantitative treatment of this case is straightforward. Let the compositions and the sets of separation factors and $H$-function roots of the presaturant and influent be $\mathbf{x}''$ and $\mathbf{x}'$, $\boldsymbol{\alpha}''_{1i}$ and $\boldsymbol{\alpha}'_{1i}$, and $\mathbf{h}''$ and $\mathbf{h}'$. The sets $\mathbf{h}''$ and $\mathbf{h}'$ then are given as the roots of

$$H(h'', \mathbf{x}'', \alpha''_{1i}) = 0 \qquad \text{and} \qquad H(h', \mathbf{x}', \boldsymbol{\alpha}'_{1i}) = 0 \tag{5.19}$$

The stationary-phase composition $\mathbf{y}''$ in equilibrium with the presaturant, on the downstream side of the boundary involving the variations of total concentration and separation factors, can be calculated by application of Eq. (3.59):

$$y''_j = \left(\alpha''_{j1} \prod_i \alpha''_{1i} \Big/ \prod_i h''_i\right) x''_j \qquad j = 1, \ldots, n \tag{5.20}$$

On the upstream side of this boundary, the stationary-phase composition is $\mathbf{y}''$, too, but the separation factors are $\alpha'_{1i}$; the $H$-function roots, $h^*_i$, on that side can thus be calculated from $\mathbf{y}''$ as the reciprocals of the roots in $1/h^*$ of

$$H(1/h^*, \mathbf{y}'', \boldsymbol{\alpha}'_{i1}) = 0 \tag{5.21}$$

Once the set $\mathbf{h}^*$ has been obtained, the complete response $\mathbf{x}(z, t)$, $\mathbf{y}(z, t)$ upstream of the boundary with variation of total concentration can be calculated with the procedures derived in Chapter 4, Section I.D, with substitution of $h^*_i$ for $h''_i$ and with separation factors $\alpha'_{1i}$. This completes the calculation of the response, since the velocity of the boundary and the composition on its downstream side are already known to be $u_0$ and $\{\mathbf{x}'', \mathbf{y}''\}$.

This treatment is readily extended to more complex cases. Any single or multiple, abrupt or gradual variations of the total concentration with attendant variations of the separation factors are always propagated at the rate of mobile-phase flow, do not involve variations of the stationary-phase composition (except for those in other boundaries they may cross), and always leave in their wake separation factors and a mobile-phase composition which differ from those in front. The $H$-function roots upstream can be calculated, as in Eq. (5.21), from the unchanged stationary-phase composition and the changed separation factors, and can then be used for computing the other parts of the response by standard procedures. The only new feature is that existing coherent boundaries in general become noncoherent when the separation factors are changed; therefore, whenever a variation of total concentration overtakes a previously coherent boundary, the latter is made noncoherent and will give rise, upon resolution, to an entire set of new coherent boundaries.

A slight complication arises if the variation of the separation factors is caused by a change of the solvent, co-ion, or other constituent not entirely excluded from the stationary phase. Equation (3.19) applies also to solvents, co-ions, etc., and gives velocities lower than $u_0$ if a concentration variation in the stationary phase is involved, as is necessarily the case for both the replacing and the replaced constituent if they are not completely excluded. With the boundary moving at a rate lower than $u_0$, Eq. (3.19) applied to the sorbable species then shows that their concentrations in the stationary phase must also vary, so that the composition on the upstream side will deviate from the presaturation composition $\mathbf{y}''$. For co-ions in ion exchange, the invasion of the stationary phase is usually so slight that the deviations from $u_0$ and $\mathbf{y}''$ as well as their consequences are negligible. For solvent changes, on the other hand, they may be sizable. The response can then be computed as follows. First, the velocity $u_f$ of the front of the new solvent is calculated with Eq. (3.19). The composition $\{\mathbf{x}^*, \mathbf{y}^*\}$ on the upstream side of this front can then be calculated from

$$\frac{x_i^* - x_i''}{y_i^* - y_i''} = u_f, \qquad \frac{y_1^* x_i^*}{x_1^* y_i^*} = \alpha_{1i}' \qquad \text{for all } i \tag{5.22}$$

The corresponding root set $\mathbf{h}^*$ is now obtained from $\mathbf{x}^*$ or $\mathbf{y}^*$, and the calculation can then proceed as in the simpler case above.

A further complication arises if Eq. (3.19) gives significantly different velocities for the advancing front of the replacing solvent and the retreating rear of the solvent being replaced. This will entail either a nonuniform volumetric flow rate or a considerable volume change of the stationary phase, resulting in expansion or shrinkage of the bed and possible changes in the void fraction. Both effects lead too far afield to be discussed in this context.

## III. Nonstoichiometric Sorption

One of the most serious limitations of the treatment in the earlier sections is the restriction to stoichiometric exchange of sorbable species between the mobile and stationary phases. Obviously, this premise is untenable for sorption systems in which the stationary phase can be loaded to different degrees and, as a rule, even be stripped completely of any solutes. Even for ion exchange, stoichiometric exchange is no more than a reasonable approximation, since exchange of counterions is nearly always accompanied, if to a small extent, by electrolyte sorption or desorption.

The problem of extending the treatment to systems with nonstoichiometric sorption will be examined in this section. This extension requires the premise of constant total concentrations in the stationary and mobile phases to be relaxed. For simplicity, however, we shall retain the premise of constant and uniform flow rate. This essentially restricts the treatment in the present section to liquid chromatography: In gas chromatography, nonstoichiometric sorption at concentrations high enough to produce interference usually entails a nonuniform flow rate and other side effects, whose discussion will be relegated to Section VI.

### A. Formal Equivalence to Stoichiometric Exchange

Fortunately, the extension to systems with nonstoichiometric sorption can be accomplished with relative ease because a formal mathematical equivalence can be established to stoichiometric exchange. The principle is simple: For the sorption system, a fictitious additional species (*dummy species*) is introduced which, in each phase, makes up at any time for the difference between the total concentration of the physically existing species and some selected mathematical constant. In this way, an $n$-component system with nonstoichiometric sorption becomes the formal equivalent of an $(n+1)$-component system with stoichiometric exchange and constant total concentrations in both phases, including the influent. The implementation of this principle is not without problems, however, in that the dummy species is not necessarily subject to the same rules and constraints as the others. In exceptional cases, entirely new phenomena may arise, as the following deductions will show.

In practical applications, the dummy species must be taken into account when the problem is formulated. For example, a column receiving influents of different total concentrations of the existing species receives, in the equivalent stoichiometric system, influents of the same total concentration but with varying proportions of the dummy species. A column in reality stripped entirely of sorbable species is, in the equivalent stoichiometric system, saturated with the dummy species. Displacement development in such a stripped column, and with species $n+1$ as the dummy, corresponds to

$$1, n+1 \quad \rightarrow \quad 2, \ldots, n+1 \quad \rightarrow \quad n+1 \tag{5.23}$$

### B. Systems with Compound Langmuir Isotherms

The extension of the classical probabilistic deduction of the Langmuir sorption isotherm for a single sorbable species [5–8] to systems with more than one such species [6] gives what is commonly called a *compound Langmuir isotherm*:

$$q_j = \frac{a_j c_j}{1 + \sum_i b_i c_i} \qquad j = 1, \ldots, n \tag{5.24}$$

where $q_i$ are the concentrations in or on the sorbent,* the $c_i$ are those in the external fluid in equilibrium with the sorbent, and the $a_i$ and $b_i$ are

* Usually, the $q_i$ are given in moles per unit weight of sorbent, but any other units can be used. The choice of units merely affects the numerical values of the $a_i$, but not their ratios.

empirical constants. To give physically meaningful positive and finite $q_i$, the constants must obey the constraints

$$a_i > 0 \quad \text{for all } i \qquad \text{and} \qquad \sum_i b_i c_i > -1 \tag{5.25}$$

The $a_i$ are a measure of the intrinsic affinities of the respective species for the sorbent, and the $b_i$ are characteristic of the nature and strength of interference produced by the species. Thus, the (constant) separation factors and the affinity sequence are dictated by the $a_i$ alone:

$$\alpha_{ij} \equiv \frac{q_i c_j}{c_i q_j} = a_i/a_j = \text{const} \tag{5.26}$$

and, with the convention of numbering the species in the order of decreasing affinity,

$$i < j \qquad \text{if} \quad a_i > a_j \tag{5.27}$$

If all $b_i$ are positive, sorption is entirely "competitive," and if any $b_i$ are negative, the respective species have a "synergistic" effect, i.e., their sorption enhances rather than reduces sorption of others (for definitions, see Chapter 2, Section III). The absolute values of the $b_i$ reflect the strengths of the competitive or synergistic effects.

### 1. *Transformation into Equivalent Stoichiometric Systems*

It would be a simple matter to set mathematical constants for the fictitious total concentrations high enough to exceed the highest totals $\Sigma_i\, c_i$ and $\Sigma_i\, q_i$ that can possibly occur, and so to transform the Langmuir system into an equivalent stoichiometric one with an additional, dummy species as described earlier in this section. In general, however, the separation factors $\alpha_{ik}$ relating to the dummy $k$ would then vary with composition. To effect a transformation into a stoichiometric system having constant separation factors of all species including the dummy, we shall need the degrees of freedom provided by the option of choosing for each species a different scale to measure amounts. No inconsistencies arise as long as, for any given species, the same scale is used in both phases, so that the material balances remain valid.

The choice of different scales for different species is in no way illogical and is in keeping with the physical reasoning on which the deduction of the isotherm is based. Molecules of different species will in general differ in the space they take up in or on the sorbent, as reflected by their $b_i$ values. The different scales about to be introduced correspond to measuring the amount of each species in terms of occupied space (or space that would be occupied if the amount were sorbed).

Since the scale to be used for any given species must be the same in both phases, the distribution coefficient is invariant to the choice of the scale and must be the same in the equivalent stoichiometric as in the Langmuir system. In the latter, with Eq. (5.24),

$$\frac{q_j}{c_j} = \frac{a_j}{1+\sum_i b_i c_i} \qquad \text{for all } j \tag{5.28}$$

This equation is of the same form as that for the distribution ratio in a system with stoichiometric exchange:

$$\frac{q_j}{c_j} = \frac{\alpha_{jk} R}{1+\sum_{i \neq k} [(\alpha_{ik}-1)x_i]} \qquad \text{for all } j \tag{5.29}$$

where the constant $R$ is the ratio of the total concentrations:

$$R \equiv \sum_i q_i / \sum_i c_i = \text{const} \tag{5.30}$$

[Equation (5.29) is readily obtained from Eq. (3.7) after elimination of an arbitrary $x_k$ with Eq. (2.5) and conversion to $q_i$ and $c_i$ with Eqs. (2.4) and (2.1).]

A comparison of their equilibrium relations (5.28) and (5.29) shows that the $n$-component Langmuir system is formally equivalent to an $(n+1)$-component stoichiometric system (having species $k$ in addition to those corresponding to the Langmuir species) if one sets

$$\begin{aligned} a_i &= R\alpha_{ik}, & b_i c_i &= (\alpha_{ik}-1)x_i & i &= 1, \dots, k-1 \\ a_i &= R\alpha_{i+1,k}, & b_i c_i &= (\alpha_{i+1,k}-1)x_{i+1} & i &= k, \dots, n \end{aligned} \tag{5.31}$$

In choosing the equivalent stoichiometric system for a given Langmuir system, one must conform to these $2n$ conditions, but one has $2n+1$ degrees of freedom, namely, of choosing $R$, the $n$ constants $\alpha_{ik}$, and the $n$ conversion factors $x_i/c_i\,(i < k)$ and $x_{i+1}/c_i\,(i > k)$. Thus, there is an infinite multitude of equivalent stoichiometric systems for every given Langmuir system. In practice, the constant $R$ can be fixed arbitrarily.

Leaving the choice of a value of $R$ for later and solving Eqs. (5.31) for the $\alpha_{ik}$ and the conversion factors, one obtains

$$\left.\begin{aligned} \alpha_{ik} &= a_i/R \\ x_i/c_i &= b_i R/(a_i-R) \end{aligned}\right\} \quad i = 1, \dots, k-1$$
$$\left.\begin{aligned} \alpha_{i+1,k} &= a_i/R \\ x_{i+1}/c_i &= b_i R/(a_i-R) \end{aligned}\right\} \quad i = k, \dots, n \tag{5.32}$$

With the requirement that the same scale for measuring amounts be used in both phases, the conversion factors for the stationary-phase concentrations then automatically become

$$\begin{aligned} y_i/q_i &= b_i/(a_i - R) \qquad i = 1, \ldots, k-1 \\ y_{i+1}/q_i &= b_i/(a_i - R) \qquad i = k, \ldots, n \end{aligned} \tag{5.33}$$

With the choice of a convenient value for $R$ and with Eqs. (5.32) and (5.33), any Langmuir system can be transformed into an equivalent stoichiometric system with constant separation factors and an additional, dummy species. In the transformation, the Langmuir species $1, \ldots, k-1$ retain their numbers whereas the species $k, \ldots, n$ become $k+1, \ldots, n+1$, species $k$ in the equivalent stoichiometric system being the dummy.

## 2. *Systems with Competitive Sorption* ($b_i > 0$ *for all species*)

Usually, if a Langmuir isotherm can at all represent the experimental sorption equilibrium, all its $b_i$ are positive, reflecting competitive sorption. For cases of this type, Eqs. (5.32) and (5.33) show that $R$ should be chosen in the interval

$$0 < R < a_n \tag{5.34}$$

to give, after transformation, positive concentrations $x_i$ and $y_i$ of all physically existing species. The exact choice of $R$ within the interval does not matter. For convenience, however, one should choose $R$ small enough that even the highest potentially occurring concentrations $c_i$, after transformation with Eqs. (5.32), still give $\Sigma_{i \neq k} x_i < 1$, so that the dummy, too, will have positive concentrations $x_k$ and $y_k$ throughout. This is not a prerequisite, however, since negative concentrations of the dummy pose no problem in the formal mathematical treatment. With $R$ chosen according to condition (5.34), the dummy species becomes that of lowest affinity, i.e., species $n+1$.

It may appear surprising that different choices of $R$ give entirely equivalent transformations. To illustrate this point, Fig. 5.3 shows two transforms of the same Langmuir system but with different values of $R$. The shaded areas in both diagrams cover the same range of Langmuir compositions **c**. In the left diagram, for the smaller value of $R$, the area is smaller but the scales of the Langmuir concentrations and the composition-path grid are correspondingly compressed (note the distances of the watershed points from $P_3$). That the shaded area in the right diagram

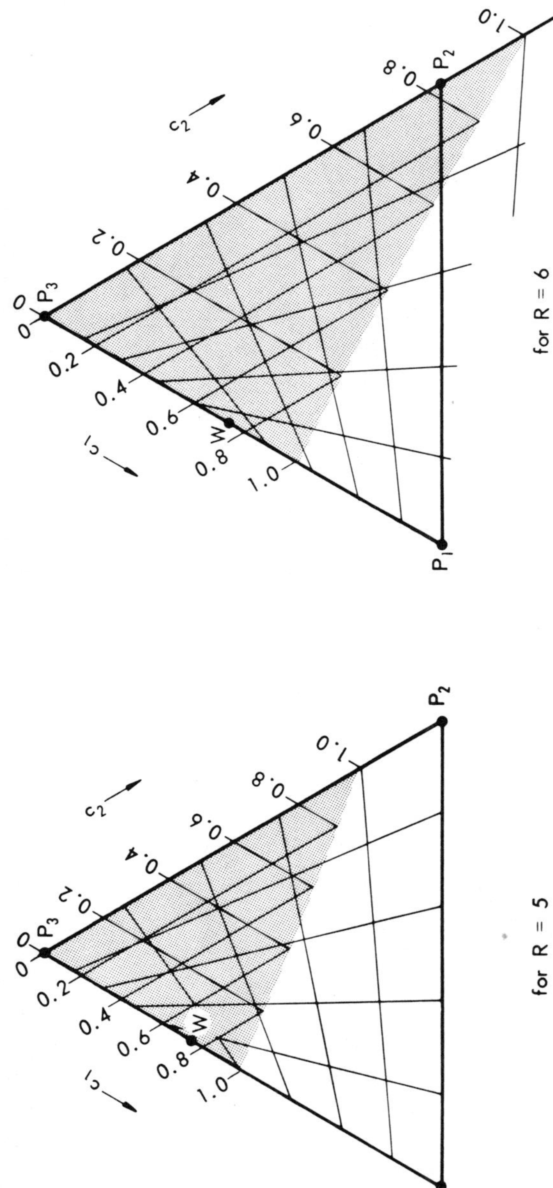


**Fig. 5.3.** Transformation of typical two-component Langmuir system with $b_1, b_2 > 0$. Mobile-phase simplex with path grid of equivalent stoichiometric systems with two different values of $R$. (For $a_1 = 25$, $a_2 = 10$, $b_1 = 2$, $b_2 = 0.8$; range $0 \leqq c_1 + c_2 \leqq 1$ is shaded.)

spills over the $P_1P_2$ border of the familiar composition triangle into the region where the dummy (species 3) has a negative concentration is of no consequence because, in this region, the path grid has no anomalies and all Langmuir concentrations remain positive.

Since a transformation into an equivalent stoichiometric system with constant separation factors and total concentrations and with positive concentrations of all species can be performed, the chromatographic response of Langmuir systems with positive $b_i$ obeys all the rules derived in the previous sections; these rules must merely be rephrased to account for the fact that the species of lowest affinity in the stoichiometric system now is a dummy without physical reality. Thus, in an $n$-component Langmuir system, coherence allows the concentrations of *all* Langmuir species to increase or decrease jointly in a coherent boundary (corresponding to an $n \mid n+1$ affinity cut in the equivalent stoichiometric system), any composition has $n$ composition-velocity eigenvalues, an abrupt influent composition change in general produces $n$ coherent boundaries, etc.

For quantitative calculations, the transformation into $x_i$ and $y_i$ of the equivalent stoichiometric system can be obviated. The Langmuir compositions **c** or **q** of the initial and operating conditions can be transformed directly into the $H$-function roots of the stoichiometric system, with which the response $\mathbf{h}(z, t)$ can be calculated as in previous sections, and which can then be directly reconverted to Langmuir compositions. The nontrivial roots are obtained as the roots in $h$, or the reciprocals of the roots in $1/h$, of

$$\begin{aligned} \sum_{i=1}^{n}\left(\frac{b_i c_i}{h a_i/a_1 - 1}\right) - 1 &= 0 \\ \sum_{i=1}^{n}\left(\frac{b_i q_i}{a_i - a_1/h}\right) - 1 &= 0 \end{aligned} \tag{5.35}$$

and the trivial roots, if any, are

$$\begin{aligned} &h_{j-1} = a_1/a_j \quad \text{or} \quad h_j = a_1/a_j \quad &&\text{if} \quad c_j = 0 \quad (1 < j \leqq n) \\ &h_1 = 1 &&\text{if} \quad c_1 = 0 \end{aligned} \tag{5.36}$$

(Note that an $n$-component Langmuir system has $n$ rather than $n-1$ $H$-function roots.) The concentrations can be calculated from the roots with

$$c_j = \prod_{i=1}^{n}\left(\frac{h_i a_j}{a_1}-1\right)\Big/ b_j \prod_{\substack{i=1\\ i\neq j}}^{n}\left(\frac{a_j}{a_i}-1\right) \qquad j = 1, \ldots, n \qquad (5.37)$$

$$q_j = \prod_{i=1}^{n}\left(\frac{a_1}{h_i}-a_j\right)\Big/ b_j \prod_{\substack{i=1\\ i\neq j}}^{n}(a_j - a_i)$$

Equations (5.35) to (5.37) can be derived from their counterparts (3.54), (3.55), (3.57), and (3.58) of the stoichiometric system, written for $n+1$ components, with elimination of the dummy concentrations $x_{n+1}$ and $y_{n+1}$ through $\Sigma_i x_i = 1$ and $\Sigma_i y_i = 1$ $(i=1,\ldots,n+1)$, and with the conversions given in Eqs. (5.32) and (5.33).

The arbitrary constant $R$ does not appear in Eqs. (5.35) to (5.37), and any Langmuir composition thus has a unique set $\mathbf{h}$, independent of $R$. Also, a condition for confining the equivalent stoichiometric system to compositions with positive dummy concentrations can now be stated. The dummy concentration $x_{n+1}$ becomes negative if $h_n$ exceeds $\alpha_{1,n+1}$. Accordingly, with Eqs. (5.32) for $\alpha_{1,n+1}$,

$$x_{n+1}(z, t) \geqq 0 \qquad \text{if} \quad 0 < R \leqq a_n/(h_n)_{\max} \qquad (5.38)$$

where $(h_n)_{\max}$ is the highest value of $h_n$ in the initial and influent conditions, and, therefore, in the entire response (see conservation properties of $H$-function roots, Chapter 3, Section V.A).

### 3. *Systems with Synergistic Sorption* ($b_i < 0$ *for one or more species*)

No systematic treatment nor even a comprehensive classification of the response behavior of Langmuir systems involving negative $b_i$ values have so far been worked out. The discussion below will therefore be confined to pointing out effects and illustrating them with simple examples of two-component systems.

Although the equivalent stoichiometric system is a crutch that can eventually be dispensed with, it will be retained initially because it allows us to draw on our familiarity with earlier parts of this treatment. For each Langmuir case, however, we need not examine more than one of the different stoichiometric systems resulting from different choices of $R$, since Eqs. (5.35) to (5.37) have established the identity of all such systems as far as $H$-function roots are concerned.

The first consequence of allowing $b_i$ values to be negative is that a transformation into a stoichiometric system with positive concentrations of all physically existing species may not be possible. Prerequisite for such a transformation is that

$$\begin{aligned} b_i &> 0 \qquad \text{for all} \quad i \leqq j \\ b_i &< 0 \qquad \text{for all} \quad i > j \end{aligned} \qquad (0 \leqq j \leqq n) \qquad (5.39)$$

and all $x_i$ and $y_i$ then are positive if $R$ is chosen in the interval $a_{j+1} < R < a_j$ ($0 < R < a_n$ if $j = n$, and $R > a_1$ if $j = 0$), as one finds with Eqs. (5.32) and (5.33).

Condition (5.39) comprises the case of positive $b_i$ of all species, covered earlier, and that of negative $b_i$ for all species. In the latter case, illustrated by the transform in Fig. 5.4, the dummy is species 1 of the stoichiometric system, and no complications arise. The constant $R$ can be chosen so

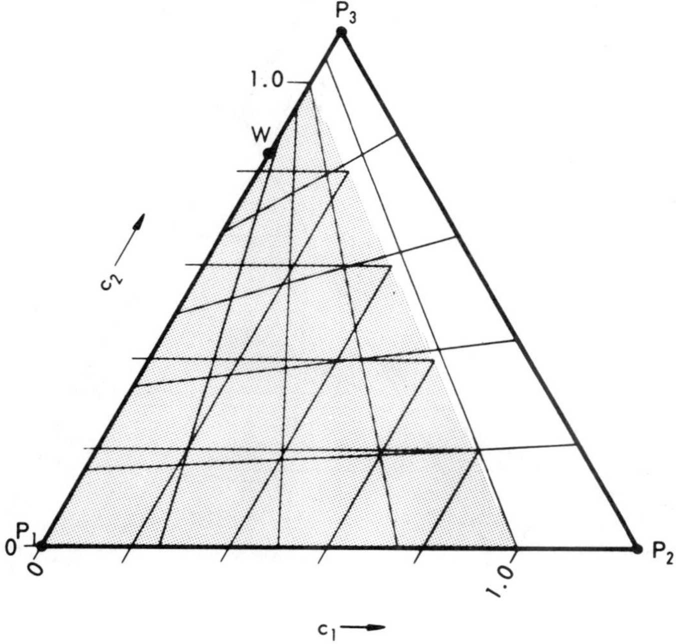


**Fig. 5.4.** Transformation of typical two-component Langmuir system with $b_1, b_2 < 0$. Mobile-phase simplex with path grid of equivalent stoichiometric system. (For $a_1 = 15$, $a_2 = 10$, $b_1 = -0.4$, $b_2 = -0.6$, $R = 30$; range $0 \leqq c_1 + c_2 \leqq 1$ is shaded.)

large that the dummy concentration is always positive. To have the stoichiometric species 1 instead of $n+1$ as the dummy, however, does affect some of the general rules. For example, the sharpening characteristics of coherent boundaries across which all Langmuir concentrations vary in the same direction are the opposite of those in cases with positive $b_i$ ($1 \mid 2$ cut instead of $n \mid n+1$ cut in the equivalent stoichiometric system).

Despite the possible transformation into a stoichiometric system with positive concentrations of all physically existing species, serious complications may arise under condition (5.39) if some but not all $b_i$ are negative. The dummy then is a species of intermediate affinity, and the leeway in choosing $R$, which is confined between two $a_i$ values, may not be sufficient to prevent the dummy concentration from becoming negative at high Langmuir concentrations. Such a case is illustrated in Fig. 5.5. Here, the

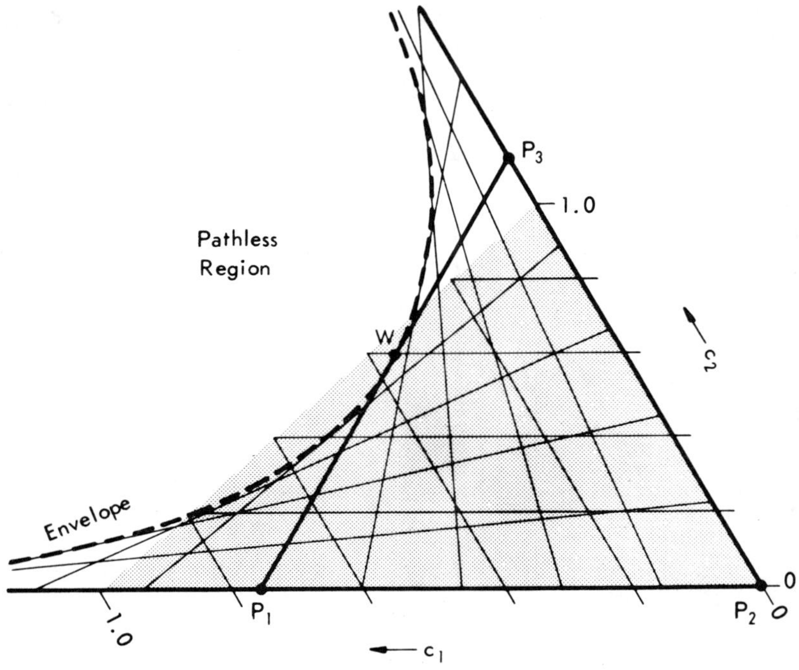


**Fig. 5.5.** Transformation of typical two-component Langmuir system with $b_1 > 0$, $b_2 < 0$, involving compositions in pathless region. Mobile-phase simplex with path grid of equivalent stoichiometric system. (For $a_1 = 24$, $a_2 = 10$, $b_1 = 0.8$, $b_2 = -0.3$, $R = 15$; range $0 \leqq c_1 + c_2 \leqq 1$ is shaded.)

dummy is species 2. In the region of negative $x_2$, the composition-path grid does not have the familar properties: In the areas near $P_1$ and near $P_3$, paths having the same variable root cross one another, and beyond the parabolic envelope of the path grid lies a pathless region in which the $H$ function has no real roots.

These properties of the path topology in the region of negative dummy concentration give rise to at least three new phenomena. First, for any composition point in the pathless region there is no route direction for which all species have the same concentration velocity (even if the dummy is not counted). Such compositions can therefore not occur in coherent boundaries nor can boundaries adjacent to plateau zones with such compositions be coherent. The physical implications of this effect will be examined in Section IV.B.1.

The second phenomenon is illustrated in Fig. 5.6. Two composition points P′ and P″ can be chosen so that there is no profile route from P′ to P″

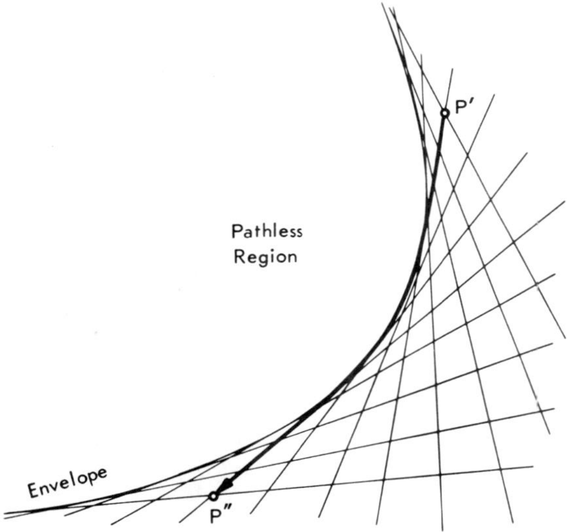


**Fig. 5.6.** Coherent composition profile route P′P″ running in part along envelope. Locations of P′ and P″ are such that $h_2' < h_1''$.

along paths in the sequence of increasing index numbers of their variable roots. The coherent route approached upon development of an arbitrary boundary between plateau zones P′ and P″ runs in part along the envelope of the path grid and involves no abrupt changes of direction and no intermediate plateau zones. This route is never completely attained, however, because the rate of development in the immediate vicinity of the envelope is nil, for the same reasons as near the watershed (see Chapter 3, Sections V.E and F.2). The example in Fig. 5.6 also reassures us that development will not shift a route across the envelope into the pathless region.

The third phenomenon may be called *metastable coherence*. As illustrated in Fig. 5.7, two composition points may be situated in such a manner that one can be reached from the other on *two* coherent profile routes which both meet the usual requirements for final, coherent routes. However, while the longer route is "stable," the more direct route tangential to the envelope is not: Constructions as in Fig. 3.32 show

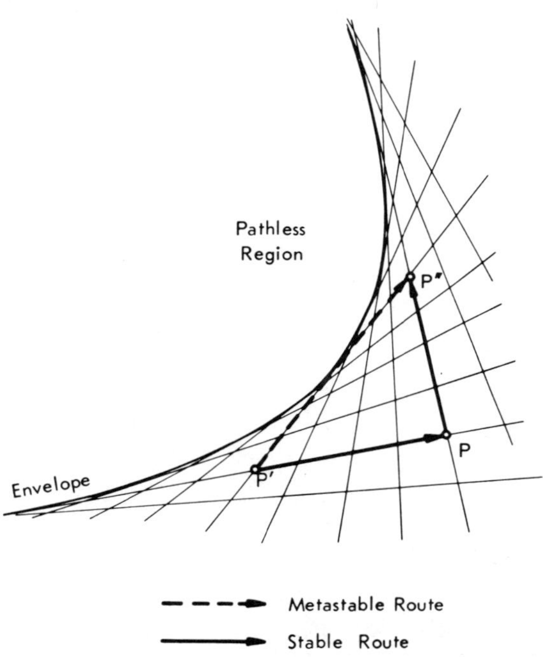


**Fig. 5.7.** Metastable coherence. Direct coherent profile route P′P″ is metastable, develops into stable route P′P P″ when disturbed.

that any disturbance in the direction away from the envelope will trigger a development changing the route to the other, stable one.

These three anomalies are not restricted to two-component systems, nor is condition (5.39) a prerequisite. They may occur whenever a species of higher affinity has a positive $b_i$ value while a species of lower affinity has a negative $b_i$ value.

A sufficient (but not necessary) condition precluding the occurrence of all three anomalies in such systems is readily stated. They can arise only if the region of negative dummy concentration is entered. As is readily shown with Eqs. (3.57), a negative $x_k$ requires either $h_{k-1} > \alpha_{1k}$ or $h_k < \alpha_{1k}$ if all other $x_i$ are positive. Therefore, negative values of $x_k$ can be precluded by choosing $R$, which equals $a_1/\alpha_{1k}$, in such a way that

$$h_{k-1} \quad < \quad a_1/R \quad < \quad h_k \tag{5.40}$$

Such a choice of $R$ is always possible if, in the initial and influent conditions,

$$(h_{k-1})_{\max} \quad < \quad (h_k)_{\min} \tag{5.41}$$

It is sufficient to state this condition for the input, because the response cannot involve root values other than those of the input conditions (see Chapter 3, Section V.A).

Fortunately, the relaxation of condition (5.39) has only minor consequences. It is true that a transformation into a stoichiometric system with positive $x_i$ and $y_i$ of all species can no longer be achieved, but this does not impair the formal mathematical treatment. A simple example is shown in Fig. 5.8, where $R$ has been arbitrarily selected in such a way that species 1 is the dummy. The area of physically meaningful compositions—with positive concentrations in the Langmuir system—then is that of positive $x_2$ and negative $x_3$. The path grid in this region has no singular features, not even a watershed. Thus, although a concentration in the equivalent stoichiometric system is negative, the response behavior is entirely normal, in some respects even more "normal" than that of a real stoichiometric system because the various exceptions related to the existence of a watershed are precluded. For example, permanent noncoherence (Chapter 3, Section V.F.2) cannot occur. Again, however, some of the general rules must be modified, particularly those related to the affinity cut. As the path grid in Fig. 5.8 shows, the concentrations of both Langmuir species increase or decrease together along the paths of both sets, and a boundary in which the two concentrations vary in opposite directions thus cannot be coherent. (Decrease of a negative concentration in the equivalent stoichiometric system amounts to increase of the respective positive concentration in the Langmuir system.)

### 4. *Langmuir Systems in the h Space*

For orientation, a general $h$ transform of composition paths for two-component Langmuir systems is shown in Fig. 5.9. The $h$ transform of the envelope is a straight line given by $h_1 = h_2$. The composition paths are "reflected" at the envelope, i.e., appear as two segments at right angles. This is a consequence of the fact that the variable root of a path changes

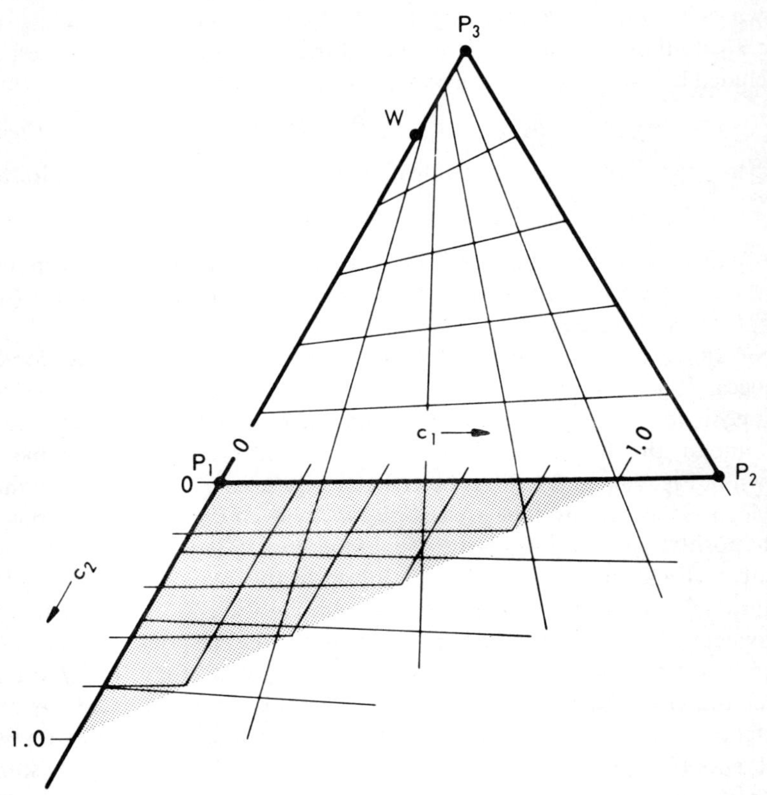


**Fig. 5.8.** Transformation of typical two-component Langmuir system with $b_1 < 0$, $b_2 > 0$. Mobile-phase simplex with path grid of equivalent stoichiometric system. (For $a_1 = 15$, $a_2 = 5$, $b_1 = -0.4$, $b_2 = 0.5$, $R = 30$; range $0 \leqq c_1 + c_2 \leqq 1$ is shaded.)

its index number where the path touches the envelope. The areas covered by the systems for which $x$-space path grids of the equivalent stoichiometric systems were shown in the earlier figures are shaded. As calculated with Eqs. (5.35) and (5.36), the location in the $(h_1, h_2)$ plane of the origin $(c_1 = 0, c_2 = 0)$ of the concentration coordinates of a given system is solely determined by the $a_i$ values, while the directions of the coordinates and the extent of the area covered depend on the signs and magnitudes of the $b_i$ values and the composition range. The systems I and II ($b_1, b_2 > 0$, and $b_1, b_2 < 0$) each have a watershed point W where a concentration co-

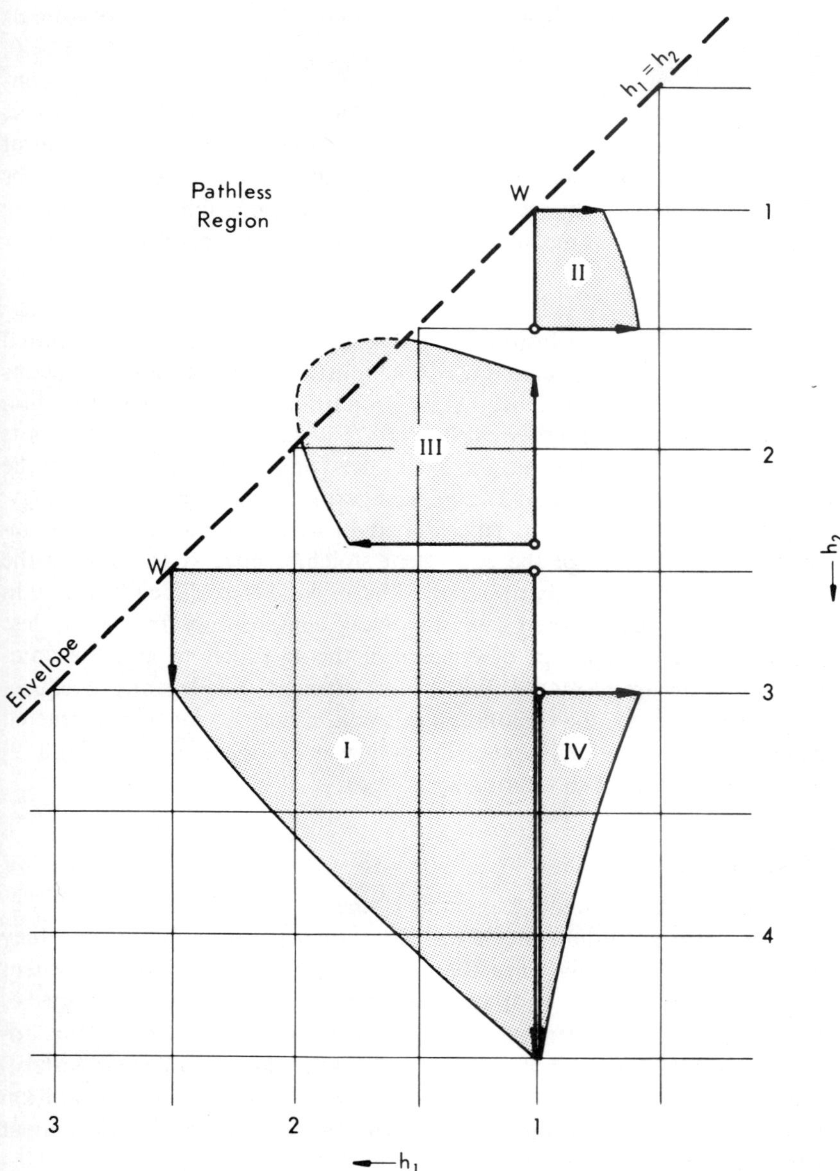


**Fig. 5.9.** *h* Space with transforms of two-component Langmuir systems. Shaded areas I, II, III, and IV correspond to composition ranges of systems shown in Figs. 5.3, 5.4, 5.5, and 5.8, respectively. Concentration coordinates are shown as heavy arrows.

ordinate is reflected at the envelope, the system III ($b_1 > 0$, $b_2 < 0$) extends across the envelope into the pathless region, and the system IV ($b_1 < 0$, $b_2 > 0$) has a path grid without watershed. The point of origin of the concentration coordinates and the area covered are invariant to the choice of $R$: When choosing different values of $R$, one merely moves the transform of the $P_1P_2P_3$ triangle of the equivalent stoichiometric system but leaves all other features unchanged. This illustrates the possibility of operating in the $h$ space without recourse to the equivalent stoichiometric system.

### 5. *Conclusions*

It can be said in conclusion that only minor modifications are required to extend the methods derived for stoichiometric systems to most systems with compound Langmuir isotherms. In particular, the $h$ transformation and the basic differential equations (3.87) in terms of $H$-function roots remain applicable, and so do the techniques for constructing routes in the composition-path grid and for deducing rules from the grid topology. While most of the general rules stated in earlier sections must be rephrased, the concept of the equivalent stoichiometric system makes the necessary changes readily apparent. The only serious difficulty arises in the fortunately exceptional cases involving compositions in the pathless region. Here, the concept of coherence, the $H$ function, and the composition paths are of no use. While more efficient and elegant procedures may yet be devised, the response in this region must so far be constructed by lengthy computations in much the same way as suggested by DeVault [9] in the earliest days of chromatographic theory.

## C. Other Systems

Systems with compound Langmuir isotherms are only one class of nonstoichiometric sorption systems with constant separation factors, but they appear to be the only ones that can be transformed into equivalent stoichiometric systems with constant separation factors of all species including the dummy. Unfortunately, if one separation factor in a stoichiometric system varies, it is of little help that the others are constant. Therefore, to discuss non-Langmuir systems with constant separation factors separately from the more general case of composition-dependent separation factors appears pointless.

The effects produced by composition-dependent separation factors will be discussed in the next section. We shall not specifically return to nonstoichiometric systems in this context. From the limited insight gained to date, one should except such systems to combine the features of the Langmuir systems in this section with those of the stoichiometric systems in the next, but not to show additional effects, because a transformation into an equivalent stoichiometric system with positive concentrations of all species (including the dummy) is always possible if constancy of the separation factors is not demanded.

### D. Comparison with Other Theories

The notion of a formal analogy between Langmuir sorption isotherms and isotherms of stoichiometric exchange or, more generally, behavior according to the mass-action law is an old one (see, for example [6, 8]) and was certainly familiar to Glueckauf, but it seems that the transformation for $n$-component systems has not previously been worked out in a quantitative manner.

Most of the chromatographic theories for systems with general equilibrium properties are applicable to stoichiometric exchange as well as to nonstoichiometric sorption and will be reviewed collectively in Section IV.C. Directly related to the material presented here, however, is Glueckauf's early work on Langmuir systems [10–13], which stands out as the most general and imaginative of all previous treatments of systems with interference. Like virtually all other authors, Glueckauf tacitly assumed entirely coherent behavior and confined himself to two- and three-component cases with relatively simple operating conditions; he pointed out specifically that his treatment covers systems with negative $b_i$ values, but was apparently not aware of the complications arising if positive and negative $b_i$ values occur side by side. With these reservations and restrictions, the present treatment of Langmuir systems is equivalent to Glueckauf's. The equivalence is not easy to trace, however, because Glueckauf neither adjusted his time variable nor used the $h$ transformation, and his equations are accordingly complex.

## IV. Composition-Dependent Separation Factors

As far as equilibrium properties are concerned, the remaining premise to be relaxed is that of composition-independent separation factors. Here, we lose what to this point has been the mainstay of the treatment: The basic differential equations (3.87) in terms of $H$-function roots are no longer valid, and the $h$ transformation therefore becomes all but useless. Accordingly, we enter a realm where most rules are deduced with conceptual arguments or from experience with numerical calculations rather than from a mathematical solution for the general case. These procedures cannot fully guarantee that all relevant effects and all possible exceptions from the normal behavior have been uncovered. Also, various questions are as yet unanswered, and there is little doubt that the efficiency and elegance of the conceptual and mathematical methods can still be improved.

In this section, systems with so-called mass-action equilibria will be singled out for a more detailed discussion, if only because they alone have seen extensive calculations [14–18] and therefore provide a logical starting point. The results will then be generalized for arbitrary equilibrium properties, within the reasonable assumptions of continuous and differentiable isotherms. The premise of stoichiometric exchange between the mobile and stationary phases will be retained in this section, since the principle of the extension to nonstoichiometric sorption has already been covered (see Section III.A).

## A. Systems with Mass-Action Equilibria

Application of the ideal mass-action law to stoichiometric exchange of arbitrary pairs of species $i$ and $j$, with stoichiometric coefficients $\nu_i$ and $\nu_j$,

$$\nu_i(i)_{\text{fluid}} + \nu_j(j)_{\text{sorbent}} \rightleftharpoons \nu_j(j)_{\text{fluid}} + \nu_i(i)_{\text{sorbent}} \tag{5.42}$$

gives

$$(q_i/c_i)^{\nu_i}(c_j/q_j)^{\nu_j} = K_{ij}^0 = \text{const} \qquad \text{for all } i \text{ and } j \tag{5.43}$$

(In ion exchange, the $\nu_i$ may be taken as the reciprocals of the valences.) In terms of equivalent fractions, here defined as

$$x_j \equiv \frac{c_j}{\nu_j}\bigg/\sum_i \frac{c_i}{\nu_i} \qquad \text{and} \qquad y_j \equiv \frac{q_j}{\nu_j}\bigg/\sum_i \frac{q_i}{\nu_i} \tag{5.44}$$

Eqs. (5.43) become

$$(y_i/x_i)^{\nu_i}(x_j/y_j)^{\nu_j} = K_{ij} \qquad \text{for all } i \text{ and } j \tag{5.45}$$

where

$$K_{ij} \equiv K_{ij}^0 \left(\frac{\sum_i (q_i/\nu_i)}{\sum_i (c_i/\nu_i)}\right)^{\nu_j - \nu_i} \qquad \text{for all } i \text{ and } j \tag{5.46}$$

While the $K_{ij}^0$ are constants within the range of validity of the ideal mass-action law, the $K_{ij}$ vary with the total equivalent concentrations $\Sigma_i\,(c_i/\nu_i)$ and $\Sigma_i\,(q_i/\nu_i)$ but are independent of the $x_i$ and $y_i$. In this section, a uniform sorbent and constant total concentration (in equivalents) of the influent will be assumed; since the exchange (5.42) does not alter the number of equivalents in either phase, the total concentrations in equivalents then are uniform and constant, and the $K_{ij}$ may be regarded as constants. Variations in the total influent concentration will add the effects discussed in Section II to those to be studied here.

The $v_i$ need not be identified as stoichiometric coefficients, but may be taken as empirical coefficients in equilibrium relations of the form of Eqs. (5.45), which alone will be referred to in the later derivations. Stoichiometric exchange equilibria with constant separation factors, with which the bulk of the present approach has dealt, can be considered as a special case of mass-action equilibria, with all $v_i$ having the same value.

### 1. *Competitive Nature of Mass-Action Equilibria*

Since we are no longer able to rely on the properties of the $H$-function roots for compact derivations, we first seek to establish that sorption under the mass-action relations (5.45) is entirely competitive, because this will facilitate the deduction of general rules (for definition of competitive sorption, see Chapter 2, Section III). As is derived below, Eqs. (5.45) give

$$\left(\frac{\partial y_j}{\partial x_k}\right)_{x_i(i \neq k,l)} < 0 \qquad \text{for all} \quad j \neq k \qquad \text{if} \quad \frac{y_k}{x_h} > \frac{y_l}{x_l} \tag{5.47}$$

that is, sorption of all other species is reduced if the concentration of a species of higher affinity, $k$, is increased at the expense of one of lower affinity, $l$. Sorption thus is competitive. It must be noted, however, that condition (5.47) is a statement about differential composition changes at an arbitrary given composition and does not preclude selectivity reversals, which will be discussed later in this section.

Condition (5.47) can be derived as follows. Equations (5.45) solved for $y_i$ and summed over all species $i$, give with $\sum_i y_i = 1$:

$$\sum_i \left[K_{ij}^{1/v_i} x_i (y_j/x_j)^{v_j/v_i}\right] - 1 = 0 \tag{5.48}$$

where $j$ is an arbitrarily chosen reference species. Implicit partial differentiation with respect to $x_k$ at constant $x_i$ $(i \neq k, l)$ yields, after rearrangement,

$$\left(\frac{\partial y_j}{\partial x_k}\right)_{x_i(i \neq k,l)} \sum_i \left[ K_{ij}^{1/v_i} \frac{v_j x_i}{v_i x_j} \left(\frac{y_j}{x_j}\right)^{(v_j/v_i)-1} \right] + \frac{y_k}{x_k} - \frac{y_l}{x_l} = 0 \tag{5.49}$$

where use has been made of the fact that $\sum_i x_i = 1$, entails

$$(\partial x_l / \partial x_k)_{x_i(i \neq k,l)} = -1 \tag{5.50}$$

Condition (5.47) follows from Eq. (5.49), since all quantities under the sum are positive.

2. *Coherence Properties*

The next step is to find the composition-velocity eigenvalues. As derived below, the eigenvalues of the adjusted velocity of an arbitrary composition with all $n$ species present can be obtained as the roots in $u$ of

$$\sum_i \frac{x_i/\nu_i}{u-u_i} = 0 \tag{5.51}$$

where $u_i$ is the adjusted species velocity [see Eq. (3.35)]. This equation has $n-1$ real roots $(u)_1, \ldots, (u)_{n-1}$, which fall into the intervals

$$u_i < (u)_i < u_{i+1} \qquad i = 1, \ldots, n-1 \tag{5.52}$$

[see the behavior of the analogous equation (A.2), discussed in Appendix I]. Accordingly, as in systems with constant separation factors, the composition has $n-1$ velocity eigenvalues, each for a different composition path.

If one or several species $k$ are absent, Eq. (5.51) has less than $n-1$ roots. The missing velocity eigenvalues are those for the paths along which the respective species $k$ appear, and are provided by

$$\lim_{x_k \to 0} u = \lim_{x_k \to 0} (x_k/y_k) = K_{jk}^{1/\nu_k}(x_j/y_j)^{\nu_j/\nu_k} \tag{5.53}$$

where $j$ can be arbitrarily chosen from the species present.

Equation (5.51) can be derived as follows. Written in logarithmic form, Eqs. (5.45) become

$$\nu_i(\ln y_i - \ln x_i) = \ln K_{ij} + \nu_j(\ln y_j - \ln x_j) \qquad \text{for all } i \text{ and } j \tag{5.54}$$

Implicit partial differentiation with respect to adjusted time at fixed location gives, after rearrangement and with Eqs. (3.33) and (3.35) for the velocities,

$$\frac{\nu_i}{x_i}(u_{x_i} - u_i)\left(\frac{\partial y_i}{\partial \tau}\right)_z = \frac{\nu_j}{x_j}(u_{x_j} - u_j)\left(\frac{\partial y_j}{\partial \tau}\right)_z \qquad \text{for all } i \text{ and } j \tag{5.55}$$

For coherence, with $u_{x_i} = u_{x_j} = u$,

$$\frac{\nu_i}{x_i}(u - u_i)\left(\frac{\partial y_i}{\partial \tau}\right)_z = \frac{\nu_j}{x_j}(u - u_j)\left(\frac{\partial y_j}{\partial \tau}\right)_z \qquad \text{for all } i \text{ and } j \tag{5.56}$$

Solved for $(\partial y_i/\partial \tau)_z$ and summed over all $i$, Eqs. (5.56) with $\Sigma_i\,(\partial y_i/\partial \tau)_z = 0$ yield Eq. (5.51). Equation (5.53) is readily obtained from Eqs. (5.56), (3.18), (3.35), and (5.45).

For systems with constant separation factors (all $\nu_i$ equal), Eq. (5.51) reduces to Eq. (3.54), as is readily shown with Eqs. (3.67) and (3.68).

A relation equivalent to Eq. (5.51) was first given by Klein *et al.* [15, 16].

A further familiar coherence property is the existence of affinity cuts: Eqs. (5.56) show immediately that the concentrations of species having velocities higher and lower than the composition velocity vary in opposite directions, while condition (5.52) guarantees the existence of a composition-velocity eigenvalue for each affinity cut.

The conditions of competitive sorption and existence of $n-1$ affinity cuts with their composition-velocity eigenvalues at any given composition do not preclude selectivity reversals, which will be discussed in Section IV.A.3, but they do ensure that, within each domain of unchanged affinity sequence in the composition space, the spatial orientation of the loci (hyperplanes and hypersurfaces) of constant $x_i$ and $y_i$ relative to one another as well as the orientation of the composition paths relative to these loci are qualitatively the same as in systems with constant separation factors. As a simple example, Fig. 5.10 shows the path grid of a representative three-component system without selectivity reversal. The topology and the qualitative features are seen to be the same as in Fig. 3.9, for a system with constant separation factors.

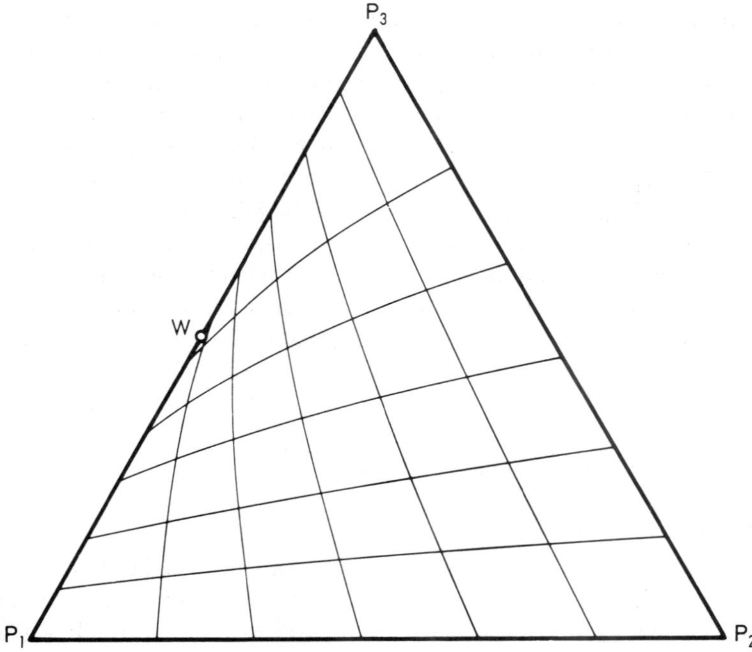


**Fig. 5.10.** Mobile-phase composition simplex with path grid of typical three-component system with mass-action equilibrium and no selectivity reversal. (For $K_{12} = 3.87$, $K_{13} = 8.06$, $\nu_1 = 2$, $\nu_2 = 1$, $\nu_3 = 1$.)

Aside from possible selectivity reversals, the properties examined so far are essentially the same as in systems with constant separation factors. There are, however, the following complications:

First, the composition paths are no longer linear, nor is their grid orthogonalized by the $h$ transformation. An algebraic expression for the paths has so far not been found, and grids as in Fig. 5.10 have been constructed by numerical integration of the differential equations

$$\left.\frac{\partial x_i}{\partial x_j}\right|_{\text{path}} = \frac{v_j x_i(u - x_j/y_j)}{v_i x_j(u - x_i/y_i)} \qquad \text{for all } i \tag{5.57}$$

with $u$ and the $y_i$ calculated at every composition with Eqs. (5.51), (5.45), and (2.5). Equation (5.57) is readily derived from Eqs. (5.56), (3.33), and (3.35).

The second complication is that pairs of points on a composition path calculated with the differential coherence condition (3.38) no longer obey the integral coherence condition (3.40). To give an unambiguous definition of the paths, we must specify that they are based on the differential, not on the integral, coherence condition. The compositions on the two sides of a sharp and self-sharpening coherent boundary then are not on the same path. Fortunately, the deviations are usually minor, so that the "differential" paths can be used in good approximation for sharp and self-sharpening boundaries, too. This complication, first noted by Glueckauf [13], has been discussed at length by Klein *et al.* [15, 16].

The third and somewhat less important complication is that so far no proof has been found for the existence, in systems with more than three components, of common surfaces and hypersurfaces of mutually intersecting paths of two or more affinity cuts (analogous to the common planes and hyperplanes in systems with constant separation factors; see Chapter 3, Section IV.C.3). The possible absence of common surfaces complicates the treatment of the transient behavior before attainment of a final pattern, as will be seen later in this section.

### 3. *Selectivity Reversals*

Owing to the fact that the species have different coefficients $v_i$, composition variations may produce reversals in the affinity sequence. Consider the equilibrium relation between two species $i$ and $j$. From Eqs. (5.45) and the definition (2.8) of the separation factors, one obtains

$$\alpha_{ij} = \text{const}\left(\frac{y_j}{x_j}\right)^{v_j/v_i - 1} \tag{5.58}$$

Unless $\nu_i$ and $\nu_j$ are equal, the distribution ratio $y_j/x_j$ will affect the separation factor $\alpha_{ij}$. As the distribution ratio varies with composition, the separation factor may vary from $\alpha_{ij} < 1$ to $\alpha_{ij} > 1$, reversing the positions of $i$ and $j$ in the affinity sequence.

With $\nu_j > \nu_i$, the exponent in Eq. (5.58) is positive, and the separation factor $\alpha_{ij}$ thus is higher where the distribution ratio is high, i.e., where there is less competition from species of high affinity. Accordingly, species with lower coefficients $\nu_i$ (higher valences in ion exchange) gain in affinity relative to others when the composition shifts toward greater predominance of low-affinity species.

In view of potential selectivity reversals, the convention of numbering the species in the sequence of decreasing affinity must be modified. We shall number the species in such a way that

$$i < j \qquad \text{if} \quad K_{ij} > 1 \tag{5.59}$$

Of a pair of arbitrary species, that with the lower number then has the higher affinity at least where all other species are absent. In a system without reversals, the new convention is equivalent to the previous one.

A selectivity reversal of two arbitrary species $j$ and $k$ will be denoted $j{:}k$. If such a reversal occurs, the composition space will consist of domains with $\alpha_{jk} < 1$ and with $\alpha_{jk} > 1$. Since the isotherm, represented by Eqs. (5.45), is continuous and differentiable, the domains must be separated by a locus of $\alpha_{jk} = 1$. As derived in Appendix IV, the equation for this locus is*

$$\sum_{i \neq j,k} \left( \frac{A_{jk} - A_{jki}}{A_{jk} - 1} \, x_i \right) - 1 = 0 \tag{5.60}$$

where

$$A_{jk} \equiv K_{jk}^{1/(\nu_j - \nu_k)} \qquad \text{and} \qquad A_{jki} \equiv K_{ij}^{1/\nu_i} A_{jk}^{\nu_j/\nu_i} \tag{5.61}$$

This equation describes an $(n-2)$-dimensional hyperplane (plane in four-component systems, straight line in three-component systems) parallel to the $P_jP_k$ edge of the $x$-space simplex. For any pair of arbitrary species $j$ and $k$, such a hyerplane exists unless $\nu_j = \nu_k$. Of course, if $\nu_j = \nu_k$, then the separation factor $\alpha_{jk}$ has a constant value different from unity, and a reversal $j{:}k$ cannot occur.

There is only one hyperplane of $\alpha_{jk} = 1$, separating a domain of $\alpha_{jk} < 1$ from one of $\alpha_{jk} > 1$. Thus, multiple reversals $j{:}k$ cannot occur.

---

* An equivalent expression has been derived independently by Tondeur [18].

For the system to have a reversal $j{:}k$, the hyperplane of $a_{jk} = 1$ must not only exist, but must intersect the simplex of physically meaningful compositions ($x_i \geqq 0$ for all $i$, $\Sigma_i\, x_i = 1$). As shown in Appendix IV, this imposes the following restrictions:

1. A reversal $1{:}i$ can occur only if $\nu_1 > \nu_i$.
2. A reversal $i{:}n$ can occur only if $\nu_i < \nu_n$.
3. A reversal $1{:}n$ cannot occur.

There are no other restrictions in principle, and an $n$-component system may thus have as many as $n(n-1)/2-1$ different reversals (for all pairs of species, except $1{:}n$). A four-component system with five reversal planes will be shown later in Fig. 5.13.

Conditions sufficient for occurrence of reversals are also readily stated (for derivations, see Appendix IV). A reversal $j{:}k$ occurs if

$$K_{ij}^{\nu_k} > K_{ik}^{\nu_j} \qquad (j < k) \tag{5.62}$$

for at least one $i \neq j, k$. Furthermore, a system has at least one reversal, $j{:}k$ or $k{:}l$, if the coefficients of three species $j$, $k$, and $l$ obey the condition

$$\nu_j,\ \nu_l \geqq 2\nu_k \qquad (j < k < l) \tag{5.63}$$

(In the singular case $K_{jk} = K_{kl}$, $\nu_j$ or $\nu_l$ should be larger than $2\nu_k$ to ensure a reversal.) Condition (5.63) does not involve the coefficients $K_{ij}$, and the widely held belief that reversals do not occur if the inequalities $K_{1n} > K_{2n} > \dots > K_{n-1,n}$ are sufficiently great thus is mistaken. For example, a three-component system with $\nu_1 = 2$, $\nu_2 = 1$, $\nu_3 = 3$ has at least one reversal, regardless of the values of the $K_{ij}$.

In three-component systems, the only reversals compatible with the foregoing rules are 1:2 and 2:3. Composition diagrams of the three possible configurations are shown in Fig. 5.11. In four-component system, the variety is much greater. The possible positions of the five potential reversal planes—for 1:2, 2:3, 3:4, 1:3, and 2:4 reversals—are shown in Fig. 5.12, and the respective conditions, derived in Appendix IV, are listed in Table 5.1. All five reversals may be realized in the same system, as illustrated in Fig. 5.13.

Despite the wealth of possible selectivity reversals in $n$-component systems, the consequences for the response behavior are less serious than one might fear, because the following rules are obeyed. First, all reversals are such that two species *adjacent* in the local affinity sequence interchange their positions in the sequence. That is, across a $j{:}k$ reversal hyperplane, the sequence changes from $\dots, j, k, \dots$ to $\dots, k, j, \dots$, not from $\dots j, i, k, \dots$

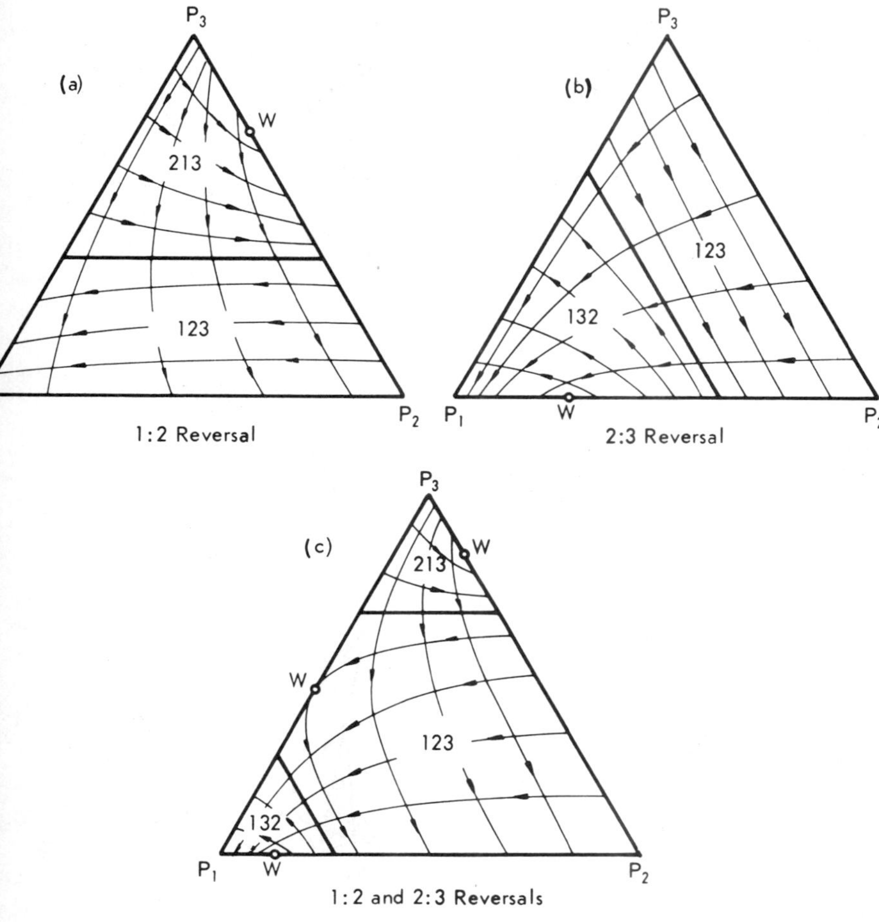


**Fig. 5.11.** Topology of mobile-phase path grids of three-component systems with mass-action equilibria and selectivity reversals (schematic). Arrows on paths point in direction of increasing composition velocity; reversal lines shown as heavy lines; numbers indicate affinity sequences in respective domains.

to ..., $k, i, j$, ..., even if $j$ and $k$ are not adjacent numbers (e.g., see Fig. 5.13). Second, the $j{:}k$ reversal hyperplane is not intersected by $j \mid k$ paths. Therefore, along any composition path even across reversal hyperplanes, the make-up of the groups of high-affinity and low-affinity species does not change and all concentrations vary monotonically. Third, sorption remains

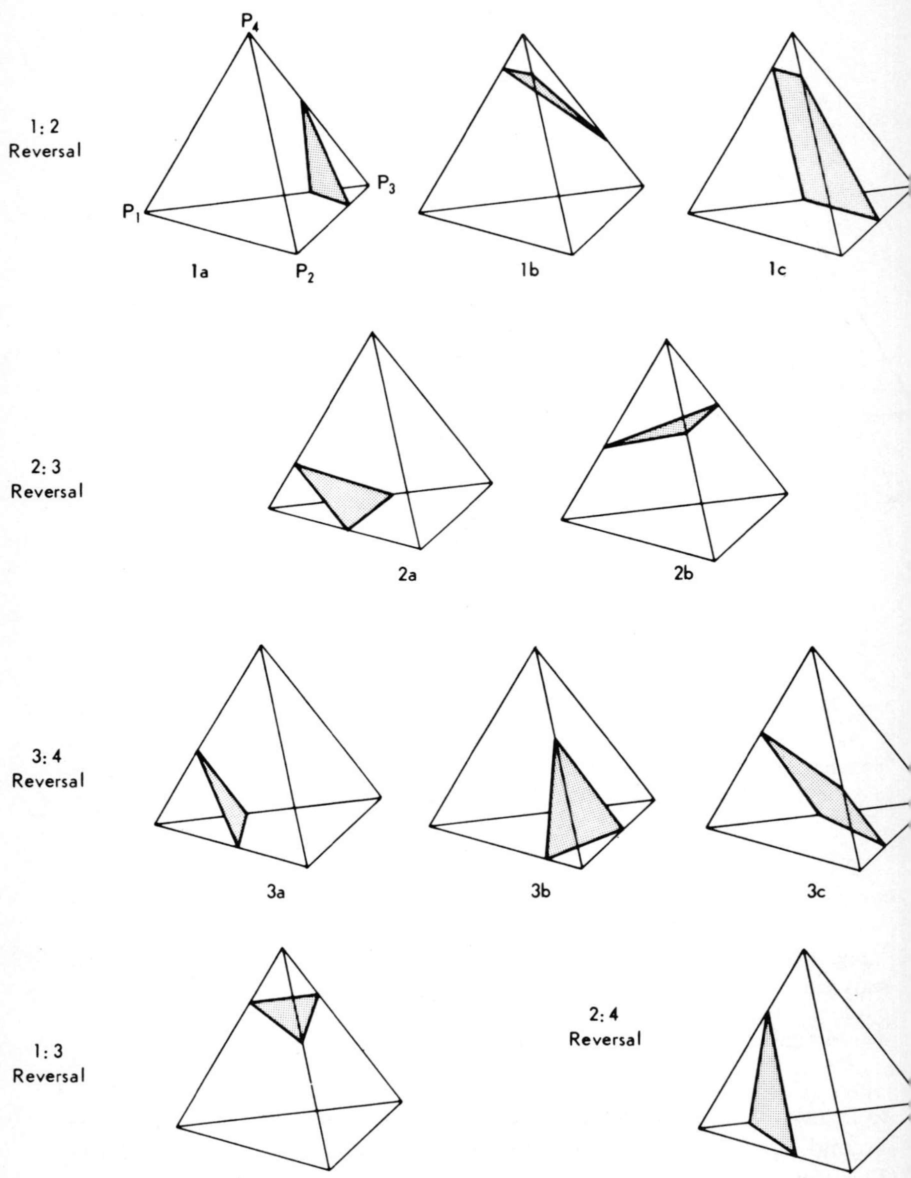


**Fig. 5.12.** Potential positions of reversal planes (shaded) in four-component systems with mass-action equilibria (schematic). Cases are numbered as in Table 5.1.

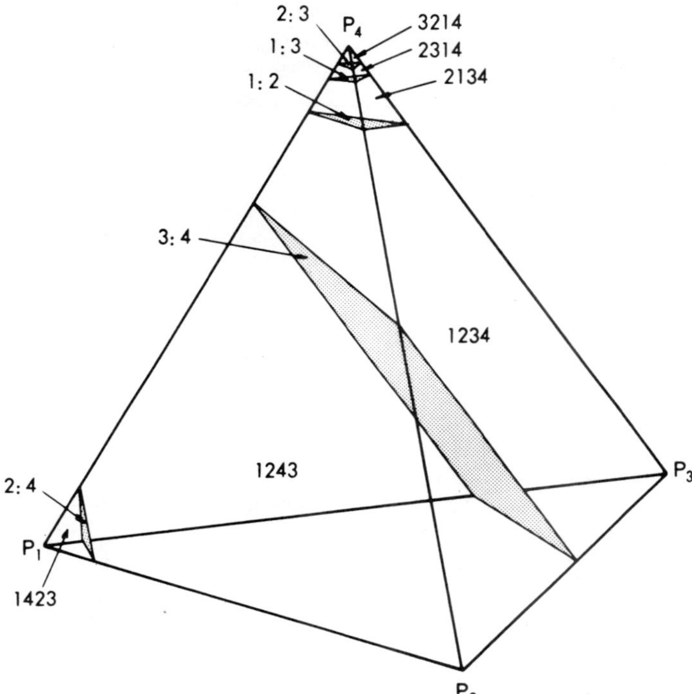


**Fig. 5.13.** Reversal planes (shaded) in mobile-phase composition simplex of four-component system with mass-action equilibrium and all five possible selectivity reversals. Four-number groups indicate affinity sequences in respective domains. (For $K_{12} = 2.35$, $K_{13} = 4.10$, $K_{14} = 8.20$, $\nu_1 = 4$, $\nu_2 = 2$, $\nu_3 = 1$, $\nu_4 = 5$.)

entirely competitive despite selectivity reversals [see condition (5.47)]. Thus, within each separate domain of invariant affinity sequence between reversal hyperplanes, the spatial orientation of the composition paths and the qualitative response behavior are the same as in systems with competitive sorption and without selectivity reversals.

For three-component systems, some of these rules are reflected in the path grids in Fig. 5.11. The 1:2 and 2:3 reversals lines are 1 | 2 and 2 | 3 path, respectively, and are not intersected by other paths of the same cut. Above the 1:2 reversal line, the path topology is typical of a system with affinity sequence 2, 1, 3 rather than 1, 2, 3 (compare with Figs. 3.9 and 5.10). Similarly, to the left of the 2:3 reversal line, the path topology is typical of a system with affinity sequence 1, 3, 2.

**Table 5.1**

CONDITIONS FOR OCCURRENCE AND POSITIONS OF REVERSAL PLANES IN FOUR-COMPONENT SYSTEMS WITH MASS-ACTION EQUILIBRIA.

| Case[a] | Reversal | Simplex edges intersected | Conditions | |
|---|---|---|---|---|
| 1a | 1:2 | $P_1P_3$, $P_2P_3$, $P_3P_4$ | $\nu_1 > \nu_2$ | $K_{13}^{\nu_2} < K_{23}^{\nu_1}, K_{14}^{\nu_2} > K_{24}^{\nu_1}$ |
| 1b | | $P_1P_4$, $P_2P_4$, $P_3P_4$ | | $K_{13}^{\nu_2} > K_{23}^{\nu_1}, K_{14}^{\nu_2} < K_{24}^{\nu_1}$ |
| 1c | | $P_1P_3$, $P_1P_4$, $P_2P_3$, $P_2P_4$ | | $K_{13}^{\nu_2} < K_{23}^{\nu_1}, K_{14}^{\nu_2} < K_{24}^{\nu_1}$ |
| 2a | 2:3 | $P_1P_2$, $P_1P_3$, $P_1P_4$ | $\nu_2 < \nu_3$ | $K_{12}^{\nu_3} > K_{13}^{\nu_2}$ |
| 2b | | $P_1P_4$, $P_2P_4$, $P_3P_4$ | $\nu_2 > \nu_3$ | $K_{34}^{\nu_2} > K_{24}^{\nu_3}$ |
| 3a | 3:4 | $P_1P_2$, $P_1P_3$, $P_1P_4$ | $\nu_3 < \nu_4$ | $K_{13}^{\nu_4} > K_{14}^{\nu_3}, K_{23}^{\nu_4} < K_{24}^{\nu_3}$ |
| 3b | | $P_1P_2$, $P_2P_3$, $P_2P_4$ | | $K_{13}^{\nu_4} < K_{14}^{\nu_3}, K_{23}^{\nu_4} > K_{24}^{\nu_3}$ |
| 3c | | $P_1P_3$, $P_1P_4$, $P_2P_3$, $P_2P_4$ | | $K_{13}^{\nu_4} > K_{14}^{\nu_3}, K_{23}^{\nu_4} > K_{24}^{\nu_3}$ |
| 4 | 1:3 | $P_1P_4$, $P_2P_4$, $P_3P_4$ | $\nu_1 > \nu_3$ | $K_{14}^{\nu_3} < K_{34}^{\nu_1}$ |
| 5 | 2:4 | $P_1P_2$, $P_1P_3$, $P_1P_4$ | $\nu_2 < \nu_4$ | $K_{12}^{\nu_4} > K_{14}^{\nu_2}$ |

[a] Denoted as in Fig. 5.12.

Two new features, however, appear in systems with more than one reversal, even if there are only three components. First, as Fig. 5.11c illustrates, three-component systems with two reversals have three watershed points instead of only one; the watershed point on the $P_1P_3$ border is unusual in that it involves composition paths that are tangential to the border, a property first described by Tondeur [18]. Second, as Fig. 5.11c also shows, it is no longer true that each path of one set intersects all paths of the other set. Analogous properties are found in $n$-component systems with more than one reversal involving the same species.

4. *Response Behavior*

The most important point is to establish that coherence is attained upon undisturbed development from arbitrary initial conditions, as in systems with constant separation factors. This can be done by an extension of the velocity and continuity considerations in Chapter 3, Section V.C, to more than three components. The inequalities on which these considerations are based are derived from the general material balance without assumptions as to equilibrium properties and thus are valid here, too. The considerations furthermore presume that all species velocities increase or decrease jointly when the composition shifts in favor of species of high and low affinity, respectively; this condition is guaranteed by competitive sorption. As stated earlier, competitive sorption also ensures that, within each domain of invariant affinity sequence, the spatial orientation of the composition paths and loci of constant $x_i$ and $y_i$ is the same as in systems with constant separation factors (and with the same affinity sequence). Under these circumstances, the considerations outlined in Chapter 3, Section V.C are applicable and suffice to establish the local trend toward coherence and the validity of qualitative constructions of transient routes as in Fig. 3.32. That selectivity reversals do not impair attainment of coherence can be shown with such constructions in path grids as in Fig. 5.11.

On most other counts, too, the response behavior turns out to be very similar to that in systems with constant separation factors. The distortion of the path grid and the nonlinearity of the paths have little effect, since the principle of constructing qualitative transient routes has remained unchanged. In particular, the rule that the final coherent pattern will consist of boundaries or pulses in the sequence of decreasing affinities of the species marking the cuts remains valid, since competitive sorption is a sufficient condition for the derivations in Chapter 3, Section IV.F, which showed plateau zones between boundaries in this sequence to lengthen. (Excepted are cases involving two boundaries with common high- or low-affinity species, in which the behavior may deviate from that shown in Fig. 3.21 and Table 3.2.)

The following is a list, possibly incomplete, of features not encountered in systems with constant separation factors:

1. While sharpening into a step, a diffuse but self-sharpening coherent boundary generates small composition variations of other affinity cuts. This is a consequence of the fact that pairs of points on a (differential) composition path do not in general obey the integral coherence condition.
2. A "crossover" of coherent boundaries or pulses (see Chapter 3, Section VI.A) may produce small composition variations of other cuts,

since the existence of common surfaces has not been proved for the general case.

3. Systems with selectivity reversals contain paths along which the composition velocity is constant. These are the paths on the reversal hyperplane ($j \mid k$ paths on $j:k$ reversal hyperplane). Coherent boundaries with routes along such paths neither sharpen nor spread (spreading through disturbances excepted).
4. In systems in which one species participates in more than one reversal, a route from one point to another along $n-1$ or less paths in the sequence required for a final coherent pattern may not exist. A single abrupt influent composition change may thus generate more than $n-1$ coherent boundaries. An example is shown in Fig. 5.14. This example also demonstrates that, in such systems, species present in both the influent and the presaturant may nevertheless be absent from intermediate plateau zones of the response.

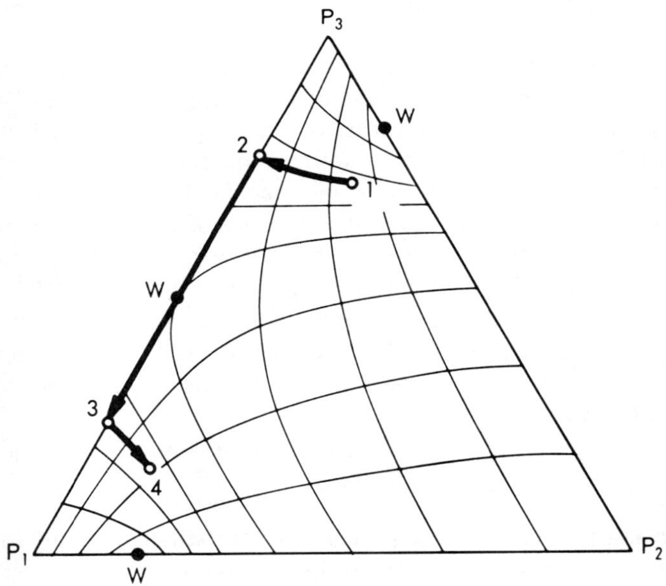


**Fig. 5.14.** Simple example of single abrupt influent composition change generating $n$ rather than $n-1$ response boundaries. Composition profile route of response in three-component system with mass-action equilibrium and two selectivity reversals (schematic). Species 2 is absent from both intermediate plateau zones 2 and 3 although being present in both influent and presaturant.

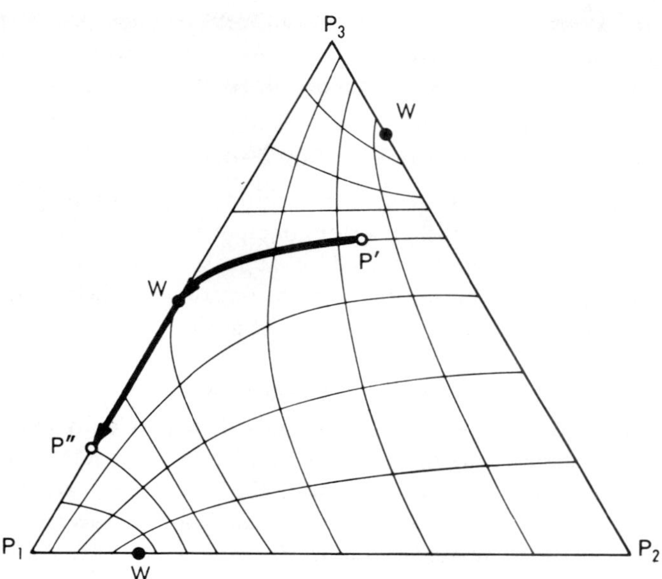


**Fig. 5.15.** Example of singular case of disappearance of a species within diffuse coherent boundary. Composition profile route of boundary in three-component system with mass-action equilibrium and two selectivity reversals (schematic).

5. In singular cases in systems with a species participating in more than one reversal, a concentration may become zero *within* rather than at the upstream or downstream end of a diffuse coherent boundary. This happens if the route follows a path which merges tangentially into a border at a watershed point. An example is shown in Fig. 5.15.*

### 5. *Quantitative Calculations*

Quantitative calculations of responses do not encounter difficulties in principle, but are very lengthy even in simple cases. Since the $h$ transformation is of no use, simultaneous equations must be solved. Also, owing to the appearance of the $v_i$ as exponents, most equations cannot be written in explicit form.

---

* Whether or not such behavior is possible has been the subject of a bitter controversy between Glueckauf [10, 19] and Offord and Weiss [20]. The matter was later examined in more detail by Baylé and Klinkenberg [21], who, however, produced the effect by assuming an unrealistic isotherm which gives infinite distribution ratios, and thus zero species velocities, at vanishing concentrations.

Calculation methods for single abrupt influent composition changes in three-component systems have been worked out by Klein *et al.* [14–16]. More complex systems have so far not been treated.

## B. Systems with Arbitrary Equilibrium Properties

In the general case of systems with arbitrary equilibrium properties, the first concern is to establish whether coherent behavior is possible and, if so, whether coherence is attained or at least approached upon undisturbed development. This question must be examined before meaningful rules for deriving coherent behavior can be given.

### 1. *Existence of Composition-Velocity Eigenvalues*

Coherent behavior is possible if, and only if, real velocity eigenvalues exist for the occurring compositions. For systems with constant separation factors and with mass-action equilibria, the eigenvalues are given by Eqs. (3.68) and (5.51), respectively, and are all real. For the general case, the problem can be formulated as follows:*

Without assumptions about equilibrium properties, the equilibrium relations for systems with stoichiometric exchange can be written in the general form

$$y_k = f_k(x_1, \ldots, x_{n-1}) \qquad k = 1, \ldots, n-1 \tag{5.64}$$

(These $n-1$ relations are sufficient, since $x_n$ amd $y_n$ can be obtained with $\Sigma_i x_i = 1$ and $\Sigma_i y_i = 1$; for systems with nonstoichiometric sorption, see Section III.A.) Total differentials $\mathrm{d}y_k$ can therefore be expressed as

$$\mathrm{d}y_k = \sum_{j=1}^{n-1} (f'_{kj}\, \mathrm{d}x_j) \qquad k = 1, \ldots, n-1 \tag{5.65}$$

where

$$f'_{kj} \equiv (\partial f_k / \partial x_j)_{x_i (i \neq j, n)} \tag{5.66}$$

Coherence requires the existence of a value $\lambda$ obeying all $n-1$ relations

$$\mathrm{d}y_k = \lambda\, \mathrm{d}x_k \qquad k = 1, \ldots, n-1 \tag{5.67}$$

With Eq. (5.65), in matrix notation,

$$\mathbf{M}\, \mathbf{dx} = \lambda\, \mathbf{dx} \tag{5.68}$$

---

* The statement of the problem follows Mangelsdorf [22]. Essentially equivalent formulations have been given by various other authors [23–25]. The matrix (5.69) already appears in the classic paper by DeVault [9], who, however, seems to have been unaware of the eigenvalue aspects.

where $\mathbf{dx}$ is a vector with $n-1$ terms $dx_i$, and $\mathbf{M}$ is a square matrix of $(n-1)$th order with the $f'_{ij}$ as elements. Equation (5.68) is a typical eigenvalue equation [26], which will in general determine a set of $n-1$ eigenvectors $\mathbf{dx}$ with associated $n-1$ eigenvalues $\lambda$. The conditions for nontrivial solutions to exist is that the determinant of the coefficients of the set of linear equations represented by Eq. (5.68) is zero [26]:

$$\begin{vmatrix} f'_{11}-\lambda & f'_{12} & \cdots & f'_{1,n-1} \\ f'_{21} & f'_{22}-\lambda & \cdots & f'_{2,n-1} \\ & \cdots & & \cdots \\ f'_{n-1,1} & f'_{n-1,2} & \cdots & f'_{n-1,n-1}-\lambda \end{vmatrix} = 0 \qquad (5.69)$$

This *characteristic equation* has $n-1$ roots $\lambda_i$, which, if real, are reciprocal adjusted composition velocities [compare Eqs. (5.67) with (3.33)].

For three-component systems, a necessary and sufficient condition for both eigenvalues to be real is readily stated [27]. Here, Eq. (5.69) becomes

$$\lambda = \tfrac{1}{2}\{f'_{11}+f'_{22}\pm[(f'_{11}+f'_{22})^2-4(f'_{11}f'_{22}-f'_{12}f'_{21})]^{1/2}\} \qquad (5.70)$$

giving real roots if, and only if,

$$(f'_{11}-f'_{22})^2+4f'_{12}f'_{21} \geqq 0 \qquad (5.71)$$

Thus, a sufficient (but not necessary) condition is that the "cross derivatives" $(\partial y_1/\partial x_2)_{x_1}$ and $(\partial y_2/\partial x_1)_{x_2}$ have the same sign. In particular, competitive sorption, for which both cross derivatives are negative, is a sufficient condition for the existence of real eigenvalues. Unfortunately, this is no longer true for systems with more than three components. For these, the treatment to date does not provide a concise statement of isotherm properties guaranteeing all eigenvalues to be real. This is one of the principal problems still to be resolved.

The physical realities behind the existence or nonexistence of real eigenvalues are illustrated by the following considerations for three-component systems.

We choose an arbitrary composition point in a domain of affinity sequence 1, 2, 3. The (adjusted) concentration velocity of the arbitrary species $j$ ($j=1$, 2, or 3) at this like at any other point depends on the concentration gradients and thus on the route direction; according to Eqs. (3.33) or (3.34), the velocity has poles for the directions of $y_j = \text{const}$,

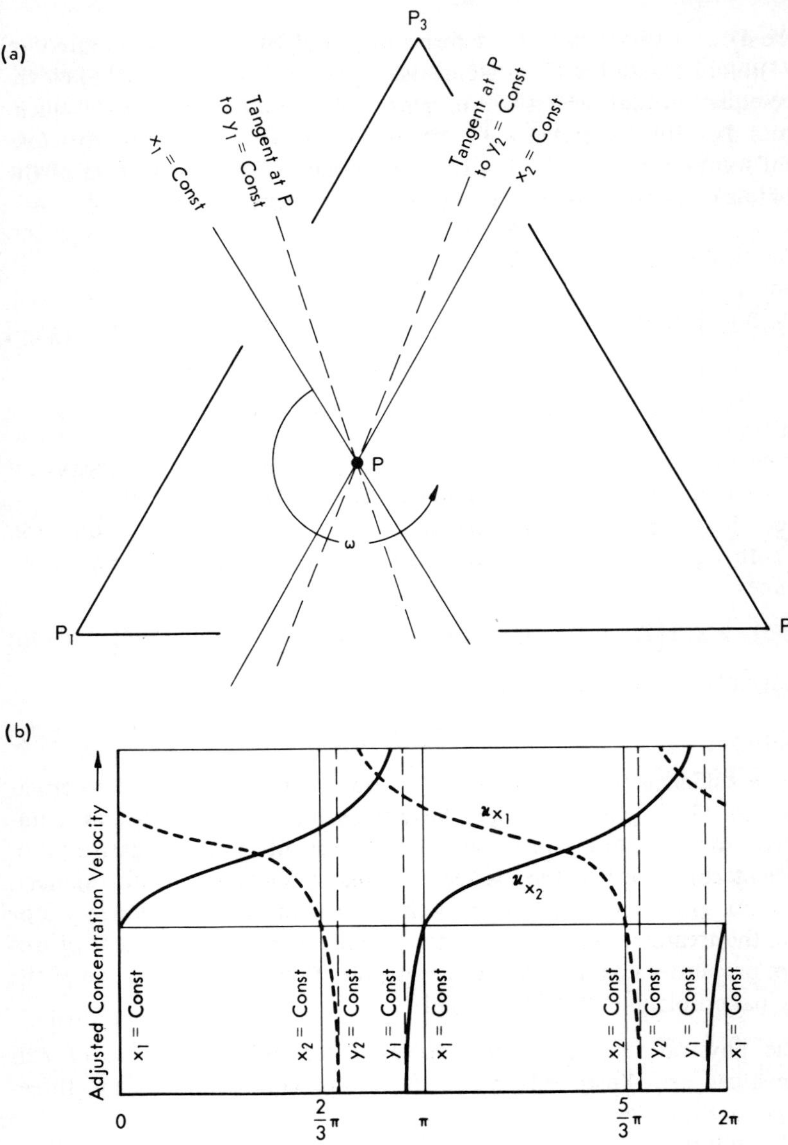


**Fig. 5.16.** Concentration velocities as function of route direction $\omega$ at given composition P in three-component system with competitive sorption (schematic). (a) Topology in x-space simplex; (b) adjusted concentration velocities of species 1 and 2 versus $\omega$.

and is zero for those of $x_j = \text{const}$. As is shown below, the adjusted concentration velocity of species 1 increases, while that of species 2 decreases, monotonically with counterclockwise rotation of the route direction if sorption is competitive. With the two velocities varying monotonically in opposite sense, each from $-\infty$ to $+\infty$ over different 180° sectors, there are necessarily two pairs of route directions for which the velocities are equal (see Fig. 5.16). The concentration velocity of the third species automatically equals the other two where these are equal (see Chapter 3, Section IV.A). The two distinct directions thus are those of composition paths, and the respective velocities are real composition-velocity eigenvalues.

*Derivation.* The route direction at a given point will be characterized by the angle $\omega$, measured counterclockwise, which the route tangent at that point forms with the line $x_1 = \text{const}$. From Eq. (5.65):

$$\left.\frac{\partial y_1}{\partial x_1}\right|_\omega = \left(\frac{\partial y_1}{\partial x_1}\right)_{x_2} + \left(\frac{\partial y_1}{\partial x_2}\right)_{x_1} \left.\frac{\partial x_2}{\partial x_1}\right|_\omega \tag{5.72}$$

After differentiation with respect to $\omega$:

$$\frac{d}{d\omega}\left[\left.\frac{\partial y_1}{\partial x_1}\right|_\omega\right] = \left(\frac{\partial y_1}{\partial x_2}\right)_{x_1} \frac{d}{d\omega}\left[\left.\frac{\partial x_2}{\partial x_1}\right|_\omega\right] \tag{5.73}$$

On the right-hand side, the cross derivative is negative if sorption is competitive, and the derivative with respect to $\omega$ is positive by virtue of the simplex topology, $\omega$ being measured in counterclockwise direction (see Fig. 5.16). The left-hand side of Eq. (5.73) thus is negative, so that

$$\frac{d}{d\omega}\left[\partial x_1/\partial y_1|_\omega\right] > 0 \tag{5.74}$$

The analogous derivation for species 2 gives

$$\frac{d}{d\omega}\left[\partial x_2/\partial y_2|_\omega\right] < 0 \tag{5.75}$$

With Eqs. (3.33) and (3.34), the partial derivatives in conditions (5.74) and (5.75) are readily identified as $u_{x_1}$ and $u_{x_2}$ if the route is a history route, and as $u_{y_1}$ and $u_{y_2}$ if the route is a profile route. Accordingly, the respective concentration velocity of species 1 increases, while that of species 2 decreases, with counterclockwise rotation of the route direction (increasing $\omega$).

The results (5.74) and (5.75) are physically plausible. As Fig. 5.17 illustrates, a composition variation involving a given increase $dx_1$ in a direction $\omega$ becomes more favorable for species 2 relative to species 3 as $\omega$ increases. For higher values of $\omega$, the greater competition because of greater predominance of the species of higher affinity leads to overall

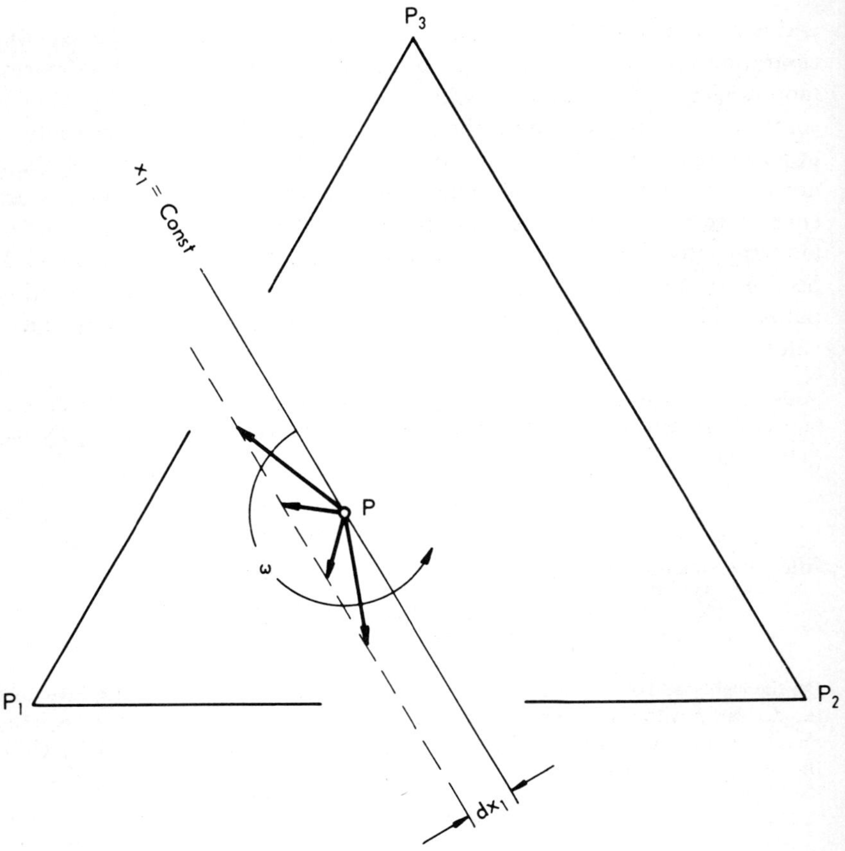


**Fig. 5.17.** Composition variations (heavy arrows) involving given increase $dx_1$ at given composition point P in three-component system.

lower distribution ratios $y_i/x_i$ and thus to a lesser increase $dy_1$, provided sorption is competitive. Analogous arguments apply for composition variations involving a given decrease $dx_1$, as well as for species 2.

From these considerations, a sufficient (but not necessary) existence condition is readily formulated: Real eigenvalues exist if the concentration velocities of two of the three species vary in opposite sense with rotation of the route direction. For competitive sorption, this condition is met. On the other hand, a necessary (but not sufficient) condition for the absence of real eigenvalues is that the concentration velocities of all three species increase or decrease jointly with rotation of the route direction. As illustrated in Fig. 5.18, directions for which the velocities are equal may or may not exist under this condition.

An example was encountered in an earlier section. In the equivalent stoichiometric system of a Langmuir case, displayed in Fig. 5.5, the concentration velocities of the physically existing species 1 and 3 vary in the same sense with rotation of the route direction at all points of the composition plane; that of the dummy, species 2, varies in the opposite sense where the dummy concentration is positive, but varies in the same sense as those of species 1 and 3 where the dummy concentration is negative. The situations at points between the envelope and the line $x_2 = 0$ and at points on the other side of the envelope are as shown in Fig. 5.18a and b, respectively.

With considerable discomfort, one can extend such considerations to systems with more than three components. While they cannot prove that competitive sorption [defined as obeying condition (5.47)] guarantees all eigenvalues to be real, they can at least show that imaginary eigenvalues will be exceptional if sorption is competitive, because, normally, one can then select different pairs of species with opposite concentration-velocity variations for different dimensions in such a manner that the hypersurfaces of pairs of equal velocities will have to intersect.

### 2. *Development toward Coherence*

Granted the existence of real composition-velocity eigenvalues for all compositions in the range considered, the general trend toward coherence upon development can be established in much the same way as for systems with simpler equilibrium properties. For differential route segments, the deformation toward eventual attainment of coherence can be proved with considerations as in Chapter 3, Section V.C. The development of routes of finite lengths toward coherence can then be shown with the following argument (which is equivalent mathematically to a finite-difference approximation). The noncoherent initial boundary is approximated by a sequence of separate, small composition steps. Being noncoherent, each of these breaks up into a set of coherent ones, attainment of coherence having been proved for differential route segments. The small coherent steps of the different sets cross one another essentially as in systems with constant separation factors (see Chapter 3, Section V.E). The only complication is that second-order perturbations are generated by each crossover if there is no common surface of the respective composition paths. However, since their development behavior is subject to the same rules as that of the initial noncoherent steps, the second-order perturbations can be shown at worst to delay, but not to prevent, attainment of coherence. The rather involved conceptual arguments are not

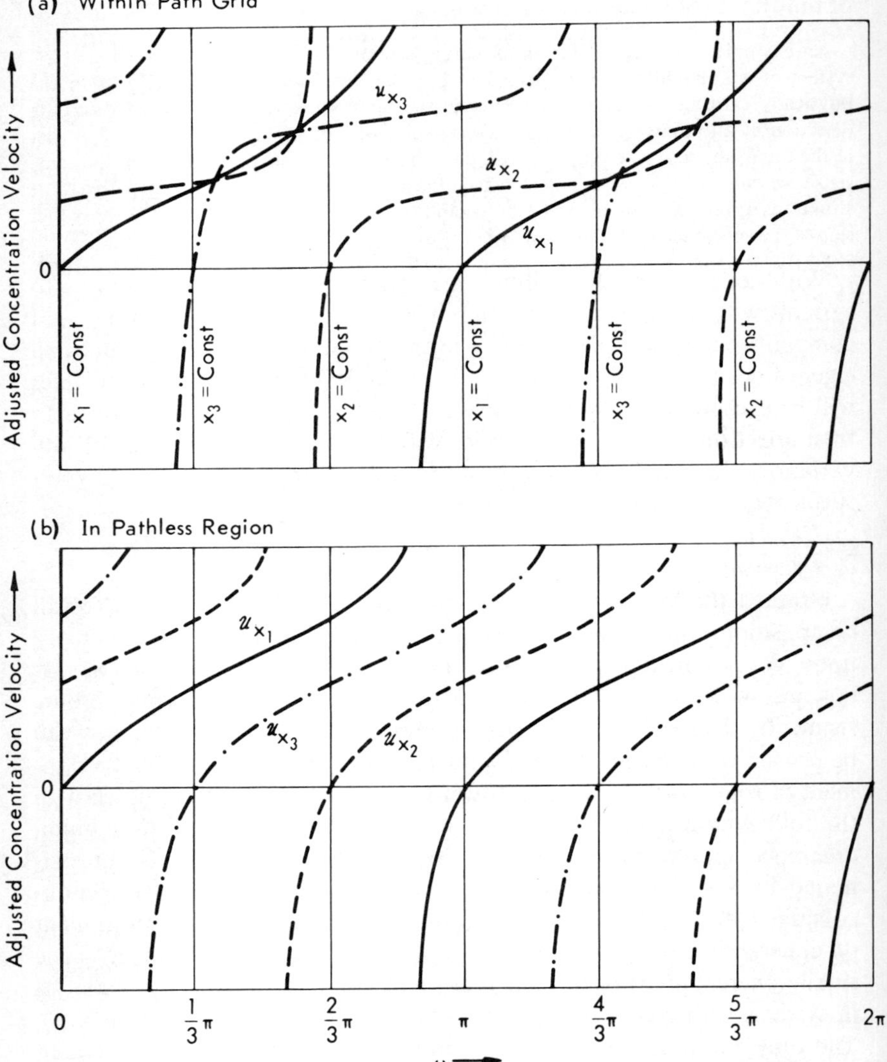


**Fig. 5.18.** Adjusted concentration velocities as function of route direction $\omega$ at given composition in three-component system with synergistic sorption under conditions such that all three velocities increase counterclockwise. (a) In path grid: velocities are equal for two route directions per 180° sector; (b) in pathless region: no route direction with equal velocities. (Schematic.)

reproduced here, because a briefer and more elegant proof can no doubt be found.

Where the composition-velocity eigenvalues are imaginary, or where a final coherent route for the specified operating conditions does not exist because some, although not all, eigenvalues are imaginary, coherence can obviously not be attained. Fortunately, such situations appear to require rather unusual equilibrium properties. One may conjecture that nonexistence of coherence is about as rare as is nonexistence of a steady state in other areas of chemical physics. The behavior of chromatographic systems under such conditions has so far remained entirely unexplored. It can at least be shown, however, that development will not shift a route from a region of real into one of imaginary eigenvalues, provided the isotherm is continuous and differentiable.

### 3. *Response Behavior*

Refraining from speculations as to exceptional conditions excluding coherence, we confine the examination of the response behavior to "well-behaved" systems—or composition ranges within systems—with $n-1$ real velocity eigenvalues for any composition.

Relatively simple rules can be stated for systems with entirely competitive sorption. The deductions in Section IV.A.4, showing that most of the basic qualitative rules for systems with constant separation factors remain valid, were based on competitive sorption rather than on the stronger condition of mass-action equilibria. Therefore, they are applicable here, too. Specifically, this is readily shown to be true for the derivations of the affinity cut and of the sequence of boundaries or pulses in final coherent patterns (with the exceptions already noted for mass-action equilibria). In general, the response behavior will thus differ little from that of mass-action systems.

Even in systems with competitive sorption, however, a few minor additional complications may arise in connection with selectivity reversals. First, a system may have a multitude of reversal hypersurfaces of the same two species, as illustrated for a simple case in Fig. 5.19. Second, the reversal hypersurfaces are no longer planar and will in general be intersected by paths of all cuts. Along a $j \mid k$ path across a $j{:}k$ reversal hypersurface, species $j$ and $k$ interchange their memberships in the groups of high- and low-affinity species; also, the composition velocity will have a maximum or minimum where the path crosses the hypersurface. Accordingly, coherent boundaries with mixed sharpening characteristics (partly

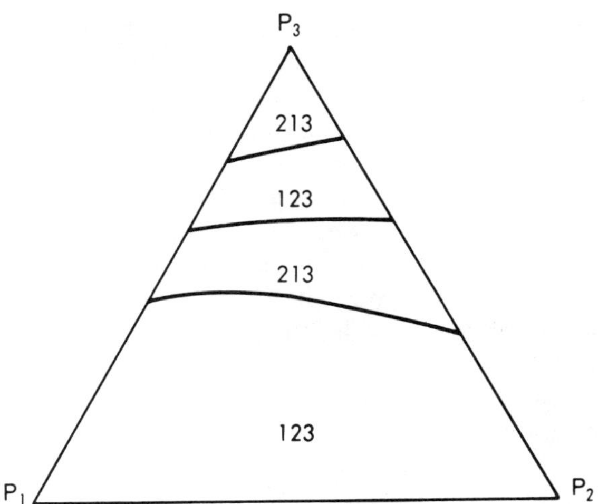


**Fig. 5.19.** Typical three-component system with multiple selectivity reversals involving same species. Domains of affinity sequences 1, 2, 3 and 2, 1, 3 in composition simplex of system with three 1:2 reversals (schematic).

self-sharpening, partly nonsharpening) and with maxima and minima of some concentrations and monotonic variations of others may occur.

No other anomalies for well-behaved systems with competitive sorption have so far been found.

Once the premise of competitive sorption is relaxed, most of the earlier derivations become inapplicable, and a variety of new phenomena can occur. Among the victims is the affinity-cut rule, based on the no longer valid premise that all species velocities increase or decrease jointly.

A systematic classification of response features in the general case has so far not been attempted. Nevertheless, the behavior in any particular case can in principle be deduced with the concepts and techniques developed in the earlier sections. Specifically, the continuity considerations in Chapter 3, Section IV.B, which led to the affinity-cut rule, and the treatment of transient routes in Chapter 3, Section V.E, can be adapted to arbitrary equilibrium properties. When applying the continuity considerations one must first identify, from the equilibrium relations, those "maverick" species whose distribution ratios, and thus velocities, do not vary in the same direction as the others when the composition shifts in favor of species of higher or lower affinities. In coherent boundaries, high- and low-affinity species can still be defined as those whose velocities are lower and higher, respectively, than the composition velocity. The behavior of the maverick species, however, then is contrary to the affinity-cut rule; for example, the concentration of a low-affinity maverick will vary in the same direction as those of the high- rather than low-affinity species. As far as transient routes are concerned, the principle of development toward a final coherent pattern with boundaries or pulses in the sequence of increasing composition velocities (or step velocities), counted in the direction of flow, remains applicable although, owing to the maverick species, the spatial orientation of the composition paths will in general be different from that in systems with competitive sorption.

Admitting entirely arbitrary equilibrium properties, the general theoretical treatment may have passed the point of diminishing returns. While the facility to predict and compute the response in any specified case still needs to be improved, the prospects of gaining further fundamental insight from considering all possible types of systems and behaviors do not appear promising.

### C. Other Theories

Attempts to formulate a treatment for arbitrary equilibrium properties have been made from the first beginnings of chromatographic theory: As early as 1943, DeVault [9] stated the basic differential equations in a general form free of assumptions as to equilibrium properties. However, neither he nor later authors were able to deduce general rules without introducing various assumptions, as is not surprising in retrospect. Almost every author, including DeVault, has taken coherent behavior for granted at least under the simple operating conditions he considered. The only notable exceptions are Baylé and Klinkenberg [21], who recognized coherent behavior as an unproved postulate, and Mangelsdorf [22], who pointed out that the composition-velocity eigenvalues are not necessarily real. (For a critique of the conclusions of Rachinskii [24] on coherence, see Chapter 4, Section I.H.)

Much of the early work in this area was directed toward development behavior in two-component systems with nonstoichiometric sorption ([10, 12, 13, 19–21, 28–30]; see also footnote to p. 313). Here, Glueckauf's approach [13] of operating with the characteristics of the differential equations (equivalent to the composition paths) must be singled out as the first great step toward making the treatment independent of specific equilibrium relations and operating conditions. His work also includes a discussion of Freundlich isotherms [30].

For two-component systems with nonstoichiometric sorption, Glueckauf [30] and later authors [21] assumed the cross derivatives $(\partial y_1/\partial x_2)_{x_1}$ and $(\partial y_2/dx_1)_{x_2}$ to have the same sign, owing to the Gibbs-Duhem equation. This convenient assumption guarantees real composition-velocity eigenvalues [see condition (5.71)] and may largely account for these having been taken for granted in the general case, too. Unfortunately, the assumption is not valid in the presence of a solvent as an additional thermodynamic component.

Other special cases treated without assumption of specific equilibrium relations include displacement development and infinitesimal influent composition pulses. For displacement development, Hagdahl *et al.* [31] gave a qualitative discussion of the consequences of selectivity reversals. For small pulses, Stalkup and Deans [23] and Mangelsdorf [22] derived the general features of the response (see also Chapter 4, Section III.F.4). The most extensive study in this general area, however, was conducted by Klein, Tondeur, and Vermeulen [15, 16], who considered responses to single abrupt influent composition changes and stated a set of rules similar to those derived in Chapter 4, Section I. Formulated largely from experience with numerical calculations, these rules have turned out to be essentially confined to mass-action systems without selectivity reversals.* This work has recently been extended by Tondeur [18], who included a detailed analysis of the consequences of selectivity reversals in mass-action systems. Some of the path grids shown in this section are based on his extension, which in other respects has drawn on the concepts developed here.

Except for recent extensions [33, 34] of the relatively simple theories for infinitesimal influent pulses, all work referred to here has assumed local equilibrium and absence of disturbances. The few and somewhat fragmentary treatments of systems involving finite composition variations with both interference and deviations from local equilibrium will be discussed in the next section.

## V. Disturbances

One of the most unrealistic of the simplifying premises, in seriousness second only to the assumptions concerning equilibrium properties, is the

* Almost identical derivations concluding that the response to an abrupt influent composition change in the general case of arbitrary equilibrium properties consists of no more than $n-1$ boundaries were given independently by Klein *et al.* [15, 16] and Zhukhovitskii ([32], see also Chapter 4, Section III.F.4). These derivations, counting equations and unknowns, overlook that certain quantities may become indeterminate, and are not valid even for mass-action systems (e.g., see Fig. 5.14). This error has meanwhile also been corrected by Tondeur [18].

postulate of local equilibrium between the mobile and the stationary phase and of complete absence of disturbances. In this respect the present treatment differs most widely from contemporary chromatographic theory, which has taken the effects of disturbances as its central topic while almost completely ignoring interference effects.

Fortunately, the principal aspects of disturbance effects are the same in systems with and without interference. The highly developed theories of column dynamics can thus be referred to, and the physical situation is well understood. The quantitative treatment of disturbances in systems with interference, however, encounters considerable difficulties. Until these are resolved, a discussion of details does not appear profitable. This section will therefore merely give an outline of the types of effects and of the extensions required to account for them.

The discussion of individual disturbances will be preceded by an examination of general features they have in common.

## A. General Features

The various effects here somewhat loosely referred to as disturbances are:

1. Deviations from local equilibrium because of finite mass-transfer resistance to sorption and desorption (*nonequilibrium*).
2. Leveling of concentration gradients through molecular diffusion in axial direction (*axial diffusion*).
3. Back-mixing of volume elements of the mobile phase because of branching and rejoining of stream paths, flow-velocity variations, etc. (*eddy dispersion*).
4. Distortions of the flow-velocity profile because of packing irregularities or wall effects (*channeling*) or hydrodynamic instability (*fingering*).
5. Volume changes of the sorbent particles constituting the bed.
6. Deviations from isothermal behavior because of release or consumption of heat of sorption, affecting equilibria and rates (*nonisothermal effects*).

### 1. *Dispersive Action*

With certain exceptions under the preceding items 4 and 6, to become apparent later, the general effect of any disturbance is dispersive, i.e., makes all concentration variations more diffuse than they would be otherwise.

It is this common property which allows the elementary theories of chromatography to operate with the so-called height equivalent to a theoretical plate (HETP), a primitive but workable concept which lumps all disturbances together to account for them jointly in terms of back-mixing on fictitious "plates," in analogy to distillation (e.g., [35–44]). This behavior may be taken as a direct consequence of the differential material balance for a species or of an upset of the orderly sequence of advancing volume elements of the mobile phase, considerations that apply regardless of interference effects. The interdependence of the equilibrium distributions of the species, here called interference, thus does not alter the general nature of the dynamic effects, here called disturbances.

With the same exceptions under items 4 and 6 as previously indicated, the effect of a disturbance is in proportion to the sharpness of the concentration variations. This is immediately apparent for the effects of warped flow-velocity profiles and actual back-mixing, the volume elements brought together at the same column level or actually mixed being of more widely differing compositions if the concentrations vary sharply. The situation is similar with respect to nonequilibrium and axial diffusion, which each contribute to the differential material balance a term involving a second-order derivative $\partial^2/\partial z \partial t$ or $\partial^2/\partial z^2$, whose magnitudes relative to the first-order terms is proportional to the respective concentration variation $\partial C_i/\partial t$ or $\partial C_i/\partial z$. This result is plausible physically: The strain on mass transfer between the mobile and the stationary phase is greater when the mobile-phase composition at a given location changes more rapidly, and axial diffusion is proportional to its driving force, the axial concentration gradient. Again, these considerations apply regardless of interference effects.

That disturbances tend to level out, but not to build up, concentration gradients indicates that they will not give rise to fundamentally different response patterns, even in systems with interference. That interference does not alter the other characteristic property of disturbances, namely, of producing effects in proportion to the sharpness of the composition variations, may serve to show that the fundamental aspects of disturbance effects in systems with and without interference are alike: In nonsharpening boundaries disturbances eventually become insignificant as boundary spreading, in the early stages assisted by disturbances, reduces all axial concentration gradients; in self-sharpening boundaries a steady state is eventually attained in which the sharpening effect of sorption equilibrium and the dispersive effects of the disturbances balance one another, so that the boundary width and the shapes of the advancing concentration profiles remain unchanged from then on. This so-called *constant pattern* of a self-sharpening boundary is a stable state that can be approached from either side, i.e., from an initially greater or lesser width of the boundary. Thus, as far as final patterns are concerned, disturbances can be ignored in nonsharpening boundaries, but must be included in any realistic treatment of self-sharpening boundaries. These effects are well known from elementary theory of nonlinear chromatography without interference (e.g., see reviews [37, 45–49]).

These considerations presume that disturbances do not block the approach to a state of coherence, impair its existence, or alter any of its properties other than the boundary widths. These assumptions are obviously justified for channeling and fingering, which are mere macroscopic distortions of the flow pattern not directly coupled with molecular events, and for nonsharpening boundaries, in which the disturbances merely assist the existing dispersive trend and eventually become insignificant. For nonequilibrium, axial diffusion, and eddy dispersion in self-sharpening boundaries the justification is not immediately apparent but is provided by an individual examination of these cases.

Since the principal consequences of disturbances are the same as in systems without interference, for which detailed theories have been developed, only a few comments qualifying or supplementing statements in earlier sections will be added here.

### 2. *Constant and Proportionate Patterns*

As mentioned earlier, coherent self-sharpening boundaries upon development approach *constant patterns*, in which all compositions travel at the same velocity, so that the boundary advances with unchanged width and profile shapes. Such a pattern is attained when the sharpening effect of sorption equilibrium and the dispersive effects of disturbances have reached a balance. In contrast, nonsharpening boundaries approach *proportionate patterns*, in which the distances between compositions are proportional to the traveled distance or elapsed time (see also Chapter 4, Section I.A). Such a pattern is attained when the boundary has spread so much that its initial relative width as well as the effects of disturbances have become negligible. Both the constant and the proportionate pattern are approached asymptotically. Distance-time diagrams and concentration profiles illustrating this behavior are shown in Fig. 5.20. These may be compared with the diagrams in Fig. 3.18, drawn in accordance with the original simplifying premises.

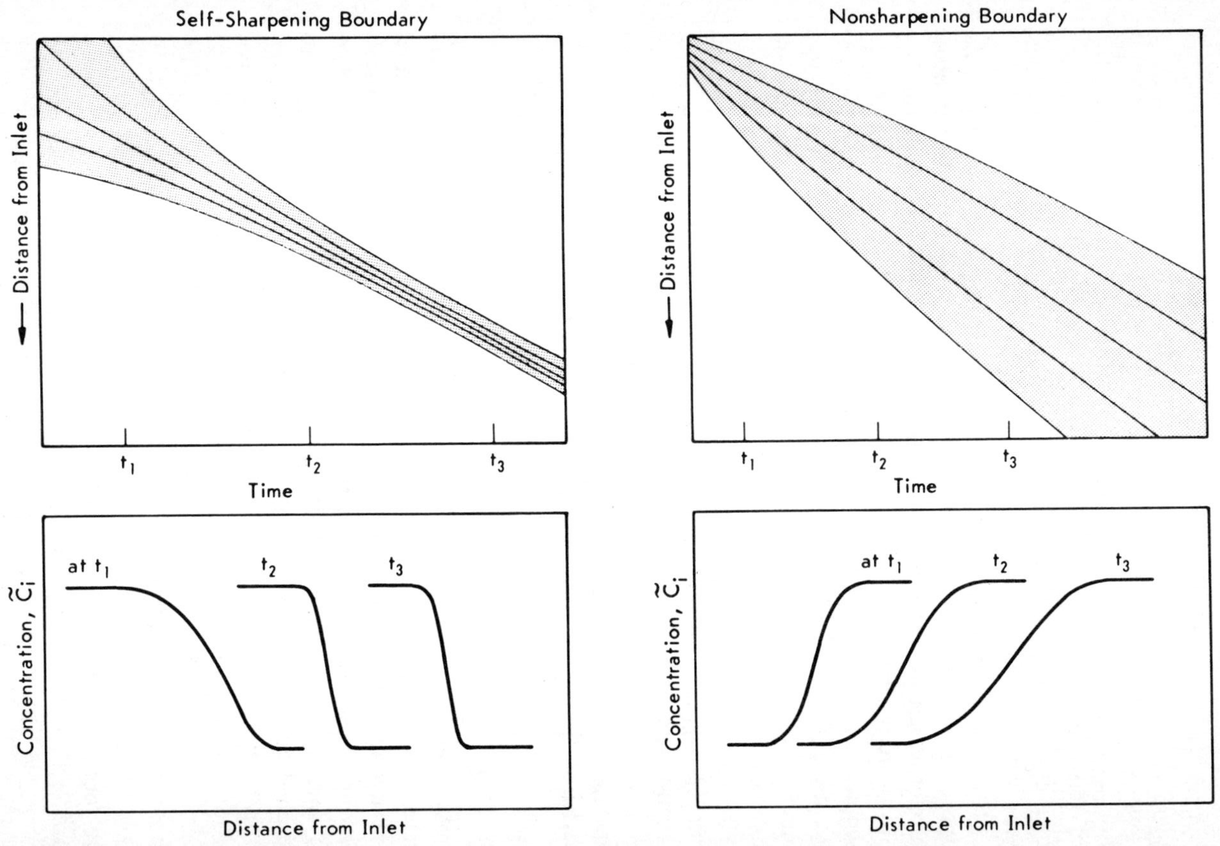


**Fig. 5.20.** Concentration trajectories and profiles for a self-sharpening and a nonsharpening boundary in presence of

Two important consequences of this behavior deserve to be singled out. First, a boundary cannot become or remain ideally sharp. Even if self-sharpening, the boundary will cease to sharpen at, or will broaden to, a finite constant-pattern width. Second, the response pattern generated by an abrupt influent composition change is not a proportionate one, since the disturbances contribute in the early but not in the later stages to the broadening of the boundaries (see Fig. 5.21; compare with Fig. 4.1).

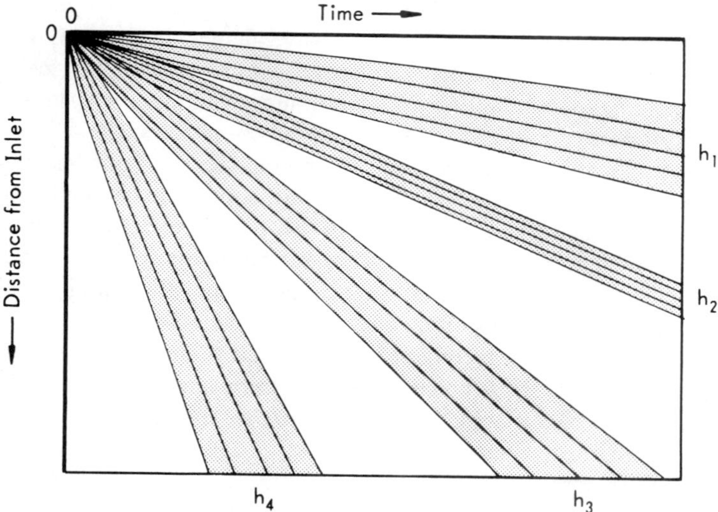


**Fig. 5.21.** Typical distance-time diagram for response to single abrupt influent composition change in presence of disturbances. Root trajectories for five-component case with one self-sharpening and three nonsharpening boundaries (schematic).

Nevertheless, the response eventually approaches a proportionate pattern when the nonsharpening boundaries have done so and when the total spread of the pattern has become very large compared with the constant widths of the self-sharpening boundaries.

The magnitude of the dispersive effect of a disturbance on a given boundary depends on the operating conditions (flow rate, temperature, etc.). Therefore, a change of operating conditions in general entails a change of the widths of constant patterns and thus may actually cause self-sharpening boundaries to spread.

### 3. *Dispersion of Infinitesimal Composition Variations. Trace-Component Systems*

Lacking the usual self-sharpening or nonsharpening tendencies of nonlinear sorption equilibria, coherent boundaries constituting infinitesimal composition variations would travel with constant widths if disturbances were entirely absent. In any real system, however, disturbances are operative and, uncombated by self-sharpening tendencies, cause the boundaries to spread. For example, the response pulses generated by a square-wave influent composition pulse of infinitesimal amplitude in a real system do not retain their square-wave shapes, as in Figs. 4.17 and 4.20, but spread and attenuate, eventually assuming Gaussian shapes.

The existing detailed theories of column dynamics in systems with linear equilibria ([50–53] and work quoted in reviews [40–49, 54]) are directly applicable to dispersion of boundaries in trace-component systems, for which interference effects were shown to be negligible (see Chapter 3, Sections III.B.1 and IV.E.3). The theories can also be applied to dispersion of infinitesimal composition variations in any other systems: Since there are virtually no self-sharpening or nonsharpening tendencies to combat or assist dispersion by disturbances, these act as in systems with linear equilibria. It may merely be more difficult to determine the relevant dynamic coefficients—rate coefficients of sorption and desorption, effective axial diffusion coefficients, etc.—because the species may affect one another's dynamic properties.

Although they spread indefinitely, boundaries constituting infinitesimal composition variations do not attain proportionate patterns. In accordance with the classical theories of chromatography, the boundary width increases in proportion to the square root, rather than to the first power, of the traveled distance or elapsed time. The rate of spreading thus decreases as the boundary advances. This behavior is readily understood: As shown earlier, the dispersive effects of the disturbances lessen as the advancing boundary becomes more diffuse.

### 4. *Resolution Behavior*

The dispersive action of disturbances in general delays resolution of boundaries or pulses or, in separations, of species. The formulas stated in earlier sections for resolution distances and times in the absence of disturbances thus give only minimum values and must be corrected for disturbance effects. To mention but one example, fully covered by conventional chromatographic theory: Resolution of the response pulses generated by an influent composition pulse of infinitesimal duration (see Chapter 4, Section III.B.1) is not instantaneous in real systems, owing to the unavoidable disturbances.

As a rule, resolution is merely delayed but can be carried to any specified degree by continued development (provided, of course, the column is long enough). This is a consequence of the fact that, under most operating conditions, the effect of sorption equilibrium by itself lets the boundaries or pulses with different affinity cuts increase their distances from one another in proportion to the traveled distance, while the dispersion that disturbances by themselves would achieve is proportional merely to the square root of that distance. In exceptional cases, however, adjacent self-sharpening boundaries of different cuts travel at the same velocity, and disturbances may then lead to a final, constant and finite overlap of the boundaries, preventing resolution that would otherwise take place. An example of practical interest is displacement development: The front and rear boundaries of a species in the final pattern may remain unresolved if the amount of that species is small (see Fig. 5.22).

It can be verified that the persistence of an overlap (or overlaps) as in Fig. 5.22 does not prevent the attainment of a constant pattern. Regardless of overlaps, the dispersive tendency of the disturbances becomes stronger as the boundaries sharpen, or weaker as the boundaries spread. A constant pattern will therefore be reached, provided the sharpening tendency of the sorption equilibria is not impaired. As it turns out, overlaps even strengthen this sharpening tendency. In the situation shown in Fig. 5.22, for example, molecules of species $l$ that have fallen behind are more efficiently displaced by species $j$ than by species $k$ alone, because species $j$ has the higher affinity; similarly, molecules of species $j$ that have got ahead are retarded by the stationary phase more

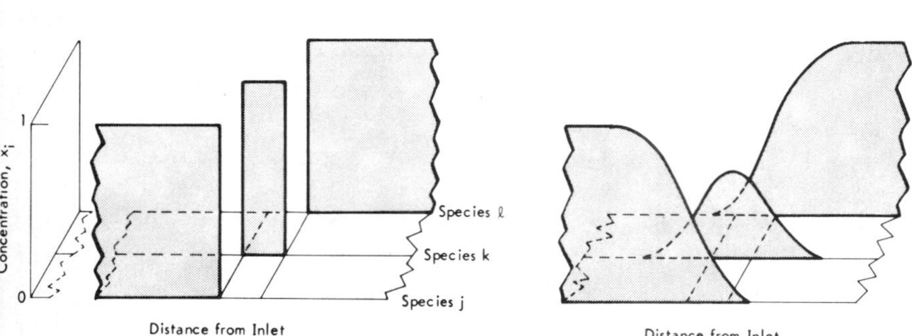


**Fig. 5.22.** Partial concentration profiles of three adjacent species in final constant pattern of displacement development in absence and presence of disturbances (schematic). Disturbances prevent resolution of front and rear boundaries of species $k$.

efficiently in the presence of species $l$ than of $k$ alone, because species $l$ has the lower affinity and therefore competes less strongly for sorption by the stationary phase. Analogous considerations apply in other cases of overlap of boundaries in a constant pattern.

### 5. *Attainment of Coherence*

Attainment of coherence may or may not coincide with completion of resolution. In elution development, for example, the response is obviously noncoherent as long as response pulses (peaks) still overlap, and becomes coherent after resolution is completed. In displacement development, on the other hand, the final constant pattern is coherent even if resolution has remained incomplete (see Fig. 5.22).

The coherence of the final, constant or proportionate patterns approached by advancing boundaries is inherent in their definitions. But attainment of coherence need not coincide with attainment of this final pattern. For example, sorption systems with a single sorbable species under equilibrium conditions are always automatically coherent, yet, finite distances and times are nevertheless required for boundaries to attain their final patterns if, say, axial diffusion is not negligible. As another example, covered by the treatment in the earlier sections, a nonsharpening boundary diffuse enough for disturbances to have become negligible may well be coherent even if, owing to a large and uneven initial spread, it is still far from attaining its eventual proportionate pattern.

In general, however, boundaries in multicomponent systems are noncoherent unless either the dispersive effects of nonequilibrium and axial diffusion have become negligible or a constant pattern has been reached. In nonsharpening boundaries, deviations from local equilibrium rule out coherence because, as shown earlier, they gradually disappear, requiring the concentrations to shift relative to one another. In self-sharpening boundaries, the extent of deviations from local equilibrium depends on the strain on mass transfer and thus varies as the boundary sharpens or spreads, again requiring the concentrations to shift relative to one another until the constant pattern is attained. Axial diffusion, on the other hand, impairs coherence through its effect on the concentration velocities: Species having different diffusion coefficients are affected to different degrees, and their concentration velocities thus undergo different variations with increasing or decreasing significance of axial diffusion as the boundary sharpens or spreads. Hence, unless the diffusion coefficients of all species happen to be the same, a boundary can be coherent only if it either has attained a constant pattern or is so diffuse that axial diffusion is negligible.

As mentioned before, the final constant or proportionate patterns are approached asymptotically. Thus, if disturbances are taken into account, coherence, too, is in general approached asymptotically rather than attained in a finite time as under the original premises.

The individual physical and mathematical features of the various disturbances, inasmuch as they relate directly to multicomponent systems with interference, will now be surveyed.

## B. Nonequilibrium

Deviations from local equilibrium occur because a finite time is required for equilibration of the stationary with the mobile phase when the composition of the latter has been varied. As noted earlier, the strain on mass transfer between the phases and, therefore, the resulting dispersive effect are greatest in sharp and fast-moving boundaries.

The differential material balances (3.16) and (3.32), derived without invoking the premise of local equilibrium, remain valid when this premise is relaxed:

$$(\partial y_i/\partial \tau)_z + (\partial x_i/\partial z)_\tau = 0 \tag{3.32}$$

for all species $i$. However, $\mathbf{y}$ now is no longer a function of $\mathbf{x}$ alone. Rather, the sorption (or desorption) rate $(\partial y_i/\partial \tau)_z$ must be supplied by a suitable mathematical model of sorption kinetics as a function of $\mathbf{x}$ and $\mathbf{y}$ and of the equilibrium relations, kinetic properties of the species and the system, and operating conditions.

The sorption or desorption rate can be written in the general form

$$(\partial y_i/\partial \tau)_z = k_i[f_i(\mathbf{x}) - y_i] \tag{5.76}$$

where $k_i$ is a first-order rate coefficient (generally a complex function of the variables and conditions), and $f_i(\mathbf{x})$ is the value that $y_i$ would assume if equilibrium with the mobile-phase composition $\mathbf{x}$ were established. (In the event of intraparticle concentration gradients in the stationary phase, $y_i$ is taken as the mean stationary-phase concentration at the location $z$; see also p. 338.) The expression in brackets thus states the distance from equilibrium with respect to species $i$ and can be considered as a "driving force" for mass transfer.

With Eq. (5.76), the material balance (3.32) can be rearranged to give

$$\left(\frac{\partial f_i(\mathbf{x})}{\partial \tau}\right)_z + \left(\frac{\partial x_i}{\partial z}\right)_\tau + \left\{\frac{\partial}{\partial \tau}\left[\frac{1}{k_i}\left(\frac{\partial x_i}{\partial z}\right)_\tau\right]\right\}_z = 0 \tag{5.77}$$

Nonequilibrium contributes the third term. Without this term—which is zero if $k_i$ is infinite—one has local equilibrium, $f_i(\mathbf{x}) = y_i$, as a comparison with Eq. (3.32) shows. Because the term is second order, its relative importance increases as the boundary sharpens or decreases as the boundary spreads.

Quantitative calculations of responses require integration over $z$ and $\tau$ of the simultaneous material balances (3.32) or (5.77) for all species, with functions $k_i$ provided by the kinetic model. No such calculations have as yet been reported.

Fortunately, the calculation of the constant pattern is much simpler. Since all concentrations in this pattern, by definition, travel at the same composition velocity $u$, they can be stated as time-independent functions of a distance coordinate $z^*$ whose origin travels at this velocity. That is, the condition for constancy of the pattern is

$$(\partial x_i/\partial \tau)_{z^*} = 0 \quad \text{and} \quad (\partial y_i/\partial \tau)_{z^*} = 0 \quad \text{for all } i \tag{5.78}$$

where

$$z^* \equiv z - u\tau \tag{5.79}$$

With the transformation (5.79) and condition (5.78), the material balance (3.32) reduces to

$$\mathrm{d}x_i/\mathrm{d}z^* - u\,\mathrm{d}y_i/\mathrm{d}z^* = 0 \tag{5.80}$$

Thus, since $u$ is constant, $x_i$ and $y_i$ vary linearly with one another. Integrating Eq. (5.80) over $z^*$, one finds†

$$\lambda_i = \omega_i \tag{5.81}$$

where

$$\lambda_i \equiv \frac{x_i(z^*) - x_i'}{x_i'' - x_i'} \quad \text{and} \quad \omega_i \equiv \frac{y_i(z^*) - y_i'}{y_i'' - y_i'} \tag{5.82}$$

(Primes and double primes refer to the upstream and downstream sides of the boundary.) From Eq. (5.76) one obtains

$$\mathrm{d}y_i/\mathrm{d}z^* = -\frac{k_i}{u}[f_i(\mathbf{x}) - y_i] \tag{5.83}$$

With Eqs. (5.81) and with the functions $k_i$ supplied by the kinetic model, the simultaneous equations (5.83) for all species can be integrated over

† This result has long been known for systems with single-component sorption and binary stoichiometric exchange [45, 55] and for the formally similar situation in distillation (e.g., [56, 57]). Equivalent derivations for multicomponent chromatographic systems have been given by Cooney and Lightfoot [58] and Clazie *et al.* [59].

$z^*$ to give the constant-pattern profile $\{\mathbf{x}(z^*), \mathbf{y}(z^*)\}$. The equations are coupled only through the dependence of the $k_i$ on the concentrations of the other species. Except for this coupling, the calculation of the pattern is the same as for a single component.

Since nonequilibrium does not alter the material-balance equation (3.16), the differential coherence condition (3.38) derived from this equation remains valid and is obeyed in the constant pattern by the non-equilibrium concentrations $C_i$ and $\bar{C}_i$. More importantly, the integral coherence condition (3.40) is also obeyed by the constant pattern, as can be shown with Eq. (5.80), $u$ having the same value for all species; the requirements that must be met by the compositions on the two sides of a coherent boundary thus are not altered by nonequilibrium, whose effect is therefore restricted to the boundary itself. Furthermore, because of Eq. (5.80) one has

$$\Delta x_i / \Delta y_i = u \tag{5.84}$$

that is, the boundary velocity, $u$, in the constant pattern is the same as under equilibrium conditions [compare Eq. (3.36)].

For a system with a single sorbable species, the route ("operating

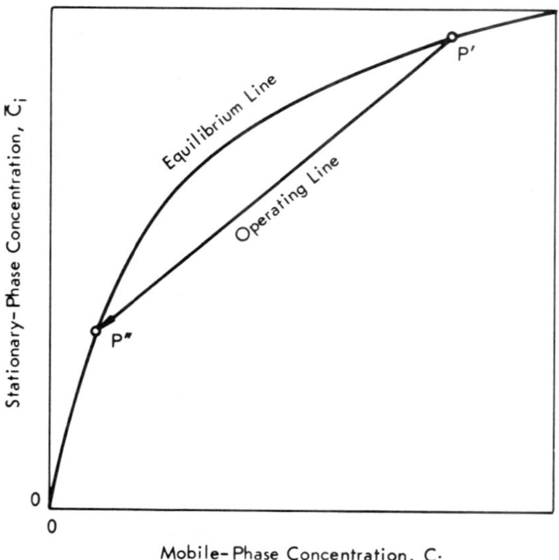


**Fig. 5.23.** Typical isotherm (equilibrium line) and profile route (operating line) for constant pattern between plateau zones P′ and P″ in system with one sorbable species (schematic).

line") and isotherm ("equilibrium line") of a constant pattern are shown in Fig. 5.23. In view of condition (5.81), the route is a straight line connecting the upstream and downstream composition points and thus is independent of the mass-transfer resistance and flow rate. To be sure, a higher mass-transfer resistance or flow rate under otherwise identical conditions results in a more diffuse constant pattern, but the balance between the sharpening and dispersive forces is nevertheless reached at the same distance from local equilibrium: For the same distance from equilibrium and thus the same "driving force," the rate of sorption or desorption is lower if the resistance is higher, but this lower rate is sufficient to conserve the constant pattern if the boundary is more diffuse, since the mobile-phase composition at any given location then varies more slowly, allowing more time for mass transfer to keep pace. Similarly, the effect of higher flow rate and greater diffuseness of the boundary compensate one another, so that the strain on mass transfer remains the same.

In multicomponent systems, the situation is analogous with respect to each separate species. Unfortunately, the route can no longer be adequately represented in $x$-space or $y$-space diagrams once the premise of equilibrium between $\mathbf{x}$ and $\mathbf{y}$ is relaxed. In view of conditions (5.81), the projections of the constant-pattern route onto the $(x_i, y_i)$ planes are straight lines; in the $(x_1, \dots x_n, y_1, \dots, y_n)$ space, however, the shape of the route depends on the relative values of the $k_i$, which determine the diffuseness of the concentration variations separately for each species. In Fig. 5.24, for example, the diffuseness is greater for species 1 than for species 2, as will be the case if $k_1 < k_2$. In principle, the sorption-desorption fluxes may be coupled in such a manner that the constant-pattern route of a boundary involves concentration maxima or minima of one or more species.

A discussion of kinetic models of sorption, ion exchange, etc., from which the rate coefficients $k_i$ are to be obtained, is beyond the scope of this book (see reviews [45–49, 54, 59–62]). It may suffice to comment that all applications to chromatography of interfering species so far have employed drastic simplifications [24, 48, 58, 59, 63], some of which are likely to be more serious in multicomponent than in single-component systems. This is particularly true for the so-called "linear driving force" approximations, which assume the rate coefficients in Eqs. (5.76) or similar rate laws to be constant; in multicomponent systems, coupling between the sorption-desorption fluxes may even be such that some species overshoot their equilibria [61, 63], so that their coefficients $k_i$ temporarily assume negative values. Also, if diffusion within the stationary

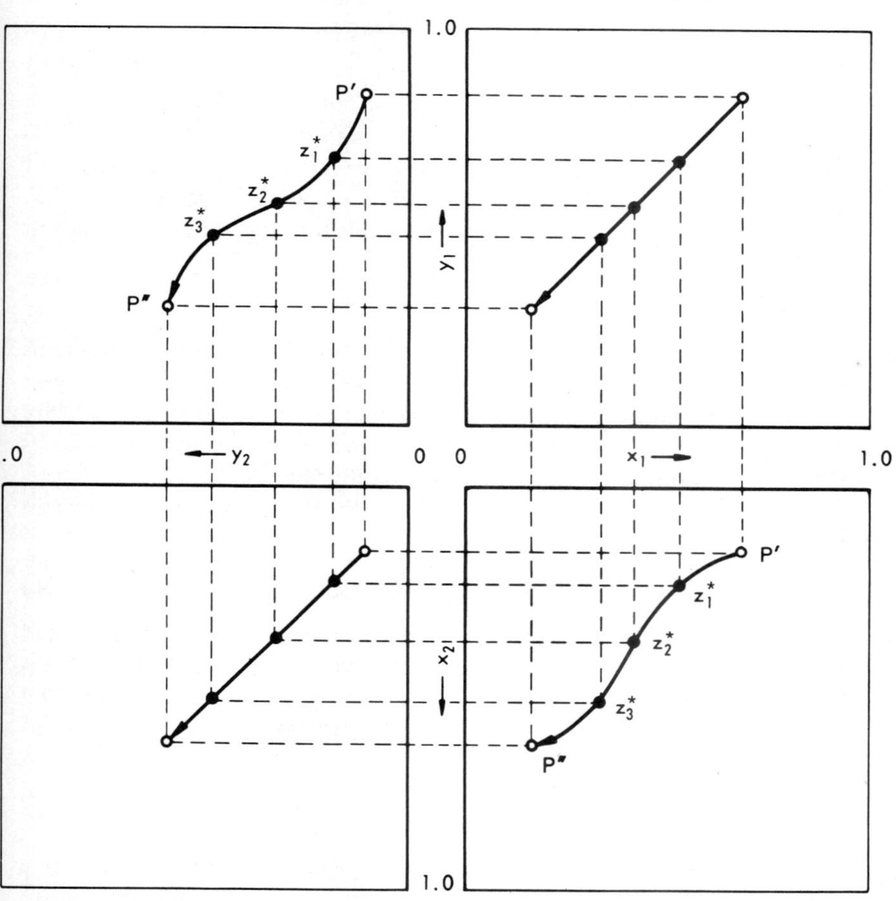


**Fig. 5.24.** Typical profile route (operating line) for constant pattern of self-sharpening boundary between plateau zones P′ and P″ in three-component system. Projections onto $(x_1, y_1)$ and $(x_2, y_2)$ planes are linear, projections onto $(x_1, x_2)$ and $(y_1, y_2)$ planes are not. Projections of compositions at three arbitrary distances $z_1^*$, $z_2^*$, $z_3^*$ are marked for orientation. (Schematic.)

phase, as an unsteady-state process, is rate-controlling, the rate cannot even be expressed as a function

$$(\partial y_i/\partial \tau)_z = f(\mathbf{x}, \mathbf{y}) \tag{5.85}$$

which does not account for the shapes of the concentration profiles in the stationary phase. These have so far been considered only in the more elaborate of the theories of linear chromatography [50, 52, 53]. It is fortunate that sorption-desorption kinetics affects only the extent of disturbances without altering the principal features of the response, so that relatively crude approximations are acceptable for most practical purposes.

## C. AXIAL DIFFUSION

Although molecular diffusion in axial direction does not cause deviations from local equilibrium, its dispersive effect is similar. The effect is strongest in sharp boundaries. However, the dependence on the boundary velocity is the opposite: Because of their longer residence times, slower boundaries are more susceptible to dispersion by axial diffusion.

If axial diffusion is significant, Eq. (3.15) for the flux of species $i$ in the direction of flow must be modified:

$$J_i = u_0 C_i - D_i(\partial C_i/\partial z)_t \tag{5.86}$$

where $D_i$ is the effective axial diffusion coefficient of species $i$ (corrected for tortuosity and partial obstruction of cross-sectional area, and potentially a function of the mobile-phase composition). The differential material balance then becomes, instead of Eq. (3.16),

$$\left(\frac{\partial(\bar{C}_i + C_i)}{\partial t}\right)_z + u_0\left(\frac{\partial C_i}{\partial z}\right)_t - \left\{\frac{\partial}{\partial z}\left[D_i\left(\frac{\partial C_i}{\partial z}\right)_t\right]\right\}_t = 0 \tag{5.87}$$

Axial diffusion contributes the third term, which is zero if the effective diffusion coefficient is nil. Because the term is second order, its relative importance increases as the boundary sharpens and decreases as the boundary spreads. (Transformation to adjusted time and normalized concentrations is not profitable at this stage because of the axial-diffusion term.)

Quantitative calculations of the response require integration over $z$ and $t$ of the simultaneous equations (5.87) for all species. Under conditions of local equilibrium, $\bar{C}_i$ can be eliminated with the equilibrium relations. Under nonequilibrium conditions, Eqs. (5.87) and (5.76) must be integrated simultaneously.

Again, the calculation of the constant pattern is much simpler. In terms of $\lambda_i$ and $\omega_i$ as defined by Eqs. (5.82), one obtains under the constant-pattern conditions (5.78)†

$$\frac{d\lambda_i}{dz^*} = \frac{u_0 - u}{D_i}(\lambda_i - \omega_i) \tag{5.88}$$

(for derivation, see below). The constant-pattern profile can be calculated by integration of the simultaneous equations (5.88) for all species. Under conditions of local equilibrium, $\omega_i$ can be eliminated with Eq. (5.82) and the equilibrium relations; under nonequilibrium conditions, Eqs. (5.88) and (5.83) for all species must be integrated simultaneously.

Equation (5.88) can be derived as follows: With Eqs. (5.78), (5.79), and (5.82), written in terms of $u$, $t$, $C_i$, and $\bar{C}_i$, Eq. (5.87) reduces to

$$(u_0 - u)\frac{d\lambda_i}{dz^*} - u\frac{\Delta\bar{C}_i}{\Delta C_i}\frac{d\omega_i}{dz^*} + \frac{d}{dz^*}\left(D_i\frac{d\lambda_i}{dz^*}\right) = 0 \tag{5.89}$$

The boundary values for integration over $z^*$, ideally at infinite distances upstream and downstream of the boundary center, are

$$\begin{aligned} z^* &\to -\infty, \quad \lambda_i = 0, \quad \omega_i = 0, \quad d\lambda_i/dz^* = 0 \\ z^* &\to +\infty, \quad \lambda_i = 1, \quad \omega_i = 1, \quad d\lambda_i/dz^* = 0 \end{aligned} \tag{5.90}$$

as follows from Eqs. (5.82) and the fact that the concentration gradients become negligible at infinite distances upstream and downstream. With the boundary conditions (5.90), integration of Eq. (5.89) over the entire boundary gives

$$\Delta\bar{C}_i/\Delta C_i = (u_0 - u)/u \tag{5.91}$$

With this substitution, integration of Eq. (5.89) over $z^*$ to an arbitrary point yields Eq. (5.88).

Axial diffusion, like nonequilibrium, neither impairs the validity of the integral coherence condition (3.40) nor alters the boundary velocity in the constant pattern, as Eq. (5.91) and its comparison with Eq. (3.19) show. This is true even if both disturbances are significant, since the premise of local equilibrium was not invoked in the derivation of Eq. (5.91). The relation between the concentration variations and concentration velocity as well as the differential coherence condition derived from it,

---

† An equivalent relation was first derived for multicomponent systems by Cooney and Lightfoot [58].

however, must be modified if axial diffusion is significant. Using Eq. (5.87) instead of (3.16), one obtains for the mobile-phase concentration velocity:

$$u_{C_i} = \frac{u_0}{1+(\partial\bar{C}_i/\partial C_i)_z - G_i} \tag{5.92}$$

where

$$G_i \equiv \left\{\frac{\partial}{\partial z}\left[D_i\left(\frac{\partial C_i}{\partial z}\right)_t\right]\right\}_t \Bigg/ \left(\frac{\partial C_i}{\partial t}\right)_z \tag{5.93}$$

The differential coherence condition, derived from Eqs. (3.37) and (5.92), then becomes

$$(\partial\bar{C}_i/\partial C_i)_z - G_i = (\partial\bar{C}_j/\partial C_j)_z - G_j \qquad \text{for all } i \text{ and } j \tag{5.94}$$

and is not normally obeyed except in the constant pattern or when the second-order terms $G_i$ and $G_j$ are negligible.

## D. Hydrodynamic Effects: Eddy Dispersion, Channeling, and Fingering

Perhaps the most unrealistic of the initial premises is that of ideal plug flow. In a bed of discrete particles of finite size, larger and smaller interstices alternate, usually in a random fashion, the stream paths are tortuous and of different lengths, involving frequent branching and rejoining, and the flow velocity varies, depending on the distance from the nearest particle surface and on the width and shape of the channel. Further complicating factors are exchange of molecules between stream paths through lateral diffusion, and possible distortions of the flow-velocity profile through gross packing irregularities or viscosity and density variations. Also, the situation is aggravated if the cross section of the column is nonuniform or its area is not large compared to the particle size.

Except for lateral diffusion between stream paths, which contributes only a correction to a correction term, all these effects act on the volume elements of the mobile phase without discriminating between the species contained. Therefore, the effects do not block or alter the molecular events leading to the establishment of chromatographic patterns, and, conversely, the hydrodynamic description is essentially independent of these events. The established theories of column hydrodynamics can therefore be directly applied to the systems discussed here.

### 1. *Eddy Dispersion*

The effect of branching and rejoining of stream paths and of intra-channel variations of the flow velocity, present even in completely regular packings, is usually called *eddy dispersion*. Its dispersive action adds to that of axial diffusion and can be accounted for in an approximate manner by the use of a correspondingly larger value of the effective axial diffusion coefficient $D_i$ in the material balance (5.87). In elementary theories [38, 41, 43, 54] the contribution of eddy diffusion to the plate height is assumed to be a constant determined solely by the particle size and packing geometry, and the contribution to $D_i$ then is proportional to the flow rate. A more detailed examination, however, reveals the dispersive contribution to be smaller at low flow rates, owing to the counteracting effect of lateral diffusion. A comprehensive treatment has been given by Giddings [51].

### 2. *Channeling*

In beds packed in a random fashion, "channels" leading through relatively loosely packed regions may exist or develop. Because of the lesser flow resistance, the flow rate through such channels is higher. Even in regular packings, such an effect appears near the wall, where the packing density differs from that in the interior of the bed. The dispersive effect of the so-called *channeling* is similar to that of eddy dispersion, except that the characteristic lateral distances of its flow-velocity variations are greater. A detailed discussion has been given by Giddings [64] (see also references in reviews [46, 47]).

### 3. *Fingering*

Less well known is a potential effect produced by hydrodynamic instability resulting from viscosity or density differences in the mobile phase. Without losing its local sharpness, a boundary between two zones may become distorted in a peculiar manner: Long and fairly uniform-sized "fingers" of the upstream zone may intrude deeply into the downstream zone.

It is obvious that a boundary between two liquids tends to be unstable if the upper liquid has the higher density. This gravity effect may be enhanced, partially offset, or outweighed by a viscosity effect; displacement by a more viscous liquid favors stability of the boundary, and displacement by a less viscous one tends to produce instability. Where an accidental disturbance has caused the boundary to bulge out in the direction of flow, the flow resistance is less and the bulge thus grows if the displacing liquid has the lower viscosity, and the opposite is true if the displacing liquid has the higher viscosity. A criterion for hydrodynamic stability can be calculated from the density and viscosity differences. The existence of the effect, better known in immiscible displacement in porous media [65, 66], has been confirmed in chromatographic operations [46, 47, 67, 68], and an adaptation of the mathematics of immiscible displacement to chromatographic boundaries has been given by Cooney [69].

The hydrodynamic viscosity-gravity effect can help to keep a boundary sharper: Although it is not by itself a sharpening force, it can counteract the dispersive effect of channeling.

## E. Volume Changes of Sorbent Particles

In operations with gels as the stationary phase, particularly with ion-exchange resins, changes of the mobile-phase composition may be accompanied by significant swelling or shrinking of the sorbent particles. As a result, the bed may expand or contract. The column performance may be affected in several ways:

1. The total concentration (ion-exchange capacity) of the stationary phase varies. This may also affect the equilibrium parameters.
2. The variation of the lengths and widths of the intraparticle diffusion paths may affect the mass-transfer coefficients.
3. The effective bed length and, for any given layer of particles, the distance from the inlet varies.
4. Swelling of the sorbent tends to reduce the fractional void volume and may clog the bed; shrinking tends to increase the fractional void volume and to produce channeling.

The effects, usually disregarded in quantitative theories of column performance, are not directly related to the number of species and to interference and can only be mentioned in passing.

## F. Nonisothermal Effects

In operations with solutes at high concentrations, heat-of-sorption effects may give rise to appreciable deviations from isothermal behavior. To include such effects in the theoretical treatment is a major enterprise, and hardly more than a promising start has so far been made.

Isothermal behavior is not specifically stipulated in the initial premises, but the assumed invariance of the equilibrium parameters usually precludes significant temperature variations. The effect of temperature variations on the equilibrium properties constitutes the major complication; the effect on kinetic properties such as sorption-desorption rates may also be appreciable, but is of lesser concern because it merely affects the extent of disturbances.

A reasonably complete description of nonisothermal operation would have to account for heat transfer to or from the environment. The temperature then varies not only with the distance from the inlet, but also with the lateral distance from the column axis. As a significant first step, however, the much simpler case of adiabatic operation is of interest. Also, the effects of heat of mixing in the mobile phase and of frictional heat generation are usually disregarded and will not be considered here.

For adiabatic operation and with the assumptions of thermal equilibrium between the mobile and the stationary phase and of axial heat transfer exclusively by mobile-phase flow, one can write the following energy balance:

$$\left(\frac{\partial(\bar{\kappa}T)}{\partial\tau}\right)_z + \left(\frac{\partial(\kappa T)}{\partial z}\right)_\tau - \sum_i \left[(\Delta H)_i \left(\frac{\partial \bar{C}_i}{\partial \tau}\right)_z\right] = 0 \tag{5.95}$$

where $\kappa$ and $\bar{\kappa}$ are the specific heat capacities of the mobile and the stationary phase (in calories per degree and unit volume of the column), and $-(\Delta H)_i$ is the molar heat of sorption of species $i$. (A more elaborate energy balance accounting for intraparticle temperature gradients and heat transfer by axial diffusion has been given by Smith [62].) The derivation of Eq. (5.95) is analogous to that of the material balance (3.32), except that the heats of sorption contribute a source-and-sink term. Through this term and the temperature dependence of the equilibrium parameters, Eq. (5.95) is coupled with the material balances.

The energy balance is formally quite similar to the material balances. It is therefore not surprising that the temperature, in many respects, acts like an additional species. For example, the response to an abrupt influent composition change under otherwise identical conditions contains one boundary more in adiabatic than in isothermal operation. In adiabatic systems, coherent boundaries in general involve variations of temperature and all concentrations. The temperature velocity

$$u_T \equiv (\partial z/\partial t)_T \tag{5.96}$$

must be included with the species velocities in the coherence criterion (3.37). However, the analogy between temperature (or enthalpy) and species is not complete, because the equations are not symmetrical with respect to their coupling.

Nonisothermal behavior may have dispersive or sharpening effects on concentration variations. In conventional chromatography, for example, it usually results in sharper front and more diffuse rear boundaries of the peaks [70]; because of the heat of sorption, the temperature at the peak maximum tends to be higher [70, 71], giving a lower distribution ratio and thus a higher species velocity at the higher concentrations.

## G. Other Theories

Relatively little work has so far been published on chromatographic theory of systems involving both interference and disturbances, and most of it has been confined to systems with no more than three components. In some of these studies [14, 72, 73], the HETP concept of classical chromatographic theory has been used, which models the combined effect of all disturbances as back-mixing on fictitious "plates," on which the mobile phase undergoes successive equilibrations. Although physically unrealistic, this model [35, 38] has proved very useful in conventional chromatography. Its adequacy for multicomponent systems, however, is questionable because of its inability to discriminate between species with different equilibrium and kinetic properties. For example, the model cannot reproduce the differences in spread of the profiles of species with different mass-transfer coefficients in the constant pattern of a self-sharpening boundary. Only its convenience in numerical calculations can at all justify this approach.

A different approach has been taken by Dranoff and Lapidus [63], who carried out numerical calculations for ternary ion exchange under nonequilibrium conditions with an extension of the Thomas model [74], which postulates the validity of a rate law of a reversible second-order reaction. This model, too, is physically unrealistic for ion exchange and most sorption processes, whose rates usually are controlled by diffusion rather than chemical reaction, but it allows at least distinctions to be made between species of different kinetic properties. In contrast to the HETP approach, for example, the model can adequately represent systems in which some species may overshoot their distribution equilibria [63] (see also Section V.B). However, the main advantage of the Thomas model, namely, its ability to provide algebraic solutions, is lost with the extension to three or more exchanging species. The cases calculated by Dranoff and Lapidus [63] are of the type $1, 2 \rightarrow 3$. Apparently unaware of the developing frontal-analysis pattern, the authors terminated their calculations short of resolution and only part-way through the profiles and histories.

Mass-transfer models for operation under nonequilibrium conditions have been employed by various authors. Rachinskii [24] has stated differential material-balance equations providing for nonequilibrium and axial-diffusion effects without specifying particular forms of the equilibrium and mass-transfer relations, and applied them to frontal analysis and displacement development (see, however, the critique in Chapter 4, Section I.H). Cooney and Lightfoot [58] derived constant-pattern material balances equivalent to Eqs. (5.81) and (5.88) and evaluated them numerically for an ion-exclusion case (two-component nonstoichiometric sorption with variable separation factors), assuming constant mass-transfer coefficients in the rate law (5.83). Clazie *et al.* [59] omitted axial diffusion, but correlated the mass-transfer coefficients with diffusion data; although the Whitman two-film concept [75–78] used in this work is unrealistic for the stationary phase, the correlations are an improvement over the assumptions of constant reaction-rate or mass-transfer coefficients. The evaluation was, again, restricted to constant patterns, here in ternary ion-exchange frontal analysis.

Cooney and Lightfoot [58, 79] have introduced the concept of "pseudobinary isotherms" for the constant pattern. These are functions $y_i = f(x_i)$ along the composition route and thus are neither equilibrium isotherms nor refer to two species, as the name may be interpreted to imply. The authors argue that the asymptotic constant-pattern solutions $\{x_i(z^*), y_i(z^*)\}$ in multicomponent systems must have the same form as in single-component sorption, because the validity of the solutions should not depend on whether $y_i = f(x_i)$ is a "pseudobinary" or "true binary isotherm". With nonequilibrium as the only disturbance, the pseudobinary isotherms are $\lambda_i = \omega_i$ [see Eq. (5.81)] and indeed facilitate the calculation. In general, however, little appears to be accomplished with the argument, inasmuch as the constant pattern is presupposed and the functions $f(x_i)$ are not known beforehand.

Evaluations of mass-transfer models have so far remained confined to constant patterns. It seems as though the preoccupation with coherence and the convenience afforded by transformations developed for, and implying, coherent behavior has delayed the treatment of other cases, which generally are noncoherent.

No work on hydrodynamic disturbances particularly oriented toward multicomponent systems appears to have been carried out, but this is not a serious omission since the effects are known to be essentially the same as in conventional systems. For adiabatic operation in multicomponent systems, a theoretical basis has recently been laid by Heerdt [80] with a treatment that assumes local thermal equilibrium and absence of disturbances and takes entirely coherent behavior for granted. In this area more than in any other can interesting new developments still be expected.

## VI. Gas-Chromatographic Effects

The theoretical principles presented in the earlier sections were derived without specifying the nature of the mobile phase and therefore are as applicable to gas as to liquid chromatography. However, the properties of gases, particularly their compressibility, lead to various effects not normally observed with liquids and make some of the simplifying premises much less acceptable for gas than for liquid chromatography. Most of all, the premises of constant total concentration and flow rate of the mobile phase must be relaxed in any realistic treatment of gas chromatography. The effects are covered by the more recent theories of gas chromatography and need be referred to only briefly.

Pressure-drop and flow-rate effects in gas chromatography are closely interrelated, but two independent causes can be identified. First, the pressure drop through the column, needed to produce gas flow, lets the total concentration and the flow rate of the gas vary with the distance from the inlet. This effect is not contingent on composition variations or sorption effects and becomes significant if the pressure difference between inlet and outlet is not small compared with the absolute pressures. Second, nonuniform gas-phase compositions produce nonuniformities of the flow rate. This effect becomes significant only if the composition variations are large, but does not require a sizable pressure drop. The two effects may be superimposed, but will be viewed separately. In conventional analytical gas chromatography, the pressure-drop effect is usually appreciable but the composition variations are too small to produce flow-rate variations. In preparative applications, on the other hand, the opposite may be true.

It will be assumed in this section that the mobile phase obeys the ideal gas laws. The inlet and exit pressures will be taken as fixed and a steady-state (or at least quasi-stationary) pressure profile will be assumed. Complications arising from nonuniform temperature and viscosity will be discussed separately. Furthermore, in contrast to the practice in the earlier sections, any nonsorbable components (e.g., carrier gas) will be counted as species and be included in summations; the total concentration $C \equiv \Sigma_i C_i$ of the gas phase thus is proportional to the local total pressure, and the $x_i \equiv C_i/C$ can be identified with the mole fractions in the gas.

## A. Pressure Drop

With the pressure decreasing along their path, the volume elements of the gas phase expand while traveling. The total concentration (in moles per unit volume) then decreases in the direction of flow, and continuity requires the flow rate to increase in this direction. The corresponding corrections in the equations for the retention volume etc. were first derived by James and Martin [81] and can be found in most texts on gas chromatography (e.g., [42, 82, 83]).

In the absence of composition variations large enough to affect the flow rate, the pressure gradient $dp/dz$ and the flow rate $u_0$ are related by

$$u_0 = -(K/\eta)\, dp/dz \tag{5.97}$$

where $\eta$ is the gas viscosity and $K$ is the permeability coefficient of the bed. Since the ideal gas laws are assumed to be valid, $u_0$ can be eliminated with

$$pu_0 = p^0 u_0{}^0 \tag{5.98}$$

where the superscripts zero refer to the inlet. From Eqs. (5.97) and (5.98) one obtains after integration over $z$ and rearrangement

$$\frac{z}{Z} = \frac{p^{0\,2} - p^2}{p^{0\,2} - p^{*2}} \quad \text{and} \quad \frac{z}{Z} = \frac{(p^0/p^*)^2 - (u_0{}^*/u_0)^2}{(p^0/p^*)^2 - 1} \tag{5.99}$$

where $Z$ is the bed length, and $p^*$ and $u_0{}^*$ are the pressure and flow rate at the exit. Equations (5.99) are readily solved to give the pressure and flow rate as functions of the distance from the inlet.

In the present treatment, the pressure-drop effect is readily accounted for if the equilibrium relations are independent of total pressure. The basic features of the response behavior then are not affected: The response $\mathbf{x}(z, \tau)$ can be calculated as in the earlier sections, and the pressure and flow-rate variations merely affect the conversion to the form $\mathbf{C}(z, t)$ through the dependence of $C$ on $p$ and of $\tau$ on $C$ and $u_0$. Of course, plateau zones must now be more narrowly defined as zones of constant $\mathbf{x}$, allowing the $C_i$ to vary with total pressure.

On the other hand, if the equilibrium relations depend on total pressure, then the composition paths can no longer be uniquely defined in terms of $\mathbf{x}$ (or $\mathbf{y}$) alone. Also, plateau zones of constant $\mathbf{x}$ then are noncoherent in that they will give rise to variations of $\mathbf{x}$ as they travel. No attempts have so far been made to resolve these difficulties, which appear to make some of the basic concepts of the present treatment inapplicable.

## B. Variation of Flow Rate with Composition

Even in a column with negligible pressure drop, the flow rate is nonuniform if there are significant composition variations involving net sorption or desorption. This effect was first reported by Bosanquet [84, 85] and has since been studied in various contexts [86–91] including multicomponent systems [89].

The physical causes of the effect are readily understood and are illustrated by a simple example in Fig. 5.25. The figure shows a pulse of a sorbable solute being eluted by pure carrier gas. With the assumptions of negligible pressure drop and gas-phase ideality, the number of moles per unit volume in the gas phase does not vary along the column, and continuity then requires the gas velocity to be higher in the region of the pulse, where a fraction of the gas volume is occupied by solute molecules. In other words, where the carrier gas overtakes the more slowly moving pulse, its velocity (and thus the velocity of the gas phase) must be higher, in much the same way as water flowing in a pipe must have a higher velocity where the pipe is constricted.

The foregoing argument is based on species velocities, which, as defined in Chapter 3, Section III.A, are different for the carrier gas and the solute. Of course, the linear velocities of the carrier-gas and solute molecules *in the gas phase* at any given place and time are the same. If one argues in terms of these velocities, one must consider the sources and sinks of the system: Net desorption at the rear and net sorption at the front of the pulse supply and withdraw solute molecules to and from the gas phase, and the velocity must then be higher in the region of the pulse because, here, the volumetric flow at any time must transfer not only the carrier gas supplied at the inlet, but also the solute vapor produced by desorption at the rear and consumed by sorption at the front of the pulse (see Fig. 5.25).

The argument is readily extended to more complex situations: At any given time, the flow rate increases in the direction of flow where net desorption occurs, and decreases where net sorption occurs. This effect is superimposed on the pressure-drop effect discussed previously. In liquid chromatography, the effect is usually negligible because, here, the net volume change of the combined mobile and stationary phases upon sorption or desorption tends to be quite small, both phases being con-

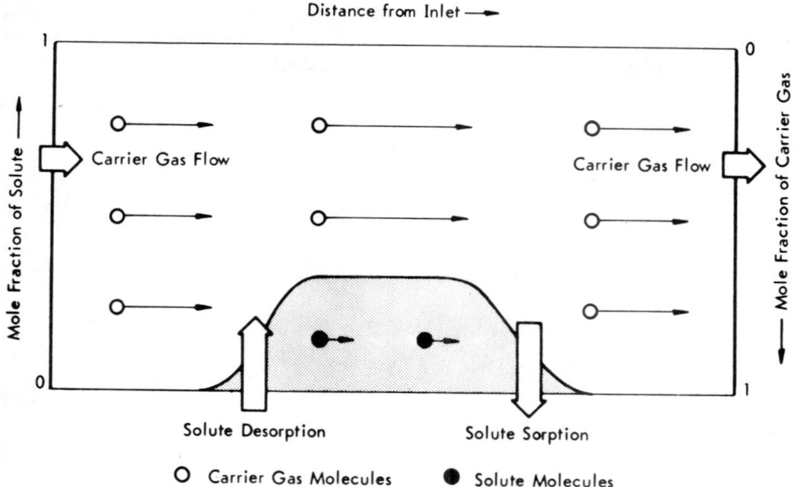


**Fig. 5.25.** Variation of flow rate with extent of sorption, illustrated by elution of pulse of sorbable species with carrier gas. Broad arrows indicate flows entering and leaving gas space in column; lengths of thin arrows indicate magnitudes of species velocities. (Schematic.)

densed to comparable degrees.

In the mathematical treatment, the following modifications of the basic equations are required in order to account for variations of the flow rate with composition at constant pressure. In Eq. (3.15), the mobile-phase flow rate $u_0$ must be considered as variable. The material balance (3.16) then becomes

$$\left(\frac{\partial(\bar{C}_i + C_i)}{\partial t}\right)_z + u_0\left(\frac{\partial C_i}{\partial z}\right)_t + C_i\left(\frac{\partial u_0}{\partial z}\right)_t = 0 \tag{5.100}$$

With the assumptions of negligible pressure drop and validity of the ideal gas laws, one has

$$C \equiv \sum_i C_i = \text{const} \tag{5.101}$$

Summation of Eq. (5.100) over all species then gives

$$(\partial u_0/\partial z)_t = -\frac{1}{C}(\partial \bar{C}/\partial t)_z \tag{5.102}$$

or, after integration over $z$,

$$u_0(z, t) = u_0^0(t) - \frac{1}{C}\int_0^z (\partial \bar{C}/\partial t)_z \, dz \tag{5.103}$$

at any given time $t$. Equations (5.102) and (5.103) reflect the previously discussed variation of the flow rate with net sorption or desorption. Furthermore, Eq. (3.13) for the concentration velocities must be revised. After replacing the last term in Eq. (5.100) by means of Eq. (5.102), one obtains with the general rule (3.17)*:

$$u_{C_i} = \frac{u_0}{1 + (\partial \bar{C}_i/\partial C_i)_z - (C_i/C)(\partial \bar{C}/\partial C_i)_z} \tag{5.104}$$

Accordingly, the differential coherence condition (3.38) must be replaced by

$$\left(\frac{\partial \bar{C}_i}{\partial C_i}\right)_z - \frac{C_i}{C}\left(\frac{\partial \bar{C}}{\partial C_i}\right)_z = \left(\frac{\partial \bar{C}_j}{\partial C_j}\right)_z - \frac{C_j}{C}\left(\frac{\partial \bar{C}}{\partial C_j}\right)_z \qquad \text{for all } i \text{ and } j \tag{5.105}$$

Aside from its effect on the coherence condition and thus on the composition paths, the nonuniformity of the flow rate affects the sharpening characteristics [84, 85, 89]. The higher flow rate at compositions with higher total concentration in the stationary phase gives rise to a sharpening tendency of the front, and spreading tendency of the rear boundaries of zones or pulses with such compositions. In systems with interference, these tendencies are usually in the same direction as the normal sharpening and spreading tendencies existing at uniform flow rate (see Chapter 3, Section IV.E). In linear chromatography, however, the other tendencies are absent and the flow-rate variation, if significant, produces peak asymmetry.

Even in systems without interference—that is, with independent sorption isotherms of the various species—the constraint of a total gas-phase concentration $C$ depending only on total pressure results in a coupling of the species. This case, with constant distribution ratios $\bar{C}_i/C_i$ of the species, is covered by Zhukhovitskii's theories [25, 32, 92] discussed in a different context (see Chapter 4, Section III.F.4).

## C. Viscosity and Temperature Effects

The effects of pressure drop and variation of flow rate with composition may further be complicated by viscosity and temperature variations. Such variations may, of course, occur in liquid chromatography, too, but

* An equivalent relation for coherent composition variations was first derived by Peterson and Helfferich [89].

they are more serious in gas chromatography, where the viscosity directly affects the flow rate, the heat capacity and thermal conductivity of the mobile phase are much smaller, and the total gas-phase concentration depends on the temperature.

The effect of nonuniform viscosity in conventional chromatography has been examined by Haarhoff and van der Linde [93] and Dyson and Littlewood [94]. Here, usually, the viscosity in a peak is higher than in regions of pure carrier gas, and the pressure drop then is larger across the peaks [see Eq. (5.97)]. Higher pressure within peaks has also been postulated [95], but this hypothesis is not compatible with Eq. (5.97) and has been abandoned [96]. Temperature effects, discussed earlier (see Section V.F), are of particular interest in gas chromatography in view of the wide-spread use of temperature programming (for a discussion of theoretical aspects, see [97]).

No attempts to incorporate viscosity and temperature effects into a quantitative theory of gas chromatography with interference have so far been made.

## VII. Review and Perspective

The discussions in this chapter have touched on a great variety of facets. As far as quantitative theory is concerned, complete and satisfactory extensions have been achieved in only a few instances, while in most others the treatment has not gone beyond formulating the problems and pointing out directions in which improvements are still needed. These shortcomings are not surprising, since even the single-component theories of chromatography to date are less than satisfactory in their quantitative treatment of many of the aspects discussed here. The chief purpose of the exposition, however, has been to show that the basic concepts essentially survive the relaxation of the simplifying premises under which they were first introduced and illustrated. Only this result can and does vindicate the initial use of these premises, which otherwise would have to be judged unacceptably confining.

The topics discussed can be grouped into four categories, relating to (1) operating conditions and column properties, (2) equilibrium properties, (3) dynamic behavior, and (4) nature of the mobile phase. There are wide differences between the categories in the impact of the relaxation of premises on the theoretical treatment. Variable operating conditions and nonuniform column properties (category 1) are relatively easily incorporated into the quantitative treatment by the use of generalized independent variables, and do not produce new basic features. Relaxation of the equilibrium premises of stoichiometric exchange and constant separation factors (category 2) significantly alters the results, although not the substance, of the treatment and calls for modifications of many rules derived in earlier sections, such as the affinity-cut rule. This reflects the fact that, as in other fields of chromatography, the equilibrium properties are the most important single factor determining the response behavior. Disturbances through nonideal dynamic behavior (category 3), while causing serious inconveniences in the quantitative treatment, have relatively little effect on the response behavior. In general, the individual boundaries become more diffuse, but their sequence, nature, and velocities and the plateau-zone compositions remain essentially unaffected. That a simple equilibrium theory correctly gives all these principal features of often quite complex response patterns establishes its usefulness and once again shows that the response behavior is chiefly dictated by equilibrium properties. The brief survey of mobile-phase effects (category 4) allows no definite conclusions, but draws attention to the gap that still exists between the theories of gas chromatography at high concentrations and those of multicomponent chromatography assuming uniform mass flow.

As already pointed out, the principal concepts of the present approach remain essentially unaffected by the relaxation of premises. Coherence is not normally impaired, but is approached asymptotically rather than attained in a finite time if disturbances are significant. Only under exceptional circumstances—e.g., with compositions in the pathless region—does the principle of approach to coherence become inapplicable, and the response may then be expected to exhibit unusual features not covered by any existing theory. Even in such cases, however, the response can be deduced step by step in terms of species velocities and concentration velocities. Another generally applicable principle, newly introduced for the extensions in this chapter, is that of the equivalent stoichiometric system, which establishes the formal mathematical equivalence between $n$-component systems with nonstoichiometric sorption and $(n+1)$-component systems with stoichiometric exchange. In particular, systems with compound Langmuir isotherms are mathematically equivalent to the previously treated systems with stoichiometric exchange and constant separation factors.

Some of the tools extensively used in the previous chapters do not fare so well. One of the main and early victims is the $h$ transformation, which is still applicable to Langmuir systems but becomes practically useless in the more general case of equilibria with variable separation factors. At least for systems with more than three components, this is a complication in principle rather than merely a matter of mathematical technique: Unless the composition paths are arranged in common surfaces and hypersurfaces (analogous to the common planes and hyperplanes, see Chapter 3, Section IV.C.3), an orthogonalization of the path grid is impossible in principle, and no set of composition variables having the conservation and coherence properties of the $H$-function roots then exists. Thus, while generalized composition variables $h_i$ could be defined for systems with arbitrarily complex equilibria as the "natural coordinates"—i.e., as suitably normalized distances along the composition paths—their usefulness would not approach that of the $H$-function roots under the initial premises and would probably not be commensurate to the effort of calculating them.

In conclusion, it may be said that a general theory with satisfactory coverage of all facets and eventualities does not exist at this time, but that the approach of generalizing step by step a simple theory containing the essential elements of interference, as outlined here, appears feasible.

## REFERENCES


1. F. G. Helfferich, *Theory of Chromatography*. 1. *Basis of a Generalized Theory for Multicomponent Systems with Interdependent Nonlinear Isotherms*, Shell Development Company, Technical Report 315–64, Emeryville, California, 1964 (limited distribution, out of print).
2. H. G. Cassidy, *Fundamentals of Chromatography* (Techniques of Organic Chemistry, Vol. X, A. Weissberger, ed.), Interscience, New York, 1957, Chap. IV.
3. R. S. Alm, R. J. P. Williams, and A. Tiselius, *Acta Chem. Scand.*, **6**, 826 (1952).
4. E. C. Freiling, *J. Am. Chem. Soc.*, **77**, 2067 (1955).
5. I. Langmuir, *J. Am. Chem. Soc.*, **38**, 2221 (1916); **40**, 1361 (1918).
6. G. M. Schwab, in *Ergebnisse der exakten Naturwissenschaften*, Vol. 7, Julius Springer, Berlin, 1928, p. 276.
7. J. W. McBain, *The Sorption of Gases and Vapours by Solids*, George Routledge, London, 1932, p. 467.
8. H. Dunken, *Z. Physik. Chem.*, **A 187**, 105 (1940).
9. D. DeVault, *J. Am. Chem. Soc.*, **65**, 532 (1943).
10. E. Glückauf, *Nature*, **156**, 205 (1945).
11. E. Glückauf, *Proc. Roy. Soc. (London)*, **A 186**, 35 (1946).
12. J. I. Coates and E. Glueckauf, *J. Chem. Soc.*, **1947**, 1308.
13. E. Glueckauf, *Discussions Faraday Soc.*, 7, 12 (1949).
14. P. J. Pandya, G. Klein, and T. Vermeulen, Sea Water Conversion Laboratory, University of California, Report 65–2, 1965.
15. G. Klein, D. Tondeur, and T. Vermeulen, Sea Water Conversion Laboratory, University of California, Report 65–3, 1965.
16. G. Klein, D. Tondeur, and T. Vermeulen, *Ind. Eng. Chem. Fundamentals*, **6**, 339 (1967).

17. D. Tondeur, *Chim. Ind., Génie Chim.*, **100**, 1058 (1968).
18. D. Tondeur, personal communications, 1968; thesis, Nancy (France), 1969.
19. E. Glueckauf, *Discussions Faraday Soc.*, **7**, 50, 53 (1949).
20. A. C. Offord and J. Weiss, *Nature*, **156**, 570 (1945); *Discussions Faraday Soc.*, **7**, 51 (1949).
21. G. G. Baylé and A. Klinkenberg, *Rec. Trav. Chim.*, **73**, 1037 (1954).
22. P. C. Mangelsdorf, Jr., *Anal Chem.*, **38**, 1540 (1966).
23. F. I. Stalkup and H. A. Deans, *A. I. Ch. E. J.*, **9**, 106 (1963).
24. V. V. Rachinskii, *The General Theory of Sorption Dynamics and Chromatography*, English translation, Consultants Bureau, New York, 1965, Chaps. 4 and 6.
25. A. A. Zhukhovitskii, M. L. Sazonov, A. F. Shlyakhov, and V. P. Shvartsman, *Zh. Fiz. Khim.*, **41**, 2640 (1967).
26. N. R. Amundson, *Mathematical Methods in Chemical Engineering*, Prentice-Hall, Englewood Cliffs, 1966, Chap. 5.
27. R. R. Meyer, personal communication, 1968.
28. A. C. Offord and J. Weiss, *Nature*, **155**, 725 (1945); *Discussions Faraday Soc.*, **7**, 26 (1949).
29. E. Glückauf, *Nature*, **156**, 748 (1945).
30. E. Glueckauf, *J. Chem. Soc.*, **1947**, 1321.
31. L. Hagdahl, R. J. P. Williams, and A. Tiselius, *Arkiv Kemi*, **4**, 193 (1952).
32. A. A. Zhukhovitskii, M. L. Sazonov, A. F. Shlyakhov, and A. I. Karymova, *Zavodsk. Lab.*, **31**, 1048 (1965).
33. C. G. Collins and H. A. Deans, *A. I. Ch. E. J.*, **14**, 25 (1968).
34. C. A. Barrere and H. A. Deans, *A. I. Ch. E. J.*, **14**, 280 (1968).
35. A. J. P. Martin and R. L. M. Synge, *Biochem. J.*, **35**, 1358 (1941).
36. S. W. Mayer and E. R. Tompkins, *J. Am. Chem. Soc.*, **69**, 2866 (1947).
37. E. Glueckauf, in *Ion Exchange and its Applications*, Society of Chemical Industry, London, 1955, p. 34.
38. J. J. van Deemter, F. J. Zuiderweg, and A. Klinkenberg, *Chem. Eng. Sci.*, **5**, 271 (1956).
39. A. I. M. Keulemans, *Gas Chromatography*, 2nd ed., Reinhold, New York, 1959, Chap. 4.
40. D. Ambrose and B. A. Ambrose, *Gas Chromatography*, George Newnes, London, 1961, pp. 7–10.
41. S. Dal Nogare and R. S. Juvet, Jr., *Gas-Liquid Chromatography*, J. Wiley, New York, 1962, Chap. 5.
42. A. B. Littlewood, *Gas Chromatography*, Academic Press, New York, 1962, Chap. 2.
43. H. Purnell, *Gas Chromatography*, J. Wiley, New York, 1962, Chap. 8.
44. R. A. Keller and J. C. Giddings, in *Chromatography* (E. Heftmann, ed.), 2nd ed., Reinhold, New York, 1967, Chap. 6.
45. T. Vermeulen, in *Advances in Chemical Engineering* (T. B. Drew and J. W. Hoopes, Jr., eds.), Vol. 2, Academic Press, New York, 1958, p. 147.
46. F. Helfferich, *Angew. Chem. Intern. Ed. Engl.*, **1**, 440 (1962); *Chem. Ing.-Tech.*, **34**, 269 (1962).
47. F. Helfferich, *Ion Exchange*, McGraw-Hill, New York, 1962, Chap. 9.
48. E. N. Lightfoot, R. J. Sanchez-Palma, and D. O. Edwards, in *New Chemical Engineering Separation Techniques* (H. M. Schoen, ed.), Wiley-Interscience, New York, 1962, Chap. 2.

49. N. K. Hiester, T. Vermeulen, and G. Klein, in *Chemical Engineers' Handbook* (R. H. Perry, C. H. Chilton, and S. D. Kirkpatrick, eds.), 4th ed., McGraw-Hill, New York, 1963, Sec. 16.
50. J. B. Rosen, *J. Chem. Phys.*, **20**, 387 (1952); *Ind. Eng. Chem.*, **46**, 1590 (1954).
51. J. C. Giddings, *Dynamics of Chromatography*, Part I, Marcel Dekker, New York, 1965.
52. O. Grubner, in *Advances in Chromatography* (J. C. Giddings and R. A. Keller, eds.), Vol. 6, Marcel Dekker, New York, 1968, p. 173.
53. P. Schneider and J. M. Smith, *A. I. Ch. E. J.*, **14**, 762, 886 (1968).
54. R. F. Baddour and J. R. Valbert, in *Modern Chemical Engineering* (A. Acrivos, ed.), Vol. 1, Reinhold, New York, 1963, p. 487.
55. T. Vermeulen and N. K. Hiester, *J. Chem. Phys.*, **22**, 96 (1954).
56. C. S. Robinson and E. R. Gilliland, *Elements of Fractional Distillation*, 4th ed., McGraw-Hill, New York, 1950, Chaps. 7 and 9.
57. J. A. Gerster, in *Chemical Engineers' Handbook* (R. H. Perry, C. H. Chilton, and S. D. Kirkpatrick, eds.), 4th ed., McGraw-Hill, New York, 1963, Sec. 13.
58. D. O. Cooney and E. N. Lightfoot, *Ind. Eng. Chem. Process Design Develop.* **5**, 25 (1966).
59. R. N. Clazie, G. Klein, and T. Vermeulen, Sea Water Conversion Laboratory, University of California, Report 67–4, 1967.
60. R. B. Bird, W. E. Stewart, and E. N. Lightfoot, *Transport Phenomena*, J. Wiley, New York, 1960, Chap. 2.
61. F. Helfferich, in *Ion Exchange* (J. A. Marinsky, ed.), Vol. 1, Marcel Dekker, New York, 1966, Chap. 2.
62. J. M. Smith, in *Adsorption From Aqueous Solution* (Advances in Chemistry Series No. 79, R. F. Gould, ed.), American Chemical Society, 1968, p. 8.
63. J. S. Dranoff and L. Lapidus, *Ind. Eng. Chem.*, **50**, 1648 (1958); **53**, 71 (1961).
64. Ref. 51, Chap. 5.
65. R. L. Chuoke, P. van Meurs, and C. van der Poel, *Trans. AIME*, **216**, 188 (1959).
66. A. E. Scheidegger, *The Physics of Flow Through Porous Media*, 2nd ed., University of Toronto Press, Toronto, 1960, p. 202.
67. S. Hill, *Chem. Eng. Sci.*, **1**, 247 (1952).
68. F. Helfferich, in *Advances in Chromatography* (J. C. Giddings and R. A. Keller, eds.), Vol. 1, Marcel Dekker, New York, 1965, Chap. 1, p. 46.
69. D. O. Cooney, *Ind. Eng. Chem. Fundamentals*, **5**, 426 (1966).
70. R. P. W. Scott, *Anal. Chem.*, **35**, 481 (1963).
71. J. Peters and C. B. Euston, *Anal. Chem.*, **37**, 657 (1965).
72. R. S. Henly, A. Rose, and R. F. Sweeny, *Anal. Chem.*, **36**, 744 (1964).
73. J. Shankar and T. R. Bhat, *J. Chromatog.*, **9**, 461 (1962).
74. H. C. Thomas, *J. Am. Chem. Soc.*, **66**, 1664 (1944).
75. W. G. Whitman, *Chem. Met. Eng.*, **29**, 146 (1923).
76. W. K. Lewis and W. G. Whitman, *Ind. Eng. Chem.*, **16**, 1215 (1924).
77. T. K. Sherwood and R. L. Pigford, *Absorption and Extraction*, 2nd ed., McGraw-Hill, New York, 1952, Chap. 3.
78. R. E. Treybal, *Liquid Extraction*, McGraw-Hill, New York, 2nd ed., 1963, Chap. 5.
79. D. O. Cooney and E. N. Lightfoot, *Ind. Eng. Chem. Fundamentals*, **4**, 233 (1965).
80. E. D. Heerdt, Thesis, University of Minnesota, 1969.
81. A. T. James and A. J. P. Martin, *Biochem. J.*, **50**, 679 (1952).

82. Ref. 2, Chap. V.
83. Ref. 39, Chap. 5.
84. C. H. Bosanquet and G. O. Morgan, in *Vapour Phase Chromatography* (D. H. Desty, ed.), Academic Press, New York, 1957, p. 35.
85. C. H. Bosanquet, in *Gas Chromatography 1958* (D. H. Desty, ed.), Academic Press, New York, 1958, p. 107.
86. G. Schay, *Theoretische Grundlagen der Gaschromatographie,* VEB Deutscher Verlag der Wissenschaften, Berlin, 1960, Chaps. 1 and 4.
87. Ref. 42, Secs. 2.10 and 5.6.
88. M. J. E. Golay, *Nature*, **202,** 489 (1964).
89. D. L. Peterson and F. Helfferich, *J. Phys. Chem.*, **69**, 1283 (1965).
90. J. R. Conder, in *Progress in Gas Chromatography* (Advances in Analytical Chemistry and Instrumentation, Vol. 6, J. H. Purnell, ed.), Wiley-Interscience, New York, 1968, p. 209.
91. J. R. Conder and J. H. Purnell, *Trans. Faraday Soc.*, **64**, 1505, 3100 (1968).
92. A. A. Zhukhovitskii, N. M. Turkeltaub, L. A. Malyasova, A. F. Shlyakhov, V. V. Naumova, and T. I. Pogrebnaya, *Zavodsk. Lab.*, **29**, 1162 (1963).
93. P. C. Haarhoff and H. J. van der Linde, *Anal. Chem.*, **37**, 1742 (1965).
94. N. Dyson and A. B. Littlewood, *Anal. Chem.*, **39**, 638 (1967).
95. R. P. W. Scott, *Anal. Chem.*, **36**, 1455 (1964).
96. R. P. W. Scott, *Anal. Chem.*, **37**, 1764 (1965).
97. Ref. 41, Chap. 15.

# 6

# ASSESSMENT AND OUTLOOK

In conclusion, an attempt will be made to view the substance of this book in a broader context by exploring its implications for chromatographic theory and for physics of dynamic systems in general. At an early stage of a new development, such a view is necessarily subjective and conjectural; here, it is meant as a challenge to the reader to question, to disagree, and to proceed along his own lines where we have left off.

The immediate goal of this book was to provide a comprehensive, if rudimentary, treatment of chromatographic behavior of interfering species. This we believe to have achieved. The logical and consistent manner in which all the well-known chromatographic techniques find their proper place in the general theoretical framework inspires confidence in the theory's ability to represent the principal physical features. What makes the theory even more convincing is that agreement with experimental facts is achieved not by rationalization *a posteriori*, but as an inescapable and almost incidental result of deductions from first principles. Yet, as the chapter on relaxation of premises has made evident, much detail still needs to be filled in before a degree of sophistication comparable to that of the advanced theories of column dynamics will have been attained. Such refinements, however, might better await a more thorough confirmation of the principal aspects. Here, literally hundreds of experiments suggest themselves as critical tests of the theory's predictive power, since we have in no way shied from making predictions just as freely on purely theoretical grounds in the many instances where experimental evidence was lacking. For example, a verification of the prominent features of the response pattern predicted for single-component influent pulses, shown in Fig. 4.19, would go a long way toward establishing the reliability of the theory.

Aside from the concept of coherence and the $h$ transformation, to which we shall return shortly, the main contribution to chromatographic theory has been the rephrasing of the classical material-balance argument in terms of species velocities and concentration velocities. Parallel to the mathematical derivations, on which the presentation has relied for rigor, conceptual reasoning on this basis has provided insight into the physical causes and effects as well as predictive power on a more general level than can be achieved with mathematical solutions, which in each case remain contingent on the form of the equilibrium relations. To give only three examples, the conceptual arguments help to understand why and how a noncoherent composition variation breaks up into coherent ones (see Chapter 3, Section V.C), explain various seemingly paradoxical situations such as development without resolution (Chapter 4, Section IV.D), and have led to the design of the tracer-pulse as an alternative to the more unwieldy concentration-pulse technique (Chapter 4, Section III.E). The potential of this approach has barely been tapped, and one may expect the conception of new techniques and applications to spring from the more general realization that migration of boundaries and pulses in chromatography is a wave rather than a mass-transfer phenomenon.

The mathematical and conceptual developments have also shed new light on the interrelations between chromatography and other disciplines. The formal equivalence of the fundamental differential equations in chromatography and physics of compressible fluids and, in particular, the analogy between self-sharpening boundaries and shock waves have been pointed out (see Chapter 3, Sections III.B.3 and IV.E). Similarities to optics are apparent not only in the significance of coherence, but also in the interference phenomena of boundaries, which resemble diffraction of waves. The fact that a single input signal (i.e., composition change at the inlet) in a chromatographic column produces in general $n$ response signals* which are propagated at different velocities can be related to information

* This statement refers to systems with arbitrary equilibrium properties. The constraint of stoichiometric exchange reduces the number of response signals to $n-1$.

theory. Perhaps the most intriguing analogy is to distillation, invoked on an essentially intuitive basis in the earliest days of chromatographic theory by Martin and Synge, debated ever since, and now established on a more fundamental level through the formal equivalence of the description of chromatography in terms of *H*-function roots with Underwood's stripping-factor treatment of multicomponent fractionation (for references, see Chapter 1).

As already mentioned in the introductory remarks, various other multicomponent problems besides distillation have been, or can be, efficiently handled with the same or very similar tools. In particular, the *h* transformation and the concept of coherence prove to be of key importance for multicomponent systems in general. This result of the present study may well turn out to be more significant than the contribution to chromatography, which, after all, can command only a limited general interest even if defined as generously as here.

In this broader context, it seems as though the *h* transformation could be used to the same advantage as here wherever interference in a multicomponent system is "indiscriminate," that is, if the species affect one another merely by sharing, to fixed relative degrees, in the performance of a given function. Such a function may be, for example, the occupancy of sorption sites as in the present treatment, the occupancy of vapor space as in Zhukhovitskii's theory of gas chromatography and Underwood's treatment of multicomponent fractionation, or the transport of electric current as in Dole's theory of electrophoresis, and interference is "indiscriminate" if the respective interaction coefficients—separation factors in chromatography, relative volatilities in distillation, mobility ratios in electrophoresis—are independent of composition. The potential of this approach can now be appreciated in retrospect: It would have been possible to derive all relevant equations and rules stated in Chapter 4 directly and with ease from the transformation and proofs in Appendices I and II. In disciplines other than chromatography, only a small fraction of this potential has so far been put to use, since the respective treatments have confined themselves to the steady state or state of coherence.

Unfortunately, there seems to be no easy way of extending the compact treatment based on the *h* transformation to systems with other than indiscriminate interference. For this reason, the present approach has been based instead on the more general principle of coherence and has used the *h* transformation only as a convenient but dispensable tool.

The most general result of the present study is undoubtedly the principle of development toward what has been called coherence. Previously hardly recognized except in its narrow sense as used in optics, coherence has emerged as a fundamental concept applicable to dynamic systems in general. It is unlikely that the full implications of this concept have as yet been realized, and a general definition attempted at this stage may therefore require later revisions. With this reservation, we would define coherence in its general sense as *a state in which the values of the dependent variables coexisting at any given place and time continue to coexist, either remaining stationary or moving jointly, i.e., in the same direction and with the same velocity*; the motion may be with time in physical space, as in chromatography, fluid mechanics, or optics, or may be in any other coordinate system of independent variables. The significance of coherence is that, in general, a system will tend toward this state if not disturbed. Under the broad definition given above, equilibrium and steady state are coherent states, and their attainment is a special case of development toward coherence. If the boundary conditions of the system preclude attainment of equilibrium or a steady state, as is usually the case in chromatography, development toward coherence will still take place. Under such conditions, a coherent unsteady state is generally attained if the boundary values are constant. But even if these values vary, the deformation of the composition route (or its equivalent in a system with other dependent variables) nevertheless obeys locally the principle of development toward coherence, thus being amenable to simple constructions of the type illustrated in Fig. 3.32. We believe to have gone far toward exhausting the possibilities of this new concept in chromatography, but still expect fruitful results from its application to other areas of physics.

The trend toward coherence may seem surprising at first, but is really not more so than that toward equilibrium or a steady state, with which coherence has much in common. This similarity is, perhaps, the reason why so general a principle as coherence seems to have remained hidden for so long: In most systems of broad interest, such as batch operation or continuous fractionation, the boundary conditions are such that equilibrium or a steady state is attained, and little may then be gained from a distinction between such a state and coherence. Therefore, we believe the most promising applications of the new concepts to be to nonsteady-state phenomena, be they coherent or not.

# APPENDIX

## I. *H*-Function Roots and *h* Transformation

Derivations of the equations, conditions, and rules stated without proof in Chapter 3, Sections IV.C and D, are supplied in this Appendix.

### 1. *H-Function Roots*

By definition, the $h_i$ are the roots in $h$ of

$$\sum_i \Big[\prod_{j \neq i} (h-\alpha_{1j})x_i\Big] = 0 \tag{A.1}$$

The $H$ function, on the left-hand side of this equation, is a polynomial of degree $(n-1)$ and thus has $n-1$ roots. It will first be shown that for all physically meaningful compositions (i.e., $x_i \geqq 0$ for all $i$) the $n-1$ roots are all real and fall into the intervals

$$\alpha_{1i} \leqq h_i \leqq \alpha_{1,i+1} \qquad (i = 1, \ldots, n-1) \tag{3.50}$$

*Compositions with all Species Present.* Dividing Eq. (A.1) by $\Pi_j\,(h-\alpha_{1j})$ one obtains

$$\sum_i \frac{x_i}{h-\alpha_{1i}} = 0 \tag{A.2}$$

For given $\mathbf{x}$ with $x_i > 0$ for all $i$, this equation has $n$ poles at $h = \alpha_{1i}$ $(i = 1, \ldots, n)$ and is otherwise single-valued and finite. The slope

$$\left[\frac{\partial}{\partial h}\left(\sum_i \frac{x_i}{h - \alpha_{1i}}\right)\right]_{\mathbf{x}} = -\sum_i \frac{x_i}{(h - \alpha_{1i})^2} \tag{A.3}$$

is negative for all values of $h$. Equation (A.2) thus has $n - 1$ real roots in $h$, one between each pair of adjacent poles:

$$\alpha_{1i} < h_i < \alpha_{1,i+1} \tag{A.4}$$

(see Fig. A.1a). No root equals a separation factor. The division of Eq. (A.1) by $\Pi_j\,(h - \alpha_{1j})$ thus does not affect the roots, and the roots of Eqs. (A.1) and (A.2) therefore are identical. The $n - 1$ roots of Eq. (A.1) thus are all accounted for as real, obey condition (A.4) and therefore (3.50), and can be calculated from Eq. (A.2).

*Compositions with Absent Species.* For given $\mathbf{x}$ with $x_k = 0$ for $m$ different species $k$ $(1 \leqq m < n)$, Eq. (A.1) has $m$ so-called *trivial roots*, one at each $h = \alpha_{1k}$, since the term or terms involving $x_k$ are zero and all others contain $h - \alpha_{1k}$ as a factor. The left-hand side of Eq. (A.2) is indeterminate at these values of $h$.

For all other values of $h$, the term or terms involving $x_k$ in Eq. (A.2) are zero. The nontrivial roots of Eq. (A.1) thus are those of

$$\sum_{i \neq k} \frac{x_i}{h - \alpha_{1i}} = 0 \tag{A.5}$$

The same argument as for Eq. (A.2) shows Eq. (A.5) to have $n$-$m$-1 real roots between its $n$-$m$ poles at $h = \alpha_{1i}$ $(i \neq k)$ (see Fig. A.1b). Together with the $m$ trivial roots, all $n - 1$ roots of Eq. (A.1) thus are again accounted for as real.

Comparing condition (A.4) with the intervals in which the nontrivial roots must lie, one finds that, for each $x_k \to 0$, a nontrivial root from either interval $\alpha_{1,k-1} < h < \alpha_{1k}$ or $\alpha_{1k} < h < \alpha_{1,k+1}$ (from $\alpha_{11} < h < \alpha_{12}$ if $x_1 \to 0$, and from $\alpha_{1,n-1} < h < \alpha_{1n}$ if $x_n \to 0$) becomes a trivial root $h = \alpha_{1k}$. Accordingly, with the roots indexed in the sequence of increasing numerical values, the trivial root equalling $\alpha_{1k}$ acquires the index number $k - 1$ or $k$, depending on whether the interval that lost its nontrivial root is $\alpha_{1,k-1} < h < \alpha_{1k}$ or $\alpha_{1k} < h < \alpha_{1,k+1}$. Also, the nontrivial roots keep

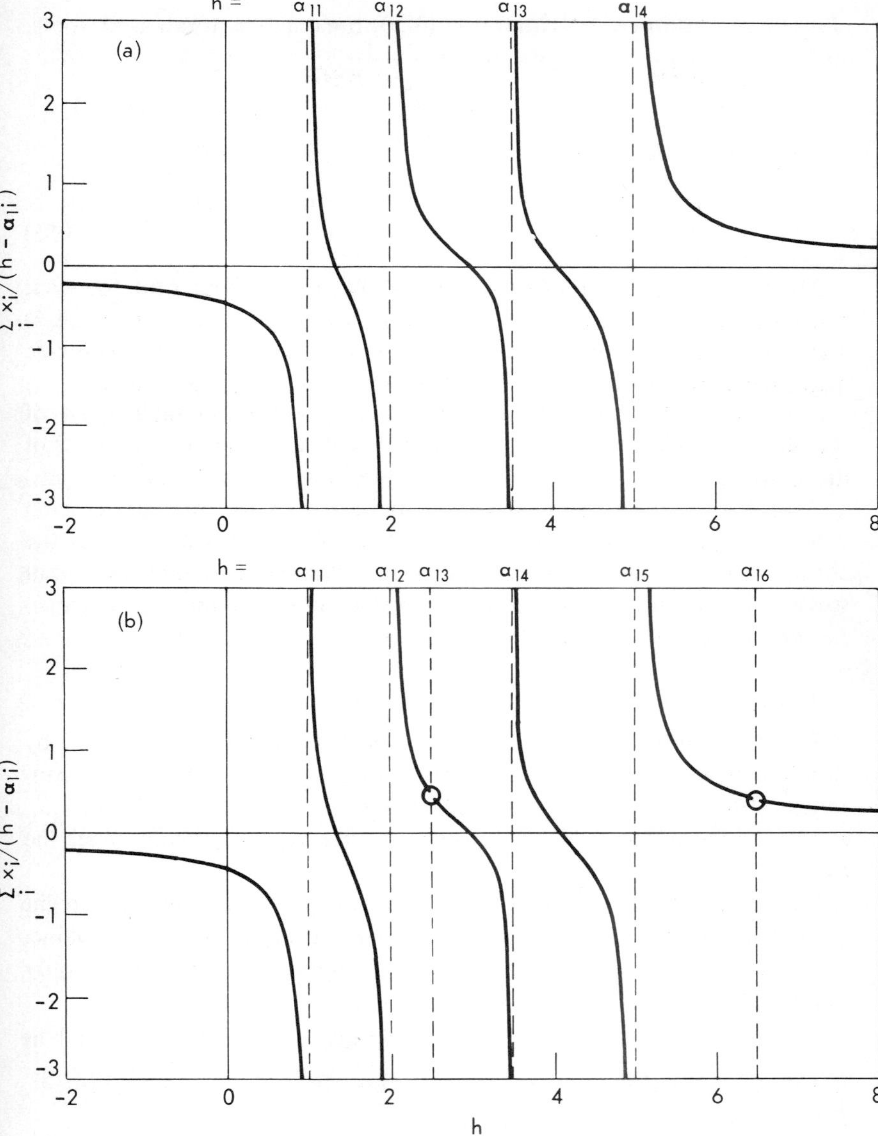


**Fig. A.1.** Function $F(h) \equiv \Sigma_i\,[x_i/(h-\alpha_{1i})]$ for given compositions (a) in four-component system, (b) in six-component system with two species absent. Circles in lower diagram mark points where function is indeterminate. [For (a) $\alpha_{12}=2$, $\alpha_{13}=3.5$, $\alpha_{14}=5$, $x_1=0.2$, $x_2=0.3$, $x_3=0.1$, $x_4=0.4$; (b) $\alpha_{12}=2$, $\alpha_{13}=3.5$, $\alpha_{14}=4$, $\alpha_{15}=5$, $\alpha_{16}=6.5$, $x_1=0.2$, $x_2=0.3$, $x_3=0$, $x_4=0.1$, $x_5=0.4$, $x_6=0$.]

obeying condition (A.4). Condition (3.50) thus applies to all $n-1$ roots, and

$$\begin{aligned} &h_{k-1} = \alpha_{1k} \quad \text{or} \quad h_k = \alpha_{1k} && \text{if } x_k = 0 \quad (k \neq 1, n) \\ &h_1 = \alpha_{11} = 1 && \text{if } x_1 = 0 \\ &h_{n-1} = \alpha_{1n} && \text{if } x_n = 0 \end{aligned} \tag{3.55}$$

Although condition (A.4) does not allow a nontrivial root to equal a separation factor $\alpha_{1i}$ of a species present, a nontrivial root of Eq. (A.5) may equal a separation factor $\alpha_{1k}$ of an absent species $k$ and thus coincide with a trivial root of Eq. (A.1). In this singular case, $h_{k-1} = h_k = \alpha_{1k}$.

For Fig. A.1b, the concentrations $x_i$ and separation factors $\alpha_{1i}$ of the present species $i = 1, 2, 4$, and 5 were taken to be the same as those of the species $i = 1, 2, 3$, and 4 in Fig. A.1a. The function $F(h) \equiv \Sigma_i\, x_i/(h-\alpha_{1i})$ in Fig. A.1b then is identical with that in Fig. A.1a, except for being indeterminate where $h$ equals the separation factors $\alpha_{13}$ and $\alpha_{16}$ of the absent species. Thus, as the comparison illustrates, to count an absent species $k$ as a component of the system does not alter the function $F(h)$ except for making it indeterminate at $h = \alpha_{1k}$.

## 2. *Calculation of* **h** *from* **x**

The $H$-function roots for given **x** can be calculated from Eq. (A.1) by standard procedures for obtaining roots of polynomials. It is more convenient, however, to calculate the nontrivial roots from Eq. (A.2) or (A.5), equate the trivial roots, if any, to the respective $\alpha_{1k}$, and then number the roots according to condition (3.50).

For compositions with no more than three species present, convenient explicit solutions of Eq. (A.5) for the nontrivial root or roots are available. If only two species $j$ and $k$ (not necessarily 1 and 2) are present, the only nontrivial root is

$$h = \alpha_{1k}x_j + \alpha_{1j}x_k \tag{A.6}$$

If only three species $j$, $k$, and $l$ (not necessarily 1, 2, and 3) are present, the two nontrivial roots are given by

$$h_{1,2} = \tfrac{1}{2}[A \mp (A^2 - 4B)^{1/2}] \tag{A.7}$$

where

$$A \equiv (\alpha_{1k} + \alpha_{1l})x_j + (\alpha_{1j} + \alpha_{1l})x_k + (\alpha_{1j} + \alpha_{1k})x_l$$

$$B \equiv \alpha_{1k}\alpha_{1l}x_j + \alpha_{1j}\alpha_{1l}x_k + \alpha_{1j}\alpha_{1k}x_l$$

### 3. *Calculation of* **x** *from* **h**

For a given **x**, each root value $h_l$ in Eq. (A.8) below is, by definition, a solution of Eq. (A.1):

$$\sum_i \left[ \prod_{j \neq i} (h_l - \alpha_{1j}) x_i \right] = 0 \qquad (l = 1, \ldots, n-1) \tag{A.8}$$

This set of $n-1$ equations is linear in the $x_i$ and can be solved explicitly for these as follows.*

The polynomial (A.1), with constant $\alpha_{1i}$ and **x**, is of degree $(n-1)$ in $h$ with constant coefficients. Its roots are called $h_1, \ldots, h_{n-1}$, and the coefficient of the term in $h^{n-1}$ is $\Sigma_i\, x_i = 1$ [see Eq. (2.5)]. The polynomial $\Pi_i\,(h-h_i)$ is also of degree $(n-1)$ with constant coefficients, has the same roots and the same coefficient of $h^{n-1}$, and is therefore identical with the polynomial (A.1):

$$\sum_j \left[ \prod_{i \neq j} (h - \alpha_{1i}) x_j \right] \equiv \prod_i (h - h_i) \tag{A.9}$$

(On the left-hand side, indices $i$ and $j$ have been interchanged to avoid later conflict.) The equality holds for all values of $h$. Specifically, for any $h = \alpha_{1j}$:

$$x_j \prod_{i \neq j} (\alpha_{1j} - \alpha_{1i}) = \prod_i (\alpha_{1j} - h_i) \tag{A.10}$$

where $j$ may be $1, \ldots, n$. Solved for $x_j$:

$$x_j = \prod_i (h_i - \alpha_{1j}) \Big/ \prod_{i \neq j} (\alpha_{1i} - \alpha_{1j}) \qquad \text{for all } j \tag{3.57}$$

With this set of equations, **x** can be calculated from **h**. All $n$ unknown $x_j$ are obtained from the $n-1$ equations (A.8), but the constraint $\Sigma_i\, x_i = 1$ has been used as the $n$th relation.

### 4. *Identity of* **h**(**x**) *and* **h**(**y**) *in Equilibrium*

The $H$-function roots in terms of **y** are given as the reciprocals of the roots in $1/h$ of

$$\sum_i \left[ \prod_{j \neq i} \left( \frac{1}{h} - \alpha_{j1} \right) y_i \right] = 0 \tag{A.11}$$

It will now be shown that these $h_i$ are identical with those calculated from **x** with Eq. (A.1) if **x** and **y** are in equilibrium.

Equation (A.11) has a trivial root at each $1/h = \alpha_{k1}$, and thus $h = \alpha_{1k}$,

---

* The authors are indebted to Dr. R. R. Meyer for having suggested this compact proof of Eqs. (3.57).

if $y_k = 0$ for one or several species $k$. The $h_i$ provided by the trivial roots, if any, thus are identical with those obtained from Eq. (A.1) with $x_k = 0$.

To prove that the $h_i$ provided by the nontrivial roots are also identical, one divides Eq. (A.11) by $\Pi_j\,(1/h-\alpha_{j1})$:

$$\sum_i \frac{y_i}{1/h-\alpha_{i1}} = 0 \tag{A.12}$$

Noting that $1/\alpha_{i1} = \alpha_{1i}$ and that, for **x** and **y** in equilibrium,

$$\alpha_{1i} = y_1 x_i / x_1 y_i \tag{A.13}$$

[see definition (2.8) of $\alpha_{ij}$], one can rearrange the left-hand side of Eq. (A.12):

$$\sum_i \frac{y_i}{1/h-\alpha_{i1}} = -\frac{hy_1}{x_1}\sum_i \frac{x_i}{h-\alpha_{1i}} \qquad \text{(in equilibrium)} \tag{A.14}$$

The sum on the right-hand side of Eq. (A.14) is the same as that in Eq. (A.2), and $y_1/x_1 \neq 0$, even for $y_1 \to 0$. The reciprocals of the roots in $1/h$ of Eq. (A.12) therefore are identical with the roots in $h$ of Eq. (A.2). [The right-hand side of Eq. (A.14) is zero for $h = 0$, too, but this does not correspond to a root in $1/h$.] The identity of **h**(**x**) and **h**(**y**) for **x** and **y** in equilibrium thus is proved.

5. *Calculation of* **h** *from* **y**

The $h_i$ can be calculated from given **y** in the same way as from given **x** (see Appendix I.2). The most convenient procedure is to calculate the reciprocals of the nontrivial roots from Eq. (A.12) with terms involving absent species, if any, omitted, equate one $h_i$ to $\alpha_{1k}$ for each absent species $k$, and then number the $h_i$ according to condition (3.50).

The relations analogous to Eqs. (A.6) and (A.7) are, for binary compositions:

$$h = 1/(\alpha_{k1} y_j + \alpha_{j1} y_k) \tag{A.15}$$

and for ternary compositions:

$$h_{1,2} = \frac{2}{\bar{A} \pm (\bar{A}^2 - 4\bar{B})^{1/2}} \tag{A.16}$$

where

$$\bar{A} \equiv (\alpha_{k1}+\alpha_{l1})y_j + (\alpha_{j1}+\alpha_{l1})y_k + (\alpha_{j1}+\alpha_{k1})y_l$$

$$\bar{B} \equiv \alpha_{k1}\alpha_{l1}y_j + \alpha_{j1}\alpha_{l1}y_k + \alpha_{j1}\alpha_{k1}y_l$$

6. *Calculation of* **y** *from* **h**

The composition **y** can be calculated from given **h** in the same manner as **x** from **h** (see Appendix I.3). A simpler symmetry argument, however, is sufficient. Comparison of Eq. (A.11) with Eq. (A.1) shows that the relation between the $y_i$, $1/h_i$, and $\alpha_{i1}$ is identical with that between the $x_i$, $h_i$, and $\alpha_{1i}$. The appropriate substitutions in Eq. (3.57) give

$$y_j = \prod_i \left(\frac{1}{h_i} - \alpha_{j1}\right) \Big/ \prod_{i \neq j} (\alpha_{i1} - \alpha_{j1}) \qquad \text{for all } j \tag{3.58}$$

7. *Differential Coherence Condition, Existence of Distinct Composition Paths, and Proof of Orthogonalization*

This section supplies proof that the differential coherence condition is obeyed only along distinct composition paths, and that the grid of these paths is orthogonal in the $h$ space. For this purpose, the derivative $\partial x_j/\partial y_j \mid_{\mathbf{h}(\eta)}$ for an arbitrary species $j$ is calculated for a route $\mathbf{h}(\eta)$ of arbitrary local direction in the $h$ space. The differential coherence condition, which requires this derivative to have the same value for all species, is then invoked to discern the route directions permitted by coherence.

Along a fixed, arbitrarily selected route in the $h$ space, **h** is a function of a single variable $\eta$, where $\eta$ may be distance from the inlet at a fixed time (profile route) or time at a fixed distance (history route) or any function of time and distance. The $x_i$ and $y_i$ along the route $\mathbf{h}(\eta)$ can also be expressed as functions of $\eta$, since they are related to **h** by Eqs. (3.57) and (3.58). The general expressions for the derivatives $\partial x_j/\partial\eta$ and $\partial y_j/\partial\eta$ along the route for an arbitrary species $j$ at any composition point are

$$\left.\frac{\partial x_j}{\partial \eta}\right|_{\mathbf{h}(\eta)} = \sum_{k=1}^{n-1} \left[ \left(\frac{\partial x_j}{\partial h_k}\right)_{h_i(i \neq k)} \frac{\mathrm{d}h_k}{\mathrm{d}\eta} \right] \tag{A.17}$$

$$\left.\frac{\partial y_j}{\partial \eta}\right|_{\mathbf{h}(\eta)} = \sum_{k=1}^{n-1} \left[ \left(\frac{\partial y_j}{\partial h_k}\right)_{h_i(i \neq k)} \frac{\mathrm{d}h_k}{\mathrm{d}\eta} \right] \tag{A.18}$$

Partial differentiation of Eqs. (3.57) and (3.58) gives

$$\left(\frac{\partial x_j}{\partial h_k}\right)_{h_i(i \neq k)} = \frac{\prod_{i \neq k} (h_i - \alpha_{1j})}{\prod_{i \neq j} (\alpha_{1i} - \alpha_{1j})} = \frac{x_j}{h_k - \alpha_{1j}} \tag{A.19}$$

$$\left(\frac{\partial y_j}{\partial h_k}\right)_{h_i(i \neq k)} = \frac{\prod_i \alpha_{1i} \prod_{i \neq k} (h_i - \alpha_{1j})}{h_k \prod_i h_i \prod_{i \neq j} (\alpha_{1i} - \alpha_{1j})} = \frac{x_j \prod_i \alpha_{1i}}{h_k(h_k - \alpha_{1j}) \prod_i h_i} \tag{A.20}$$

Dividing Eq. (A.17) by (A.18) one obtains, after substitution with Eqs. (A.19) and (A.20)

$$\left.\frac{\partial x_j}{\partial y_j}\right|_{\mathbf{h}(\eta)} = \prod_i h_i \sum_i \left(\frac{dh_i/d\eta}{h_i - \alpha_{1j}}\right) \Big/ \left[\prod_i \alpha_{1i} \sum_i \left(\frac{dh_i/d\eta}{h_i(h_i - \alpha_{1j})}\right)\right] \tag{A.21}$$

By definition, the route is a composition path if and only if the differential coherence condition (3.38) is obeyed, i.e., if $\partial x_j/\partial y_j|_{\mathbf{h}(\eta)}$ has the same value for all species at any composition point. The products in Eq. (A.21) are the same for all species $j$, but the sums, which involve $\alpha_{1j}$, are not. Thus, the differential coherence condition is obeyed only if all but one of the terms of each sum are zero, so that the factors involving $\alpha_{1j}$ cancel. This requires

$$dh_i = 0 \qquad \text{for all } i \neq k \tag{3.56}$$

where $k$ may be $1, \ldots, n-1$. This restriction, which states the differential coherence condition in terms of $H$-function roots, shows that coherence permits composition paths to have only those directions for which all but one $h_i$ are constant. The paths thus are orthogonal in the $h$ space.

## 8. *Composition-Velocity Eigenvalues*

The composition velocity exists only in coherent composition variations, i.e., along composition paths. For these, Eq. (A.21) is subject to the restriction (3.56), under which it reduces to

$$\left(\frac{\partial x_j}{\partial y_j}\right)_{h_i(i\neq k)} = h_k \prod_i h_i \prod_i \alpha_{i1} \qquad (k = 1, \ldots, n-1) \tag{A.22}$$

Since $\partial x_j/\partial y_j|_{\text{path}}$ equals the adjusted composition velocity for the path direction, the set of Eqs. (A.22) gives the $n-1$ eigenvalues, one for each path direction, of the adjusted velocity of a fixed composition, as stated in Eqs. (3.68).

## 9. *Composition Paths, Common Planes, and Common Hyperplanes*

In the $h$ space, the linearity of the composition paths and the existence of common planes and common hyperplanes are obvious consequences of the orthogonality of the path grid. That the linearity and planar character are conserved upon transformation into the $x$ or $y$ space is readily shown as follows:

By definition, a $j$-hyperplane in the $h$ space is an $(n-2)$-dimensional common hyperplane of mutually intersecting paths with variable roots of all index numbers other than $j$, and thus is a locus of constant $h_j$, given by

$$H(h_j, \mathbf{x}, \boldsymbol{\alpha}_{1i}) = 0, \qquad H(1/h_j, \mathbf{y}, \boldsymbol{\alpha}_{i1}) = 0 \tag{3.60}$$

Being linear in the $x_i$ or $y_i$, Eqs. (3.60) describe hyperplanes in the $x$ space or $y$ space. The $j$-hyperplanes thus remain planar upon transformation. This proof does not cover the $j$-hyperplanes with $h_j = \alpha_{1j}$ or $h_j = \alpha_{1,j+1}$, for which Eqs. (3.60) are indeterminate; these hyperplanes, however, are borders $x_j = 0$ or $x_{j+1} = 0$ and therefore are also planar in the $x$ space and $y$ space.

Common hyperplanes of lower order and common planes are intersections of two or more $j$-hyperplanes (with different $j$). As intersections of hyperplanes, they are also planar in the $x$ space and $y$ space. For the same reason, composition paths are linear in the $x$ space and $y$ space.

Borders with watersheds, although "folded" in the $h$ space, are obviously planar in the $x$ or $y$ space. Composition paths crossing watersheds, also folded in the $h$ space, are intersections of common hyperplanes with borders and thus are linear (i.e., not folded) in the $x$ space and $y$ space.

### 10. *Watersheds*

The existence of watersheds on all borders $x_k = 0$ $(k = 2, \ldots, n-1)$ can be proved as follows. As shown in Appendix I.1:

$$\begin{aligned}
&h_{k-1} = \alpha_{1k} \quad \text{or} \quad h_k = \alpha_{1k} && \text{if } x_k = 0 \quad (k \neq 1, n) \\
&h_1 = \alpha_{11} = 1 && \text{if } x_1 = 0 \\
&h_{n-1} = \alpha_{1n} && \text{if } x_n = 0
\end{aligned} \tag{3.55}$$

The single $x$-space border $x_k = 0$ $(k \neq 1, n)$ thus is transformed into two $h$-space borders $h_{k-1} = \alpha_{1k}$ and $h_k = \alpha_{1k}$. Being hyperplanes of constant values of one root, both are intersected only by the paths along which this root varies. Accordingly, $h_{k-1} = \alpha_{1k}$ and $h_k = \alpha_{1k}$ are the transforms of the anchor-point regions of the paths with variable $h_{k-1}$ and $h_k$, respectively. The existence of separate anchor-point regions of the paths of the two sets on $x_k = 0$ proves the existence of the "watershed," defined as separating these regions. The $h$ transform of the watershed is the intersection of the two $h$-space borders:

$$h_{k-1} = h_k = \alpha_{1k} \qquad (k \neq 1, n) \tag{3.62}$$

Equation (3.62) is readily translated into $x$ or $y$ coordinates. For both $h_{k-1}$ and $h_k$ to equal $\alpha_{1k}$, a nontrivial root of Eq. (A.1) must coincide with the trivial one at $\alpha_{1k}$. Therefore, Eq. (A.5) must have a root $h=\alpha_{1k}$ (see Appendix I.1). Substituting $\alpha_{1k}$ for $h$ in Eq. (A.5) and noting that $\alpha_{1i}/\alpha_{1k}=\alpha_{ki}$, one finds

$$h_{k-1} = h_k = \alpha_{1k} \qquad \text{if} \ \sum_{i\neq k} \frac{x_i}{1-\alpha_{ki}} = 0, \qquad x_k = 0 \tag{A.23}$$

and for the $y$ space, by virtue of symmetry (see Appendix I.6),

$$h_{k-1} = h_k = \alpha_{1k} \qquad \text{if} \ \sum_{i\neq k} \frac{y_i}{1-\alpha_{ik}} = 0, \qquad y_k = 0 \tag{A.24}$$

Equations (3.63) in Section IV.D.4 of Chapter 3 are taken from conditions (A.23) and (A.24).

As an intersection of two $h$-space borders with constant $h_{k-1}$ and $h_k$, the watershed on $x_k=0$ (or $y_k=0$) is a common hyperplane of paths with variable roots of all index numbers other than $k-1$ and $k$, and is intersected only by paths with variable $h_{k-1}$ and $h_k$. As a common surface of linear paths of $n-3$ sets, the $(n-3)$-dimensional watershed is a hyperplane in the $x$ or $y$ space, too. The two paths with variable $h_{k-1}$ and $h_k$ intersecting the watershed at the same point are in a plane, and, in the $x$ or $y$ space, both are intersections of that plane with the border and therefore constitute a single, linear path. Involving variations of both $h_{k-1}$ and $h_k$, this path can be identified as having an affinity cut bracketing the absent species $k$. The watershed thus is intersected only by paths having such a bracketing cut.

The criterion (3.64) for distinguishing on which side of the watershed a composition $\mathbf{x}$ or $\mathbf{y}$ is located can be proved as follows: As shown in Appendix I.1, the nontrivial roots for a composition with $x_k=0$ are those of Eq. (A.5) and obey condition (A.4). The function $\Sigma_{i\neq k}\, x_i/(h-\alpha_{1i})$ in Eq. (A.5) has negative slope between its poles at $h=\alpha_{1,k-1}$ and $h=\alpha_{1,k+1}$ (see Appendix I.1) and, therefore, is positive at $h=\alpha_{1k}$ if it has a root in $\alpha_{1k}<h<\alpha_{1,k+1}$, and is negative at $h=\alpha_{1k}$ if it has a root in $\alpha_{1,k-1}<h<\alpha_{1k}$ (see Fig. A.1b). In the first case, condition (A.4) identifies the root as $h_k$, so that the trivial root equalling $\alpha_{1k}$ must be $h_{k-1}$. In the second case, the root of Eq. (A.5) is identified as $h_{k-1}$, so that the trivial root must be $h_k$. Accordingly, after substitution of $\alpha_{1k}$ for $h$ under the sum, and with $\alpha_{1i}/\alpha_{1k}=\alpha_{ki}$,

$$\begin{aligned} h_{k-1} &= \alpha_{1k} \qquad \text{if} \ \sum_{i\neq k} \frac{x_i}{1-\alpha_{ki}} > 0, \qquad x_k = 0 \\ h_k &= \alpha_{1k} \qquad \text{if} \ \sum_{i\neq k} \frac{x_i}{1-\alpha_{ki}} < 0, \qquad x_k = 0 \end{aligned} \tag{A.25}$$

The derivation for the $y$ space is analogous, except that the slope of $\Sigma_{i\neq k}\, y_i/(1/h-\alpha_{i1})$, being negative with respect to $1/h$, is positive with respect to $h$. The result is

$$\begin{aligned} h_{k-1} &= \alpha_{1k} \quad \text{if} \quad \sum_{i\neq k} \frac{y_i}{1-\alpha_{ik}} < 0, \qquad y_k = 0 \\ h_k &= \alpha_{1k} \quad \text{if} \quad \sum_{i\neq k} \frac{y_i}{1-\alpha_{ik}} > 0, \qquad y_k = 0 \end{aligned} \tag{A.26}$$

The criterion (3.64) in Chapter 3, Section IV.D.4 is compiled from conditions (A.25) and (A.26).

### 11. *Rule of Equal Intercept Ratios*

The "rule of equal intercept ratios" (see Chapter 3, Sections IV.C. 2–3) states that any composition path on a common plane divides opposite borders of this plane into segments which are in the same ratio to one another; on triangular common planes, the border segments between corner points and watershed point count as separate borders for the purpose of this rule.

For common planes in the $x$ space, the rule can be derived as follows. Any common plane in the $h$ space is bounded by four composition paths. Along any path, only one $h_i$ varies. Equations (3.57) show that, with this restriction, all $x_i$ vary linearly with the respective variable $h_i$. Hence, on any path, composition points which are equidistant in the $h$ space are also equidistant in the $x$ space. Therefore, intercept ratios on the paths bounding a given common plane are not altered by transformation from the $h$ space into the $x$ space. The rule of equal intercept ratios, valid in the $h$ space because the paths are orthogonal, thus is conserved upon transformation into the $x$ space.

The proof for common planes in the $y$ space is analogous. Equations (3.58) show that the $y_i$ along any path vary linearly with the respective variable $1/h_i$. Hence, on any path, equidistant points in the $1/h$ space (with rectangular coordinates $1/h_i$) are also equidistant in the $y$ space, and intercept ratios are the same in both. The rule of equal intercept ratios, valid in the $1/h$ space because the paths are orthogonal, thus is valid in the $y$ space also.

### 12. *Integral Coherence Condition and Step Velocities*

It will be shown that the integral coherence condition for a sharp composition step is met if, and only if, the compositions on both sides of the step fall on the same composition path. The derivation also yields the adjusted step velocity of a coherent step.

For a step between compositions $\{\mathbf{x}', \mathbf{y}'\}$ and $\{\mathbf{x}'', \mathbf{y}''\}$, the integral coherence condition (3.40) requires the ratio $(x_i' - x_i'')/(y_i' - y_i'')$ to have the same value for all species. For an arbitrary species $j$ one obtains with Eqs. (3.57) and (3.58):

$$\frac{\Delta x_j}{\Delta y_j} \equiv \frac{x_j' - x_j''}{y_j' - y_j''} = \prod_{i \neq j} \alpha_{i1} \frac{\prod_i (h_i' - \alpha_{1j}) - \prod_i (h_i'' - \alpha_{1j})}{\prod_i [(h_i' - \alpha_{1j})/h_i'] - \prod_i [(h_i'' - \alpha_{1j})/h_i'']} \tag{A.27}$$

The right-hand side of this expression will have the same value for all species if, and only if, all features distinguishing $j$ from other species disappear. This is the case if the products in the numerator as well as in the denominator have all factors but one in common. A sufficient but not necessary condition for this to be so is

$$h_i' = h_i'' \qquad \text{for all } i \neq k \tag{A.28}$$

where $k$ can be $1, \ldots, n-1$. With this constraint, Eq. (A.27) reduces to

$$\frac{\Delta x_j}{\Delta y_j} = h_k' h_k'' \prod_{i \neq k} h_i \prod_i \alpha_{i1} \tag{A.29}$$

(no primes or double primes shown for the constant roots), and the compositions on the two sides of the step are on the same path with variable $h_k$.

The requirement that the right-hand side of Eq. (A.27) carry no feature distinguishing species $j$ from others can be met in another way, since condition (3.50) allows roots with different but adjacent index numbers to have the same value. With condition (3.50) as constraint, a sufficient and necessary set of conditions for equality of the ratios $\Delta x_j/\Delta y_j$ of all species present is

$$\begin{aligned} h_i' &= h_i'' && \text{for all } i < k \text{ and } i \geqq l \\ h_{i-1}' &= h_i'' = \alpha_{1i} \quad \text{or} \quad h_i' = h_{i-1}'' = \alpha_{1i} && \text{for all } i \text{ with } k < i < l \end{aligned} \tag{A.30}$$

where $1 \leqq k < l \leqq n$. [Condition (A.28) is contained as the special case $l = k + 1$.] With this restriction, Eq. (A.27) reduces to

$$\frac{\Delta x_j}{\Delta y_j} = h'h'' \prod_{i=1}^{k-1} h_i \prod_{i=l}^{n-1} h_i \prod_{i=1}^{k} \alpha_{i1} \prod_{i=l}^{n} \alpha_{i1} \tag{A.31}$$

where $h' = h'_k$, $h'' = h''_{l-1}$, or $h' = h'_{l-1}$, $h'' = h''_k$. Under condition (A.30), species $k+1, \ldots, l-1$ are absent from both sides of the step, since roots on both sides equal the respective $\alpha_{1i}$. If species and roots are renumbered to the exclusion of the absent species $k+1, \ldots, l-1$ and trivial roots due to them, all roots except $h_k$ are found to have the same value on both sides of the step, and Eq. (A.31) reduces to Eq. (A.29). Accordingly, under conditions (A.30), too, $\mathbf{x}'$ and $\mathbf{x}''$ are on the same composition path, in this case a path with affinity cut bracketing the absent species (see also Table 3.1).

According to Eq. (3.36), the adjusted step velocity is given by the ratio $\Delta x_i/\Delta y_i$. For a coherent step, Eq. (A.29) or (A.31) thus gives the adjusted step velocity, as stated in Eqs. (3.72) and (3.71) in Section IV.D.7 of Chapter 3.

## II. Dynamic Behavior of $H$-Function Roots

The set of basic differential equations (3.87) for the $H$-function roots as a function of distance and time is derived in this Appendix.

Equations (A.1) and (A.11) defining the $h_i$, as well as Eqs. (A.2) and (A.12) for calculating the nontrivial roots, involve the concentrations of all $n$ species. However, in view of

$$\sum_i x_i = 1 \qquad \text{and} \qquad \sum_i y_i = 1 \tag{2.5}$$

only $n-1$ $x_i$ or $y_i$ can be varied independently. It will be more convenient for the derivation below to express the $h_i$ by equations involving independent concentrations only. For this purpose, $x_n$ is eliminated from Eq. (A.2) with Eq. (2.5):

$$\sum_i \frac{x_i}{h-\alpha_{1i}} = \frac{1}{\alpha_{1n}-h}\left[\sum_{i=1}^{n-1}\left(\frac{\alpha_{1n}-\alpha_{1i}}{h-\alpha_{1i}}\, x_i\right)-1\right] = 0 \tag{A.32}$$

Since $1/(\alpha_{1n}-h) \neq 0$ for all finite values of $h$, Eq. (A.32) requires, for finite $h$,

$$\sum_{i=1}^{n-1}\left(\frac{\alpha_{1n}-\alpha_{1i}}{h-\alpha_{1i}}\, x_i\right)-1 = 0 \tag{A.33}$$

Similarly, Eq. (A.12) with Eq. (2.5), after rearrangement, leads to the requirement

$$\sum_{i=1}^{n-1}\left(\frac{h(\alpha_{1n}-\alpha_{1i})}{\alpha_{1n}(h-\alpha_{1i})}\, y_i\right)-1 = 0 \tag{A.34}$$

For any given composition $\{\mathbf{x}, \mathbf{y}\}$ in equilibrium, each $h_i$ obeys Eqs. (A.33) and (A.34). (This includes the trivial roots, if any, for which the left-hand sides of the equations are indeterminate.) Thus, for the arbitrary root $h_k$:

$$\sum_{i=1}^{n-1}\left(\frac{\alpha_{1n}-\alpha_{1i}}{h_k-\alpha_{1i}}\,x_i\right)-1 = 0 \tag{A.35}$$

$$\sum_{i=1}^{n-1}\left(\frac{h_k(\alpha_{1n}-\alpha_{1i})}{\alpha_{1n}(h_k-\alpha_{1i})}\,y_i\right)-1 = 0 \tag{A.36}$$

Equations (A.35) and (A.36) give $h_k$ as an implicit function of the $n-1$ independent $x_1, \ldots, x_{n-1}$ or $y_1, \ldots, y_{n-1}$ and thus allow the partial derivatives of $h_k$ with respect to $z$ and $\tau$ to be written as

$$\left(\frac{\partial h_k}{\partial z}\right)_\tau = \sum_{j=1}^{n-1}\left[\left(\frac{\partial h_k}{\partial x_j}\right)_{x_i(i\neq j,n)}\left(\frac{\partial x_j}{\partial z}\right)_\tau\right] \tag{A.37}$$

$$\left(\frac{\partial h_k}{\partial \tau}\right)_z = \sum_{j=1}^{n-1}\left[\left(\frac{\partial h_k}{\partial y_j}\right)_{y_i(i\neq j,n)}\left(\frac{\partial y_j}{\partial \tau}\right)_z\right] \tag{A.38}$$

[The summation extends only over the $n-1$ independent concentration variables. The $n$th, dependent variable $x_n$ (or $y_n$) is allowed to vary with each $x_j$ (or $y_j$), as required by the constraint $\Sigma_i\, x_i = 1$ (or $\Sigma_i\, y_i = 1$) if the other independent concentrations are held constant.]

The derivatives $\partial h_k/\partial x_j$ in Eq. (A.37) are obtained by implicit partial differentiation of Eq. (A.35) with respect to an arbitrary $x_j$ at constant $x_i\,(i \neq j, n)$:

$$-\left(\frac{\partial h_k}{\partial x_j}\right)_{x_i(i\neq j,n)}\sum_{i=1}^{n-1}\left[\frac{\alpha_{1n}-\alpha_{1i}}{(h_k-\alpha_{1i})^2}\,x_i\right] + \frac{\alpha_{1n}-\alpha_{1j}}{h_k-\alpha_{1j}} = 0 \tag{A.39}$$

After rearrangement:

$$\left(\frac{\partial h_k}{\partial x_j}\right)_{x_i(i\neq j,n)} = \frac{\alpha_{1n}-\alpha_{1j}}{h_k-\alpha_{1j}}\Bigg/\sum_{i=1}^{n-1}\left[\frac{\alpha_{1n}-\alpha_{1i}}{(h_k-\alpha_{1i})^2}\,x_i\right] \tag{A.40}$$

Incidentally, this expression can be shown to reduce to

$$\left(\frac{\partial h_k}{\partial x_j}\right)_{x_i(i\neq j,n)} = (\alpha_{1n}-\alpha_{1j})\prod_{i\neq j}(\alpha_{1i}-h_k)\Bigg/\prod_{i\neq k}(h_i-h_k) \tag{A.41}$$

The derivatives $\partial h_k/\partial y_j$ in Eq. (A.38) are obtained by implicit partial differentiation of Eq. (A.36) with respect to an arbitrary $y_j$ at constant $y_i$ $(i \neq j, n)$:

$$-\left(\frac{\partial h_k}{\partial y_j}\right)_{y_i(i\neq j,n)} \sum_{i=1}^{n-1}\left[\frac{(\alpha_{1n}-\alpha_{1i})\alpha_{1i}}{(h_k-\alpha_{1i})^2\alpha_{1n}}\, y_i\right]+\frac{(\alpha_{1n}-\alpha_{1j})h_k}{(h_k-\alpha_{1j})\alpha_{1n}} = 0 \qquad \text{(A.42)}$$

After rearrangement:

$$\left(\frac{\partial h_k}{\partial y_j}\right)_{y_i(i\neq j,n)} = \frac{(\alpha_{1n}-\alpha_{1j})h_k}{h_k-\alpha_{1j}} \Big/ \sum_{i=1}^{n-1}\left[\frac{(\alpha_{1n}-\alpha_{1i})\alpha_{1i}}{(h_k-\alpha_{1i})^2}\, y_i\right] \qquad \text{(A.43)}$$

The $y_i$ can be expressed in terms of the $x_i$ by means of Eq. (3.59):

$$\left(\frac{\partial h_k}{\partial y_j}\right)_{y_i(i\neq j,n)} = \frac{(\alpha_{1n}-\alpha_{1j})h_k \prod_i h_i}{(h_k-\alpha_{1j}) \prod_i \alpha_{1i}} \Big/ \sum_{i=1}^{n-1}\left[\frac{\alpha_{1n}-\alpha_{1i}}{(h_k-\alpha_{1i})^2}\, x_i\right] \qquad \text{(A.44)}$$

Comparison with Eq. (A.40) shows that

$$\left(\frac{\partial h_k}{\partial y_j}\right)_{y_i(i\neq j,n)} = h_k \prod_i h_i \prod_i \alpha_{i1} \left(\frac{\partial h_k}{\partial x_j}\right)_{x_i(i\neq j,n)} \qquad \text{(A.45)}$$

In Eq. (A.38), $(\partial y_j/\partial \tau)_z$ can be eliminated with

$$(\partial y_j/\partial \tau)_z + (\partial x_j/\partial z)_\tau = 0 \qquad \text{(3.32)}$$

and $(\partial h_k/\partial y_j)_{y_i(i\neq j,n)}$ can be replaced by means of Eq. (A.45). With these substitutions, Eq. (A.38) becomes

$$\left(\frac{\partial h_k}{\partial \tau}\right)_z = -h_k \prod_i h_i \prod_i \alpha_{i1} \sum_{j=1}^{n-1}\left[\left(\frac{\partial h_k}{\partial x_j}\right)_{x_i(i\neq j,n)} \left(\frac{\partial x_j}{\partial z}\right)_\tau\right] \qquad \text{(A.46)}$$

with Eq. (A.37), and for all roots $h_k$:

$$\left(\frac{\partial h_k}{\partial \tau}\right)_z = -h_k \prod_i h_i \prod_i \alpha_{i1} \left(\frac{\partial h_k}{\partial z}\right)_\tau \qquad (k = 1, \ldots, n-1) \qquad \text{(3.87)}$$

## III. Responses to Square-Wave Influent Composition Pulses in Three-Component Systems

The quantitative calculation of responses $\mathbf{h}(z, \tau)$ to arbitrary square-wave influent composition pulses is described in this Appendix. A derivation of the relations for distances and times required for resolution and flat-top disappearance, listed in Table 4.3, is included.

*Classification and Formulas.* Responses of four types may be generated, distinguished by the sharpening characteristics of the flanks of the two response pulses. Distance-time diagrams for the four possible types are shown in Fig. A.2. The inequality conditions for the *H*-function roots of the base influent, $h_1'$ and $h_2'$, and influent pulse, $h_1^0$ and $h_2^0$, are given for each case (see also Table 4.3).

In all four cases, the "triangle formula" will be used for calculating

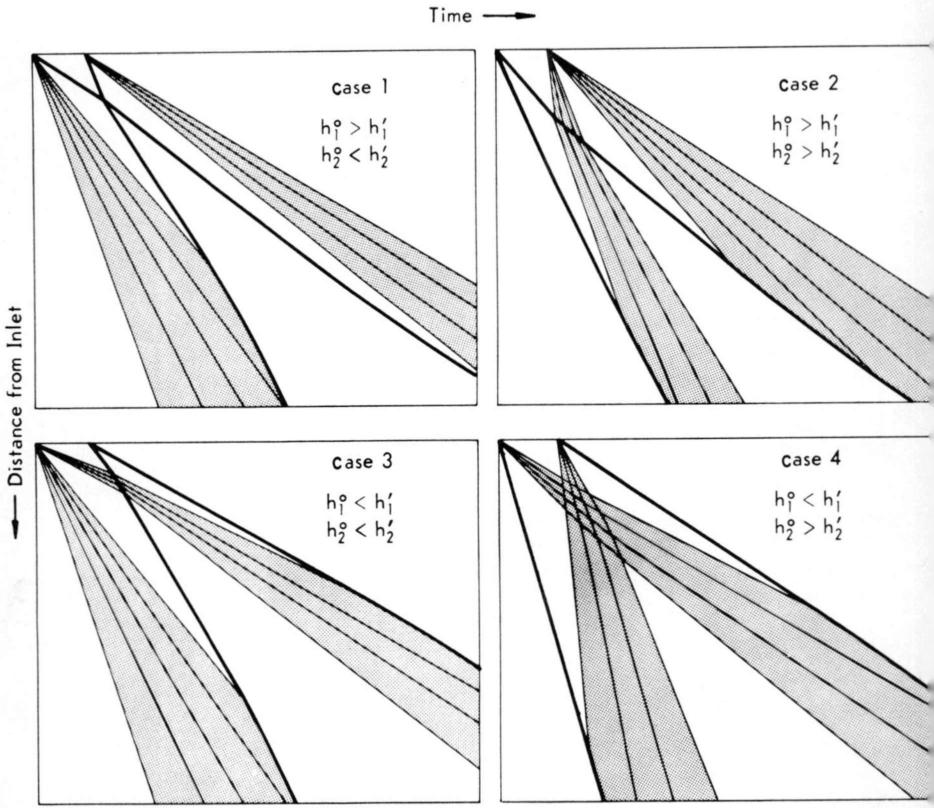


**Fig. A.2.** Schematic distance-time diagrams of the four types of responses to square-wave influent composition pulses in three-component systems. Diffuse pulse flanks are shaded, sharp pulse flanks shown as heavy lines.

the point $(z, \tau)$ of intersection of two linear trajectories with slopes (velocities) $u_A$ and $u_B$ and points of origin $(z_A, \tau_A)$ and $(z_B, \tau_B)$:

$$\tau = \frac{u_B \tau_B - u_A \tau_A - z_B + z_A}{u_B - u_A}$$
$$z = z_A + u_A(\tau - \tau_A) \qquad \text{or} \qquad z = z_B + u_B(\tau - \tau_B) \tag{4.95}$$

(See Fig. A.3; for derivation, see Chapter 4, Section IV.B.4.)

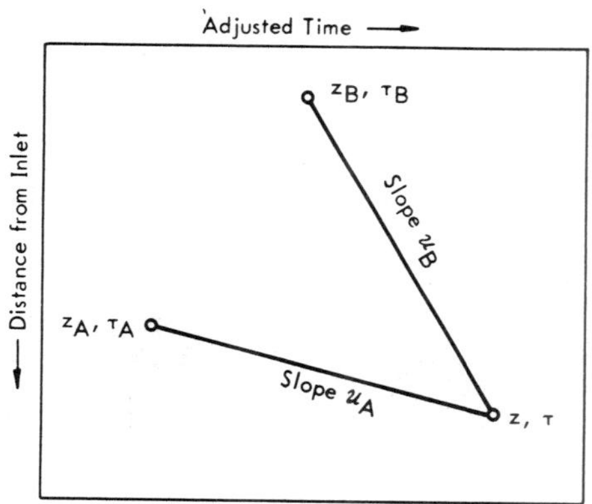


**Fig. A.3.** Notation of quantities used in triangle formula (4.95).

For three-component systems, Eqs. (3.68) and (3.71) give for the adjusted composition velocities $u$ in diffuse flanks and for the adjusted step velocities $u_\Delta$ of sharp flanks:

$$\left.\begin{aligned} u &= h_1^2 h_2/\alpha \\ u_\Delta &= h_1' h_1'' h_2/\alpha \end{aligned}\right\} \quad \text{if variable root is } h_1$$
$$\left.\begin{aligned} u &= h_1 h_2^2/\alpha \\ u_\Delta &= h_2' h_2'' h_1/\alpha \end{aligned}\right\} \quad \text{if variable root is } h_2 \tag{A.47}$$

where

$$\alpha \equiv \alpha_{11}\alpha_{12}\alpha_{13} \tag{A.48}$$

and where primes and double primes refer to the upstream and downstream sides of the step. To assist in substituting the appropriate root values in Eqs. (A.47), the detailed distance-time diagrams in Figs. A.4 to A.7 will list the root values in each plateau zone.

A further standard procedure will be the calculation of a linear trajectory in the form $z(\tau)$ from its known slope (adjusted velocity) $u$ and point of origin $(z_0, \tau_0)$. Since

$$u = (z - z_0)/(\tau - \tau_0) \tag{A.49}$$

the equation for the trajectory is

$$z(\tau) = z_0 + u(\tau - \tau_0) \tag{A.50}$$

The derivation for case 1 will be given in detail. Those for the others will only be outlined, since essentially the same methods are used.

As it turns out in all four cases, the variables $z$, $\tau$ and $\Delta\tau$, and $\alpha$ always occur in such combinations that the response, when expressed as $\mathbf{h}(\alpha z/\Delta\tau, \tau/\Delta\tau)$, is independent of the numerical values of $\alpha$ and $\Delta\tau$. Coordinate scales $\alpha z/\Delta\tau$ and $\tau/\Delta\tau$ are therefore used in the generalized distance-time diagrams in Figs. A.4 to A.7.

1. *Case* 1 ($h_1^0 > h_1'$, $h_2^0 < h_2'$): *two sharp pulse flanks cross* (see Fig. A.4)

*Points of Intersection and Flat-Top Disappearance.* The point of intersection $(z_{1/2}, \tau_{1/2})$ of the two sharp pulse flanks is calculated first. Comparison with Fig. A.3 shows that application of the triangle formula requires the following substitutions in Eqs. (4.95):

$$z_A = 0, \qquad \tau_A = 0$$
$$z_B = 0, \qquad \tau_B = \Delta\tau \tag{A.51}$$

($\Delta\tau$ = duration of influent pulse). For the velocities of the intersecting pulse flanks, $u_A$ and $u_B$ in Eqs. (4.95), Eqs. (A.47) with the root values shown in Fig. A.4 give

$$u_A = h_1' h_1^0 h_2^0/\alpha, \qquad u_B = h_1^0 h_2' h_2^0/\alpha \tag{A.52}$$

With these substitutions, Eqs. (4.95) give for the coordinates of the point of intersection

$$z_{1/2} = \frac{h_1' h_1^0 h_2' h_2^0}{(h_2' - h_1')\alpha} \Delta\tau, \qquad \tau_{1/2} = \frac{h_2'}{h_2' - h_1'} \Delta\tau \tag{A.53}$$

The point $(z_1, \tau_1)$ of disappearance of the flat top of the slower response pulse is calculated next. Here, the substitutions required for application of Eqs. (4.95) are

$$z_A = z_{1/2}, \qquad \tau_A = \tau_{1/2}$$
$$z_B = 0, \qquad \tau_B = \Delta\tau \tag{A.54}$$

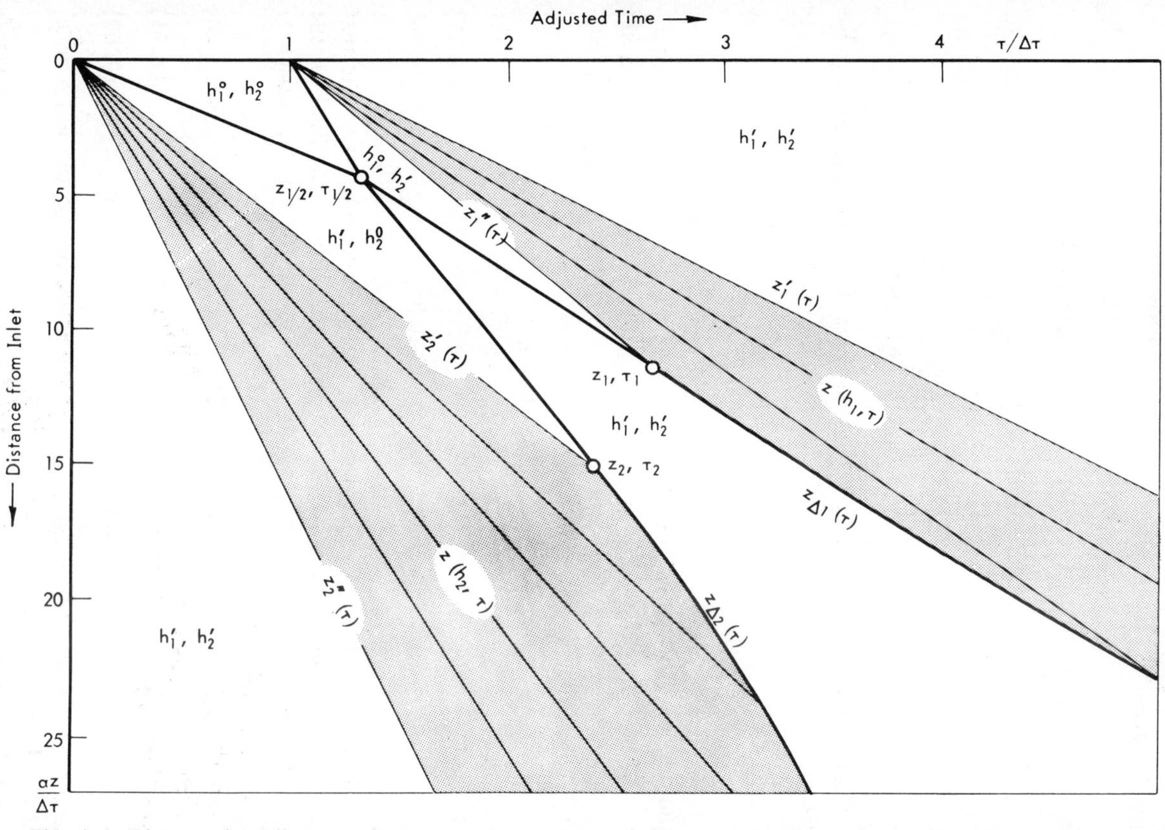


**Fig. A.4.** Distance-time diagram of response to square-wave influent composition pulse in three-component system with $h_1{}^\circ > h_1{}'$, $h_2{}^\circ < h_2{}'$ (case 1). (For $h_1{}' = 1$, $h_2{}' = 4$, $h_1{}^\circ = 1.3$, $h_2{}^\circ = 2.5$.)

For the velocities of the trajectories intersecting at $(z_1, \tau_1)$, appearing as $u_A$ and $u_B$ in Eqs. (4.95), Eqs. (A.47) with the root values listed in Fig. A.4 give

$$u_A = h_1' h_1^0 h_2'/\alpha, \qquad u_B = h_1^{0^2} h_2'/\alpha \tag{A.55}$$

With the substitutions (A.53) to (A.55), Eqs. (4.95) yield

$$\begin{aligned} z_1 &= \frac{h_1' h_1^{0^2} h_2'(h_2^0 - h_1')}{(h_2' - h_1')(h_1^0 - h_1')\alpha} \Delta\tau \\ \tau_1 &= \left(1 + \frac{h_1'(h_2^0 - h_1')}{(h_2' - h_1')(h_1^0 - h_1')}\right) \Delta\tau \end{aligned} \tag{A.56}$$

Similarly, for the point $(z_2, \tau_2)$ at which the flat top of the faster pulse disappears, the appropriate substitutions in Eqs. (4.95) are

$$\begin{aligned} z_A &= 0, \qquad & \tau_A &= 0 \\ z_B &= z_{1/2}, \qquad & \tau_B &= \tau_{1/2} \end{aligned} \tag{A.57}$$

and the slopes of the intersecting trajectories are

$$u_A = h_1' h_2^{0^2}/\alpha, \qquad u_B = h_1' h_2' h_2^0/\alpha \tag{A.58}$$

With these substitutions, Eqs. (4.95) give

$$\begin{aligned} z_2 &= \frac{h_1' h_2' h_2^{0^2}(h_2' - h_1^0)}{(h_2' - h_1')(h_2' - h_2^0)\alpha} \Delta\tau \\ \tau_2 &= \frac{h_2'(h_2' - h_1^0)}{(h_2' - h_1')(h_2' - h_2^0)} \Delta\tau \end{aligned} \tag{A.59}$$

*Trajectories.* For the trajectories $z_1'(\tau)$, $z_1''(\tau)$, and $z(h_1, \tau)$ of the rear end, the front end, and given $h_1$ values of the diffuse rear flank of the slower response pulse, all originating at the point $(0, \Delta\tau)$, the trajectory formula (A.50) with the root values listed in Fig. A.4 yields

$$z_1'(\tau) = h_1'^2 h_2'(\tau - \Delta\tau)/\alpha \tag{A.60}$$

$$z_1''(\tau) = h_1^{0^2} h_2'(\tau - \Delta\tau)/\alpha \qquad (z \leqq z_1) \tag{A.61}$$

$$z(h_1, \tau) = h_1^2 h_2'(\tau - \Delta\tau)/\alpha \qquad (z \leqq z_{\Delta 1}(\tau)) \tag{A.62}$$

where $z_{\Delta 1}(\tau)$, still to be calculated, is the trajectory of the pulse crest, at which the $h_1$ trajectories end (see Fig. A.4).

Similarly, for the trajectories $z_2'(\tau)$, $z_2''(\tau)$, and $z(h_2, \tau)$ of the rear end, the front end, and given $h_2$ values of the diffuse front flank of the faster response pulse, all originating at the point (0, 0), the formula (A.50) gives

$$z_2'(\tau) = h_1' h_2^{0\,2} \tau/\alpha \qquad (z \leqq z_2) \tag{A.63}$$

$$z_2''(\tau) = h_1' h_2'^{\,2} \tau/\alpha \tag{A.64}$$

$$z(h_2, \tau) = h_1' h_2^2 \tau/\alpha \qquad (z \leqq z_{\Delta 2}(\tau)) \tag{A.65}$$

where $z_{\Delta 2}(\tau)$ is the crest trajectory, at which the $h_2$ trajectories end.

The trajectory $z_{\Delta 1}(\tau)$ of the sharp front flank of the slower pulse, after flat-top disappearance constituting the trajectory of the pulse crest, is given by

$$\begin{aligned} z_{\Delta 1}(\tau) &= h_1' h_1^0 h_2^0 \tau/\alpha && (\tau \leqq \tau_{1/2}) \\ &= \frac{h_1' h_1^0 h_2'}{\alpha}\left(\tau - \frac{h_2' - h_2^0}{h_2' - h_1'}\Delta\tau\right) && (\tau_{1/2} \leqq \tau \leqq \tau_1) \\ &= \frac{h_1' h_2'}{\alpha}\left[[h_1'(\tau - \Delta\tau)]^{1/2} + \left(\frac{(h_2^0 - h_1')(h_1^0 - h_1')}{h_2' - h_1'}\Delta\tau\right)^{1/2}\right]^2 && (\tau \geqq \tau_1) \end{aligned} \tag{A.66}$$

These three equations give the three distinct segments of the trajectory, from origin to crossover with the other sharp flank, from crossover to flat-top disappearance, and beyond flat-top disappearance. The derivation is as follows:

The first equation (A.66) is obtained directly from the trajectory formula (A.50) with the root values listed in Fig. A.4. The second equation is obtained from the trajectory formula after elimination of the origin coordinates $z_{1/2}$ and $\tau_{1/2}$ with Eqs. (A.53). The third equation can be derived as follows: The slope of the crest trajectory at any point is

$$dz_{\Delta 1}/d\tau = h_1 h_1' h_2'/\alpha \qquad (\tau \geqq \tau_1) \tag{A.67}$$

where $h_1$ is the local value of that root at the pulse crest, decreasing from $h_1 = h_1^0$ at the point of flat-top disappearance and approaching $h_1'$ as $\tau \to \infty$. A second relation between $h_1$, $z_{\Delta 1}$, and $\tau$ is provided by Eq. (A.62), valid over the entire lengths of the $h_1$ trajectories including their end points on the pulse crest, where $z = z_{\Delta 1}$. Solved for $h_1$, Eq. (A.62) gives

$$h_1 = [\alpha z/h_2'(\tau - \Delta\tau)]^{1/2} \qquad (z \leqq z_{\Delta 1}(\tau)) \tag{A.68}$$

With this substitution, at $z = z_{\Delta 1}$, Eq. (A.67) becomes

$$dz_{\Delta 1}/d\tau = h_1'[h_2' z_{\Delta 1}/\alpha(\tau - \Delta\tau)]^{1/2} \qquad (\tau \geqq \tau_1) \tag{A.69}$$

After separation of variables and integration along the crest from the point $(z_1, \tau_1)$ of flat-top disappearance to an arbitrary point $z_{\Delta 1}(\tau)$:

$$\int_{z=z_1}^{z_{\Delta 1}} \mathrm{d}z_{\Delta 1}/z_{\Delta 1}^{1/2} = h_1'(h_2'/\alpha)^{1/2} \int_{\tau=\tau_1}^{\tau} \mathrm{d}\tau/(\tau-\Delta\tau)^{1/2} \tag{A.70}$$

one obtains

$$z_{\Delta 1}^{1/2}(\tau) = z_1^{1/2} + h_1'(h_2'/\alpha)^{1/2}[(\tau-\Delta\tau)^{1/2} - (\tau_1-\Delta\tau)^{1/2}] \qquad (\tau \geqq \tau_1) \tag{A.71}$$

Elimination of $z_1$ and $\tau_1$ with Eqs. (A.56) and rearrangement yields the third equation (A.66).

The trajectory of the sharp rear flank of the faster response pulse, after flat-top disappearance also the crest trajectory, is given by

$$\begin{aligned} z_{\Delta 2}(\tau) &= h_1^0 h_2' h_2^0 (\tau-\Delta\tau)/\alpha && (\tau \leqq \tau_{1/2}) \\ &= \frac{h_1' h_2' h_2^0}{\alpha}\left(\tau - \frac{h_2'-h_1^0}{h_2'-h_1'}\,\Delta\tau\right) && (\tau_{1/2} \leqq \tau \leqq \tau_2) \\ &= \frac{h_1' h_2'}{\alpha}\left[(h_2'\tau)^{1/2} - \left(\frac{(h_2'-h_1^0)(h_2'-h_2^0)}{h_2'-h_1'}\,\Delta\tau\right)^{1/2}\right]^2 && (\tau \geqq \tau_2) \end{aligned} \tag{A.72}$$

The first and second equations are obtained from the trajectory formula (A.50), the second after elimination of $z_{1/2}$ and $\tau_{1/2}$ with Eqs. (A.53). The third equation is calculated from the equation for the slope of the crest trajectory

$$\mathrm{d}z_{\Delta 2}/\mathrm{d}\tau = h_1' h_2 h_2'/\alpha \qquad (\tau \geqq \tau_2) \tag{A.73}$$

Equation (A.65), solved for $h_2$, gives

$$h_2 = (\alpha z/h_1'\tau)^{1/2} \qquad (z \leqq z_{\Delta 2}(\tau)) \tag{A.74}$$

and is valid up to the pulse crest, where $z = z_{\Delta 2}$. With $h_2$ thus replaced, Eq. (A.73) can be integrated along the crest from $(z_2, \tau_2)$ to an arbitrary point $z_{\Delta 2}(\tau)$. The integration yields

$$z_{\Delta 2}^{1/2}(\tau) = z_2^{1/2} + h_2'(h_1'/\alpha)^{1/2}(\tau^{1/2} - \tau_2^{1/2}) \qquad (\tau \geqq \tau_2) \tag{A.75}$$

Elimination of $z_2$ and $\tau_2$ with Eqs. (A.59) and rearrangement gives the third equation (A.72).

**Table A.1**

RESPONSE $\mathbf{h}(z, \tau)$ TO INFLUENT PULSE IN THREE-COMPONENT SYSTEM WITH $h_1^0 > h_1'$, $h_2^0 < h_2'$ (CASE 1).

| | | | |
|---|---|---|---|
| $\tau \geqq 0$, | $z \geqq z_2''(\tau)$ | | |
| $\tau \geqq \Delta\tau$, | $z \leqq z_1'(\tau)$ | $h_1 = h_1'$, $h_2 = h_2'$ | Base influent |
| $\tau \geqq \tau_{1/2}$, | $z_{\Delta 1}(\tau) < z < z_{\Delta 2}(\tau)$ | | |
| $\tau \leqq \tau_2$, | $z_2'(\tau) \leqq z \leqq z_2''(\tau)$ | $h_1 = h_1'$, $h_2 = \left(\dfrac{\alpha z}{h_1'\tau}\right)^{1/2}$ | Diffuse flank of faster response pulse |
| $\tau \geqq \tau_2$, | $z_{\Delta 2}(\tau) \leqq z \leqq z_2''(\tau)$ | | |
| $\tau \leqq \tau_{1/2}$, | $z_{\Delta 1}(\tau) < z \leqq z_1'(\tau)$ | $h_1 = h_1'$, $h_2 = h_2^0$ | Flat top of faster response pulse |
| $\tau_{1/2} \leqq \tau \leqq \tau_2$, | $z_{\Delta 2}(\tau) < z \leqq z_1'(\tau)$ | | |
| $\tau \leqq \Delta\tau$, | $z < z_{\Delta 1}(\tau)$ | $h_1 = h_1^0$, $h_2 = h_2^0$ | Influent pulse |
| $\Delta\tau \leqq \tau \leqq \tau_{1/2}$, | $z_{\Delta 2}(\tau) < z < z_{\Delta 1}(\tau)$ | | |
| $\Delta\tau \leqq \tau \leqq \tau_{1/2}$, | $z_1''(\tau) \leqq z < z_{\Delta 2}(\tau)$ | $h_1 = h_1^0$, $h_2 = h_2'$ | Flat top of slower response pulse |
| $\tau_{1/2} \leqq \tau \leqq \tau_1$, | $z_1''(\tau) \leqq z < z_{\Delta 1}(\tau)$ | | |
| $\Delta\tau \leqq \tau \leqq \tau_1$, | $z_1'(\tau) \leqq z \leqq z_1''(\tau)$ | $h_1 = \left(\dfrac{\alpha z}{h_2'(\tau-\Delta\tau)}\right)^{1/2}$, $h_2 = h_2'$ | Diffuse flank of slower response pulse |
| $\tau \geqq \tau_1$, | $z_1'(\tau) < z < z_{\Delta 1}(\tau)$ | | |

It is readily verified that the equations for the crest trajectories, i.e., the third equations (A.66) and (A.72), yield

$$\begin{aligned}
(dz_{\Delta 1}/d\tau)_{\tau=\tau_1} &= h_1' h_1^0 h_2'/\alpha \\
(dz_{\Delta 1}/d\tau)_{\tau\to\infty} &= h_1'^2 h_2'/\alpha \\
(dz_{\Delta 2}/d\tau)_{\tau=\tau_2} &= h_1' h_2' h_2^0/\alpha \\
(dz_{\Delta 2}/d\tau)_{\tau\to\infty} &= h_1' h_2'^2/\alpha
\end{aligned} \tag{A.76}$$

That is, the crest velocities at the times of flat-top disappearance and the (asymptotically approached) final crest velocities are correctly obtained as the limiting values of Eqs. (A.67) and (A.73) with $h_1 = h_1^0$, $h_1 = h_1'$, $h_2 = h_2^0$, and $h_2 = h_2'$.

*Response* $\mathbf{h}(z, \tau)$. The complete response $\mathbf{h}(z, \tau)$ is now readily compiled with the aid of Fig. A.4. A listing is given in Table A.1. The equations for the variable roots in the diffuse pulse flanks are provided by Eqs. (A.68) and (A.74).

2. *Case* 2 ($h_1^0 > h_1'$, $h_2^0 > h_2'$): *diffuse rear flank of faster response pulse crosses sharp front flank of slower response pulse* (see Fig. A.5)

The point $(z_0, \tau_0)$ is calculated with the triangle formula from the points (0, 0) and (0, $\Delta\tau$):

$$z_0 = \frac{h_1' h_1^0 h_2^{0\,2}}{(h_2^0 - h_1')\alpha} \Delta\tau, \qquad \tau_0 = \frac{h_2^0}{h_2^0 - h_1'} \Delta\tau \tag{A.77}$$

The point $(z_2, \tau_2)$ calculated with the triangle formula and Eqs. (A.77) from the points (0, 0) and $(z_0, \tau_0)$, is

$$\begin{aligned}
z_2 &= \frac{h_1' h_2' h_2^{0\,2}(h_2^0 - h_1^0)}{(h_2^0 - h_1')(h_2^0 - h_2')\alpha} \Delta\tau \\
\tau_2 &= \frac{h_2^0(h_2^0 - h_1^0)}{(h_2^0 - h_1')(h_2^0 - h_2')} \Delta\tau
\end{aligned} \tag{A.78}$$

The trajectories of the rear end, front end, and given $h_1$ values of the diffuse flank of the slower pulse, all originating from (0, $\Delta\tau$), are obtained directly from the trajectory formula (A.50):

$$z_1'(\tau) = h_1'^2 h_2'(\tau - \Delta\tau)/\alpha \tag{A.79}$$

$$z_1''(\tau) = h_1^{0\,2} h_2'(\tau - \Delta\tau)/\alpha \qquad (z \leqq z_1) \tag{A.80}$$

$$z(h_1, \tau) = h_1^2 h_2'(\tau - \Delta\tau)/\alpha \qquad (z \leqq z_{\Delta 1}(\tau)) \tag{A.81}$$

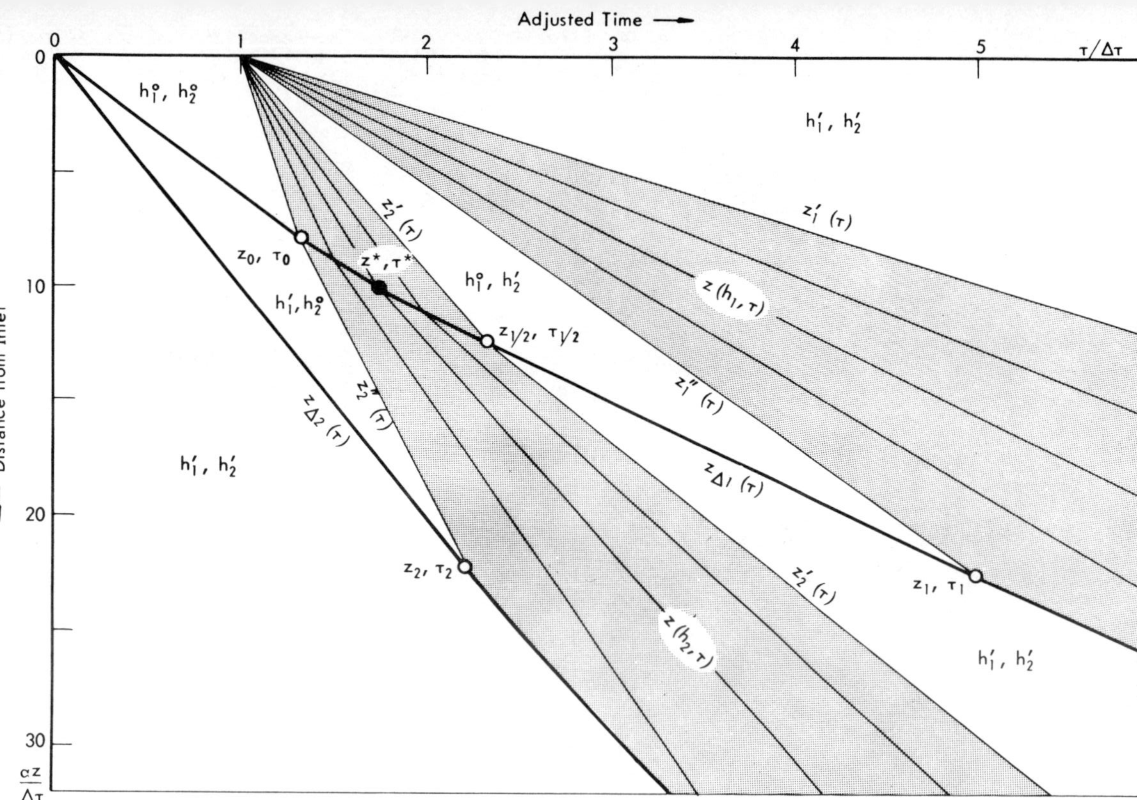


**Fig. A.5.** Distance-time diagram of response to square-wave influent composition pulse in three-component system with $h_1{}^o > h_1{}'$, $h_2{}^o > h_2{}'$ (case 2). (For $h_1{}' = 1$, $h_2{}' = 2.5$. $h_1{}^o = 1.5$, $h_2{}^o = 4$.)

The trajectories of the rear end, front end, and given $h_2$ values of the diffuse flank of the faster pulse can at this stage be given only to the crossing trajectory $z_{\Delta 1}(\tau)$:

$$z_2'(\tau) = h_1^0 h_2'^2(\tau - \Delta\tau)/\alpha \qquad (z \leqq z_{1/2}) \tag{A.82}$$

$$z_2''(\tau) = h_1^0 h_2^{0\,2}(\tau - \Delta\tau)/\alpha \qquad (z \leqq z_0) \tag{A.83}$$

$$z(h_2, \tau) = h_1^0 h_2^2(\tau - \Delta\tau)/\alpha \qquad (z \leqq z_{\Delta 1}(\tau)) \tag{A.84}$$

The trajectory of the sharp front flank of the slower pulse can now be calculated to the point $(z_{1/2}, \tau_{1/2})$:

$$\begin{aligned} z_{\Delta 1}(\tau) &= h_1' h_1^0 h_2^0 \tau/\alpha & (\tau \leqq \tau_0) \\ &= \frac{h_1' h_1^0}{\alpha} \{[h_1'(\tau - \Delta\tau)]^{1/2} + [(h_2^0 - h_1')\Delta\tau]^{1/2}\}^2 & (\tau_0 \leqq \tau \leqq \tau_{1/2}) \end{aligned} \tag{A.85}$$

The first of these equations, referring to the trajectory before crossover, is obtained directly from the trajectory formula (A.50). The second equation, referring to the trajectory during crossover with the diffuse flank of the other pulse, is calculated from the equation for the trajectory slope

$$dz_{\Delta 1}/d\tau = h_1^0 h_1' h_2/\alpha \qquad (\tau_0 \leqq \tau \leqq \tau_{1/2}) \tag{A.86}$$

after elimination of $h_2$ with

$$h_2 = [\alpha z/h_1^0(\tau - \Delta\tau)]^{1/2} \qquad (z \leqq z_{\Delta 1}(\tau)) \tag{A.87}$$

obtained from Eq. (A.84), integration along the trajectory from $(z_0, \tau_0)$ to an arbitrary point $z_{\Delta 1}(\tau)$, and elimination of $z_0$ and $\tau_0$ with Eqs. (A.77).

The point $(z_{1/2}, \tau_{1/2})$ can now be found as follows. In Eq. (A.82), $z_2'(\tau) = z_{1/2}$ if $\tau = \tau_{1/2}$; in the second equation (A.85), $z_{\Delta 1}(\tau) = z_{1/2}$ if $\tau = \tau_{1/2}$. Solving the two equations thus obtained for $z_{1/2}$ and $\tau_{1/2}$, one finds

$$\begin{aligned} z_{1/2} &= \frac{h_1' h_1^0 h_2'^2 (h_2^0 - h_1')}{(h_2' - h_1')^2 \alpha} \Delta\tau \\ \tau_{1/2} &= \left(1 + \frac{h_1'(h_2^0 - h_1')}{(h_2' - h_1')^2}\right) \Delta\tau \end{aligned} \tag{A.88}$$

The coordinates $z^*(h_2)$ and $\tau^*(h_2)$ of the intersection of a given $h_2$ trajectory with that of the sharp pulse flank (see Fig. A.5) are obtained by an analogous derivation, with Eq. (A.84) instead of (A.82) and $h_2$ instead of $h_2'$:

$$z^*(h_2) = \frac{h_1' h_1^0 h_2^2 (h_2^0 - h_1')}{(h_2 - h_1')^2 \alpha} \Delta\tau$$
$$\tau^*(h_2) = \left(1 + \frac{h_1'(h_2^0 - h_1')}{(h_2 - h_1')^2}\right) \Delta\tau \qquad \text{(A.89)}$$

The trajectories of the diffuse flank of the faster pulse after crossover, i.e., the continuations of those given in Eqs. (A.82) to (A.84), can now be calculated with the trajectory formula and Eqs. (A.88), (A.77), and (A.89):

$$z(h_2, \tau) = \frac{h_1' h_2^2}{\alpha}\left[\tau - \left(1 - \frac{(h_2^0 - h_1')(h_1^0 - h_1')}{(h_2 - h_1')^2}\right)\Delta\tau\right] \qquad \text{(A.90)}$$
$$(z_{\Delta 1}(\tau) \leqq z \leqq z_{\Delta 2}(\tau))$$

$$z_2'(\tau) = z(h_2', \tau) \qquad (z \geqq z_{1/2}) \qquad \text{(A.91)}$$

$$z_2''(\tau) = z(h_2^0, \tau) \qquad (z_0 \leqq z \leqq z_2) \qquad \text{(A.92)}$$

where the function on the right-hand sides of Eqs. (A.91) and (A.92) is that given in Eq. (A.90).

The point $(z_1, \tau_1)$ is now calculated with the triangle formula from the points $(z_{1/2}, \tau_{1/2})$ and $(0, \Delta\tau)$ and Eqs. (A.88). The result turns out to be the same as for case 1, Eqs. (A.56).

The continuation of the trajectory of the sharp flank of the slower pulse after crossover can now be calculated:

$$z_{\Delta 1}(\tau) = \frac{h_1' h_1^0 h_2'}{\alpha}\left(\tau + \frac{h_2^0 - h_1'}{h_2' - h_1'}\Delta\tau\right) \qquad (\tau_{1/2} \leqq \tau \leqq \tau_1)$$
$$= \frac{h_1' h_2'}{\alpha}\left[[h_1'(\tau - \Delta\tau)]^{1/2} + \left(\frac{(h_2^0 - h_1')(h_1^0 - h_1')}{h_2' - h_1'}\Delta\tau\right)^{1/2}\right]^2$$
$$(\tau \geqq \tau_1) \qquad \text{(A.93)}$$

The first equation is obtained with the trajectory formula and Eqs. (A.88). The second equation, referring to the crest trajectory, is calculated from the equation for the slope

$$dz_{\Delta 1}/d\tau = h_1 h_1' h_2'/\alpha \qquad (\tau \geqq \tau_1) \qquad \text{(A.94)}$$

after elimination of $h_1$ with

$$h_1 = [\alpha z/h_2'(\tau-\Delta\tau)]^{1/2} \qquad (z \leqq z_{\Delta 1}(\tau)) \tag{A.95}$$

obtained from Eq. (A.81), by integration along the crest from $(z_1, \tau_1)$ to an arbitrary point, and elimination of $z_1$ and $\tau_1$ with Eqs. (A.56).

The trajectory of the sharp flank of the faster pulse to the point of flat-top disappearance is obtained with the trajectory formula (A.50):

$$z_{\Delta 2}(\tau) = h_1' h_2' h_2^0 \tau/\alpha \qquad (\tau \leqq \tau_2) \tag{A.96}$$

No explicit equation can be given beyond this point, because Eq. (A.90) cannot be solved explicitly for $h_2$. The integration of the equation for its slope

$$\mathrm{d}z_{\Delta 2}/\mathrm{d}\tau = h_1' h_2 h_2'/\alpha \qquad (\tau \geqq \tau_2) \tag{A.97}$$

must therefore be carried out numerically.

With all trajectories, points of intersection, etc., calculated, the complete response $\mathbf{h}(z, \tau)$ is now readily compiled with Fig. A.6 in the same manner as in the previous case. The solution is explicit, except that $h_2(z, \tau)$ in the diffuse flank of the faster pulse after crossover is given as the appropriate root of Eq. (A.90).

3. *Case* 3 ($h_1^0 < h_1'$, $h_2^0 < h_2'$): *sharp rear flank of faster response pulse crosses diffuse front flank of slower response pulse* (see Fig. A.6)

This case is very similar to the previous one in that a sharp and a diffuse flank cross. In fact, as far as root values are concerned, the equations of this case can be obtained from those of the previous one by interchange of the indices 1 and 2. With respect to signs and interchange of $\tau$ and $\tau-\Delta\tau$, however, the symmetry is complex enough to make its exact derivation as lengthy as the direct calculation of the trajectories and intersections.

The equations below are ordered according to content and therefore are not in the sequence of derivation. The derivations are indicated with reference to the analogous steps in the previous case.

The points of intersection, etc., are

$$z_0 = \frac{h_1^{0\,2} h_2' h_2^0}{(h_2'-h_1^0)\alpha}\Delta\tau, \qquad \tau_0 = \frac{h_2'}{h_2'-h_1^0}\Delta\tau \tag{A.98}$$

$$z_{1/2} = \frac{h_1'^2 h_2' h_2^0 (h_2^0-h_1^0)}{(h_2'-h_1')^2\alpha}\Delta\tau, \qquad \tau_{1/2} = \frac{h_2'(h_2'-h_1^0)}{(h_2'-h_1')^2}\Delta\tau \tag{A.99}$$

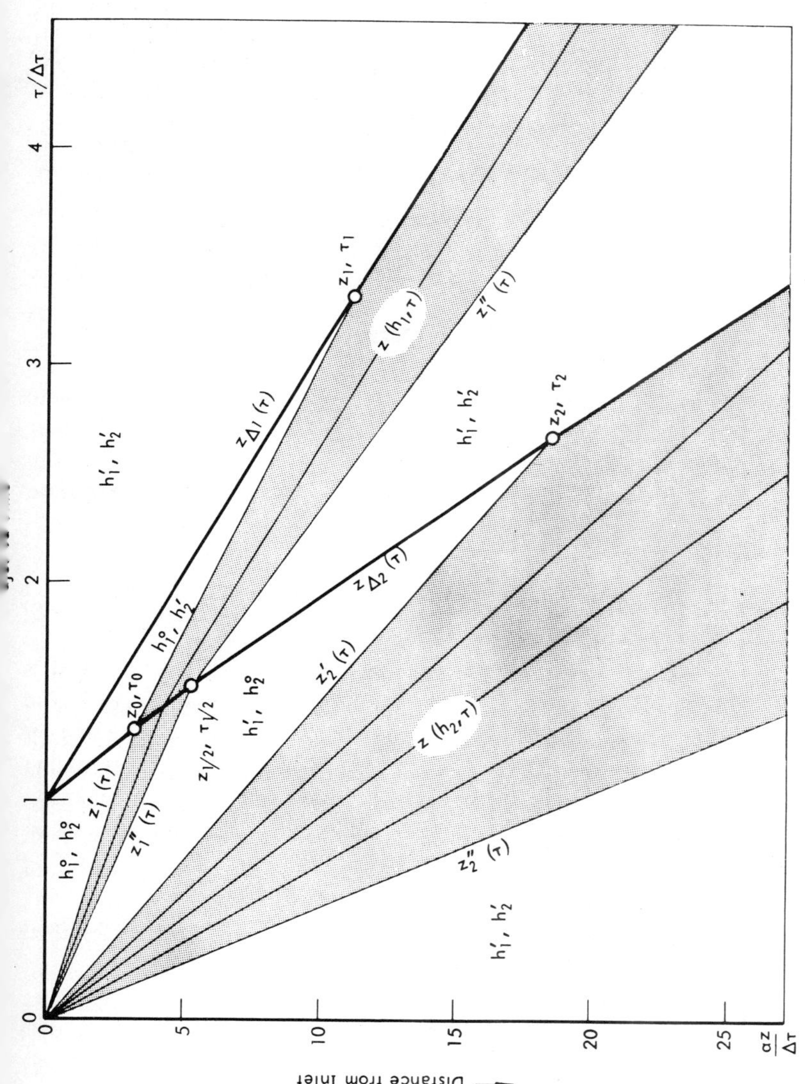


**Fig. A.6.** Distance-time diagram of response to square-wave influent composition pulse in three-component system with $h_1{}^o < h_1{}'$, $h_2{}^o < h_2{}'$ (case 3). (For $h_1{}' = 1.2$, $h_2{}' = 4$, $h_1{}^o = 1$, $h_2{}^o = 2.4$.)

$$z_1 = \frac{h_1' h_1^{0^2} h_2' (h_2^0 - h_1^0)}{(h_2' - h_1^0)(h_1' - h_1^0)\alpha} \Delta\tau$$

$$\tau_1 = \left(1 + \frac{h_1^0 (h_2^0 - h_1^0)}{(h_2' - h_1^0)(h_1' - h_1^0)}\right) \Delta\tau \tag{A.100}$$

The equations for $z_2$ and $\tau_2$ turn out to be identical with those for case 1, Eqs. (A.59). Equations (A.98) and (A.100) are calculated with the triangle formula. Equations (A.99) are obtained by solving Eq. (A.103) and the second equation (A.107) for $z$ and $\tau$ at the trajectory intersection [as for Eq. (A.88) in case 2]. The point $(z_2, \tau_2)$ can then be calculated with the triangle formula.

The trajectories for the diffuse flank of the slower pulse are

$$z(h_1, \tau) = h_1^2 h_2^0 \tau/\alpha \qquad (z \leqq z_{\Delta 2}(\tau))$$

$$= \frac{h_1^2 h_2'}{\alpha}\left(\tau - \frac{(h_2' - h_1^0)(h_2' - h_2^0)}{(h_2' - h_1)^2} \Delta\tau\right) \qquad (z_{\Delta 2}(\tau) \leqq z \leqq z_{\Delta 1}(\tau)) \tag{A.101}$$

$$z_1'(\tau) = z(h_1^0, \tau) \qquad (z \leqq z_1) \tag{A.102}$$

$$z_1''(\tau) = z(h_1', \tau) \tag{A.103}$$

where the function on the right-hand sides of Eqs. (A.102) and (A.103) is that given in Eq. (A.101). These trajectories are obtained with the formula (A.50). Before crossover, the formula can be applied directly; after crossover, the points of intersection $z^*(h_1)$, $\tau^*(h_1)$ must first be calculated. These are given by Eqs. (A.99) with $h_1$ substituted for $h_1'$ [see derivation of Eqs. (A.90) and (A.89) in case 2].

The trajectories of the diffuse flank of the faster pulse, directly obtained with the formula (A.50), are

$$z_2'(\tau) = h_1' h_2^{0^2} \tau/\alpha \qquad (z \leqq z_2) \tag{A.104}$$

$$z_2''(\tau) = h_1' h_2' \tau/\alpha \tag{A.105}$$

$$z(h_2, \tau) = h_1' h_2^2 \tau/\alpha \qquad (z \leqq z_{\Delta 2}(\tau)) \tag{A.106}$$

The trajectory of the sharp flank of the faster pulse is given by

$$
\begin{aligned}
z_{\Delta 2}(\tau) &= h_1^0 h_2' h_2^0(\tau-\Delta\tau)/\alpha && (z \leqq z_0) \\
&= \frac{h_2' h_2^0}{\alpha}\{(h_2'\tau)^{1/2}-[(h_2'-h_1^0)\Delta\tau]^{1/2}\}^2 && (z_0 \leqq z \leqq z_{1/2}) \\
&= \frac{h_1' h_2' h_2^0}{\alpha}\left(\tau-\frac{h_2'-h_1^0}{h_2'-h_1'}\Delta\tau\right) && (z_{1/2} \leqq z \leqq z_2) \\
&= \frac{h_1' h_2'}{\alpha}\left[(h_2'\tau)^{1/2}-\left(\frac{(h_2'-h_1^0)(h_2'-h_2^0)}{h_2'-h_1'}\Delta\tau\right)^{1/2}\right]^2 && (z \leqq z_2)
\end{aligned}
\tag{A.107}
$$

The first and third of these equations are obtained from the formula (A.50), the third with Eqs. (A.99) for the origin. The second and fourth equations are obtained by integration of the equation for the trajectory slope [see derivations of the second equations (A.85) and (A.93) in the previous case], with Eqs. (A.98) and (A.59) for the starting points of the integration, and with the variable root values provided by

$$h_1 = (\alpha z/h_2^0\tau)^{1/2} \qquad (z \leqq z_{\Delta 2}(\tau)) \tag{A.108}$$

$$h_2 = (\alpha z/h_1'\tau)^{1/2} \tag{A.109}$$

obtained from Eqs. (A.101) and (A.104).

The trajectory of the sharp flank of the slower pulse to the point of flat-top disappearance is obtained with the formula (A.50):

$$z_{\Delta 1}(\tau) = h_1' h_1^0 h_2'(\tau-\Delta\tau)/\alpha \qquad (\tau \leqq \tau_1) \tag{A.110}$$

Beyond this point, the equation

$$\mathrm{d}z_{\Delta 1}/\mathrm{d}\tau = h_1 h_1' h_2'/\alpha \qquad (\tau \geqq \tau_1) \tag{A.111}$$

must be integrated numerically.

The response $\mathbf{h}(z, \tau)$ is now readily compiled as in the previous cases.

## 4. *Case* 4 ($h_1^0 < h_1'$, $h_2^0 > h_2'$): *two diffuse pulse flanks cross* (see Fig. A.7)

The derivation follows similar lines as in the previous cases. The point

$$z_0 = \frac{(h_1^0 h_2^0)^2}{(h_2^0-h_1^0)\alpha}\Delta\tau, \qquad \tau_0 = \frac{h_2^0}{h_2^0-h_1^0}\Delta\tau \tag{A.112}$$

is calculated with the triangle formula, and the trajectories of the diffuse pulse flanks before crossover are obtained with the formula (A.50):

$$z_1'(\tau) = h_1^{0\,2} h_2^0 \tau/\alpha \qquad (z \leqq z_0) \tag{A.113}$$

$$z_1''(\tau) = h_1'^{\,2} h_2^0 \tau/\alpha \qquad (z \leqq z'') \tag{A.114}$$

$$z(h_1, \tau) = h_1^2 h_2^0 \tau/\alpha \qquad (z \leqq z_2''(\tau)) \tag{A.115}$$

$$z_2'(\tau) = h_1^0 h_2'^{\,2}(\tau - \Delta\tau)/\alpha \qquad (z \leqq z') \tag{A.116}$$

$$z_2''(\tau) = h_1^0 h_2^{0\,2}(\tau - \Delta\tau)/\alpha \qquad (z \leqq z_0) \tag{A.117}$$

$$z(h_2, \tau) = h_1^0 h_2^2(\tau - \Delta\tau)/\alpha \qquad (z \leqq z_1'(\tau)) \tag{A.118}$$

Equations (A.115) and (A.118), solved for $h_1$ and $h_2$, respectively, give

$$h_1 = (\alpha z/h_2^0 \tau)^{1/2} \qquad (z \leqq z_2''(\tau)) \tag{A.119}$$

$$h_2 = [\alpha z/h_1^0(\tau - \Delta\tau)]^{1/2} \qquad (z \leqq z_1'(\tau)) \tag{A.120}$$

The trajectory $z_1'(\tau)$ can now be calculated to the point of flat-top disappearance:

$$\begin{aligned} z_1'(\tau) &= \frac{h_1^{0\,2}}{\alpha}\{[h_1^0(\tau - \Delta\tau)]^{1/2} + [(h_2^0 - h_1^0)\Delta\tau]^{1/2}\}^2 \qquad (z_0 \leqq z \leqq z') \\ &= \frac{h_1^{0\,2} h_2'}{\alpha}\left(\tau + \frac{h_2^0 - h_1^0}{h_2' - h_1^0}\Delta\tau\right) \qquad (z' \leqq z \leqq z_1) \end{aligned} \tag{A.121}$$

The first equation is obtained by integration of

$$dz_1'/d\tau = h_1^{0\,2} h_2/\alpha \qquad (z_0 \leqq z \leqq z') \tag{A.122}$$

from $(z_0, \tau_0)$ along the trajectory after elimination of $h_2$ with Eq. (A.120). The second equation is obtained with the formula (A.50) after elimination of $z'$ and $\tau'$ with

$$z' = \left(\frac{h_1^0 h_2'}{h_2' - h_1^0}\right)^2 \frac{h_2^0 - h_1^0}{\alpha}\Delta\tau, \qquad \tau' = \left(1 + \frac{h_1^0(h_2^0 - h_1^0)}{(h_2' - h_1^0)^2}\right)\Delta\tau \tag{A.123}$$

obtained when Eq. (A.118) and the first equation (A.121) are solved for $z$ and $\tau$.

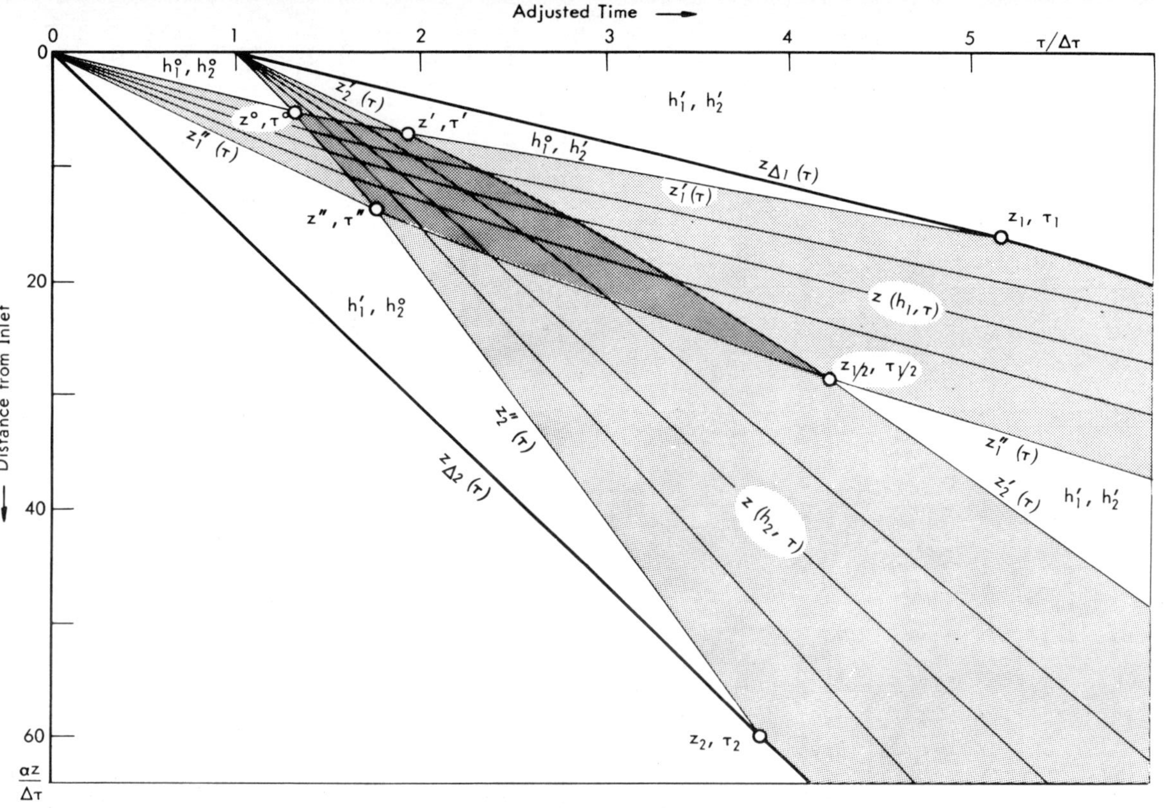


**Fig. A.7.** Distance-time diagram of response to square-wave influent composition pulse in three-component system with $h_1^o < h_1'$, $h_2^o > h_2'$ (case 4). (For $h_1' = 1.4$, $h_2' = 2.8$, $h_1^o = 1$, $h_2^o = 4$.)

The trajectory $z_2''(\tau)$, calculated in an analogous manner, is

$$z''(\tau) = \frac{h_2^{0^2}}{\alpha}\{(h_2^0\tau)^{1/2} - [(h_2^0 - h_1^0)\Delta\tau]^{1/2}\}^2 \qquad (z_0 \leqq z \leqq z'')$$

$$= \frac{h_1' h_2^{0^2}}{\alpha}\left(\tau - \frac{h_2^0 - h_1^0}{h_2^0 - h_1'}\Delta\tau\right) \qquad (z'' \leqq z \leqq z_2) \qquad \text{(A.124)}$$

where

$$z'' = \left(\frac{h_1' h_2^0}{h_2^0 - h_1'}\right)^2 \frac{h_2^0 - h_1^0}{\alpha}\Delta\tau, \qquad \tau'' = \frac{h_2^0(h_2^0 - h_1^0)}{(h_2^0 - h_1')^2}\Delta\tau \qquad \text{(A.125)}$$

is obtained by solving Eq. (A.115) and the first equation (A.124) for $z$ and $\tau$.

The points $(z_1, \tau_1)$ and $(z_2, \tau_2)$, at which the flat tops disappear, can now be calculated with the triangle formula. The results turn out to be the same as in Eqs. (A.100) and (A.78).

The trajectories of the sharp pulse flanks to the point of flat-top disappearance are obtained with the formula (A.50):

$$z_{\Delta 1}(\tau) = h_1' h_1^0 h_2'(\tau - \Delta\tau)/\alpha \qquad (z \leqq z_1) \qquad \text{(A.126)}$$

$$z_{\Delta 2}(\tau) = h_1' h_2' h_2^0 \tau/\alpha \qquad (z \leqq z_2) \qquad \text{(A.127)}$$

No analytical solutions are available for the region of noncoherence (darkly shaded in Fig. A.7), the point of resolution of the two pulses, the trajectories in the diffuse flanks after crossover, and the crest trajectories after flat-top disappearance. For the noncoherent region, the simultaneous differential equations

$$(\partial z/\partial\tau)_{h_1} = h_1^2 h_2/\alpha, \qquad (\partial z/\partial\tau)_{h_2} = h_1 h_2^2/\alpha \qquad \text{(A.128)}$$

must be integrated numerically or graphically. The further calculations and the compilation of the response $\mathbf{h}(z, \tau)$ can then proceed as in the previous cases.

## 5. *Remarks*

The methods used in the preceding derivations are more generally applicable. They were employed, for example, to calculate the responses in Figs. 3.39, 4.13, 4.24, 4.32, and 4.33. Also, the derivations are directly applicable to systems with more than three components, provided no more than two $H$-function roots vary; merely, the constant $\alpha$ has to be redefined accordingly:

$$\alpha \equiv \prod_i \alpha_{1i} \prod_{i \neq j,k} h_i \qquad \text{(A.129)}$$

if the variable roots are $h_j$ and $h_k$. Crossovers and mergers of two coherent boundaries in $n$-component systems can be calculated in this manner. Explicit analytical solutions are obtained for crossovers of two sharp boundaries in all cases, and of one sharp and one diffuse boundary if the root trajectories of the latter originate from a common point [e.g., as in Eqs. (A.101) and (A.107)]. In the general case, the calculation of crossovers or mergers of a diffuse with a sharp boundary requires integration of an ordinary differential equation with implicit argument [e.g., Eq. (A.97) with Eq. (A.90)], and that of two diffuse boundaries requires integration of two simultaneous partial differential equations [e.g., Eqs. (A.128)].

## IV. Selectivity Reversals in Systems with Mass-Action Equilibria

This Appendix provides a derivation of the general equation (5.60) for arbitrary $j{:}k$ reversal hyperplanes and of the criteria (5.62) and (5.63) for the existence of reversals.

### 1. *Reversal Hyperplanes*

We consider a reversal involving the pair of arbitrary species $j$ and $k$, with $j < k$. By definition, the system has such a reversal if there are compositions with $\alpha_{jk} > 1$ and others with $\alpha_{jk} < 1$. Since the equilibrium relations are continuous and differentiable, the domains of $\alpha_{jk} > 1$ and $\alpha_{jk} < 1$ must be separated by a locus of

$$\alpha_{jk} = 1 \tag{A.130}$$

The equation for this locus in the $x$ space is calculated as follows: From the equilibrium relations (5.45) and Eq. (A.130) one obtains

$$y_j = A_{jk}x_j, \qquad y_k = A_{jk}x_k \tag{A.131}$$

$$y_i = A_{jki}x_i \qquad \text{for all } i \neq j, k \tag{A.132}$$

where

$$A_{jk} \equiv K_{jk}^{1/(\nu_j-\nu_k)} \qquad \text{and} \qquad A_{jki} \equiv K_{ij}^{1/\nu_i}A_{jk}^{\nu_j/\nu_i} \tag{5.61}$$

In the general relation $\Sigma_i\, x_i = 1$, one can express $x_j$ and $x_k$ in terms of $y_j$ and $y_k$ by means of Eqs. (A.131), and can then eliminate $y_j$ and $y_k$ with the general relation $\Sigma_i\, y_i = 1$. Expressing the other $y_i$ in terms of $x_i$ by means of Eqs. (A.132), one obtains after rearrangement

$$\sum_{i\neq j,k}\left(\frac{A_{jk}-A_{jki}}{A_{jk}-1}\,x_i\right)-1 = 0 \tag{5.60}$$

As a linear relation between the $x_i$ $(i \neq j, k)$, this equation describes a hyperplane in the $x$ space. Since the equation does not involve $x_j$ and $x_k$, the hyperplane is parallel to the $P_jP_k$ edge of the $x$-space simplex. The hyperplane exists unless $\nu_j = \nu_k$, in which case $A_{jk}, A_{jki} \to \infty$.

## 2. *Reversal Criteria*

For the system to have a $j{:}k$ reversal, the respective reversal hyperplane must intersect the simplex of the physically meaningful compositions ($x_i \geqq 0$ for all $i$, $\Sigma_i x_i = 1$). This is the case if one pair of edges $P_jP_i$ and $P_kP_i$ are intersected. To calculate the intercepts marked by the hyperplane on these edges or their prolongations, one equates all but one $x_i$ to zero in Eq. (5.60) and solves for the nonzero $x_i$:

$$x_i = (A_{jk} - 1)/(A_{jk} - A_{jki}) \tag{A.133}$$

If the hyperplane is to intersect the edge within the range of meaningful composition, the intercept (A.133) must be in the interval

$$0 < x_i < 1 \tag{A.134}$$

In view of Eq. (A.133), this requires

$$A_{jk} > 1 > A_{jki} \tag{A.135}$$

or

$$A_{jk} < 1 < A_{jki} \tag{A.136}$$

Since $K_{jk} > 1$ by virtue of the convention of numbering species, condition (A.135) requires $\nu_j > \nu_k$, and condition (A.136) requires $\nu_j < \nu_k$, to ensure $A_{jk} > 1$ and $A_{jk} < 1$, respectively. With the definitions (5.61) and with $K_{jk} > 1$ and $K_{ij}K_{jk} = K_{ik}$ [as follows from Eqs. (5.45)], the pairs of conditions $\{\nu_j > \nu_k, A_{jki} < 1\}$ and $\{\nu_j < \nu_k, A_{jki} > 1\}$ are readily shown to reduce both to

$$K_{ij}^{\nu_k} > K_{ik}^{\nu_j} \qquad (j < k) \tag{5.62}$$

Accordingly, if this condition is obeyed, the $j{:}k$ reversal hyperplane intersects the simplex edges $P_jP_i$ and $P_kP_i$. Therefore, the system has a $j{:}k$ reversal if this condition is obeyed for any $i \neq j, k$.

It can be shown with condition (5.62) that $1{:}k$ and $k{:}n$ reversals require $\nu_1 > \nu_k$ and $\nu_k < \nu_n$, respectively, for any $k$, and that $1{:}n$ reversals cannot occur. For the $1{:}k$ reversal, one has $j = 1$ and, hence, $K_{ij} < 1$, $K_{ik}$ for any $k \neq 1$ and all $i \neq 1, k$. Condition (5.62) can then be obeyed only if $\nu_j > \nu_k$. An analogous derivation for the $k{:}n$ reversal shows that, here,

$\nu_k < \nu_n$ is required. A $1{:}n$ reversal would then require both $\nu_1 > \nu_n$ and $\nu_1 < \nu_n$ and, therefore, cannot occur.

A simple extension of this derivation shows that $1{:}k$ and $k{:}n$ reversal hyperplanes cannot intersect simplex edges $P_1P_i$ $(1 < i < k)$ and $P_iP_n$ $(k < i < n)$, respectively. For all $i$ with $1 < i < k$, one has $K_{i1} < 1$ and $K_{ik} > 1$ by virtue of the convention of numbering the species; this precludes condition (5.62) from being obeyed for $j = 1$ and such $i$, so that the respective $P_1P_i$ edges of the simplex are not intersected by the $1{:}k$ reversal hyperplane. The deduction for $k{:}n$ reversals is analogous.

That the system has at least one reversal, $j{:}k$ or $k{:}l$, if

$$\nu_j, \nu_l \geqq 2\nu_k \qquad (j < k < l) \tag{5.63}$$

for any three species $j$, $k$, and $l$ can be derived as follows. According to condition (5.62), sufficient conditions for occurrence of $j{:}k$ and $k{:}l$ reversals are

$$K_{jl}^{\nu_k} < K_{kl}^{\nu_j} \qquad \text{and} \qquad K_{jl}^{\nu_k} < K_{jk}^{\nu_l} \tag{A.137}$$

respectively (note that $K_{jl} = 1/K_{lj}$). Since $K_{jk}K_{kl} = K_{jl}$, condition (5.63) ensures that at least one of the two conditions (A.137) is met, except only for the singular case $K_{jk} = K_{kl}$, $\nu_j = \nu_l = 2\nu_k$.

### 3. *Positions of Reversal Hyperplanes*

The position of a given $j{:}k$ reversal hyperplane in the $x$-space simplex is readily calculated. The intersections of the hyperplane with any pair of simplex edges $P_jP_i$ and $P_kP_i$ are given by Eq. (A.133). The intersections, if any, with edges $P_lP_m$ $(l, m \neq j, k)$ are obtained from Eq. (5.60) with $x_i = 0$ for all $i \neq l, m$; eliminating $x_m$ with $x_m = 1 - x_l$ and solving for $x_l$, one finds

$$x_l = (A_{jkm} - 1)/(A_{jkm} - A_{jkl}) \tag{A.138}$$

The reversal planes in the four-component system shown in Fig. 5.13, for example, have been calculated with Eqs. (A.133) and (A.138).

The conditions listed in Table 5.1 for the positions of the potential reversal planes in four-component systems are a compilation of the conditions for intersection of the respective simplex edges, that is, for the respective $x_i$ in Eq. (A.133) or $x_l$ in Eq. (A.138) to be positive but smaller than unity. For any other conceivable positions of the reversal planes, the various requirements as to the relative magnitudes of the $\nu_i$ and the $K_{ij}$ are found to be incompatible.

# GLOSSARY OF SYMBOLS

The accompanying table lists the symbols, definitions, and dimensions of the principal quantities. Quantities used only once are not included and are fully defined where appearing in the text. The dimensions, stated in the last column, may or may not be used as units; in particular, in some applications, concentrations are preferably expressed in terms of equivalents rather than moles.

The following conventions as to notation have been adopted:

*Indices.* Indices normally refer to species or are index numbers of $H$-function roots. Indices 1, 2, 3, ... refer to specific species or roots (for numerical sequence of species, see Chapter 2, Section III). Indices $i, j, k$, ... are used to distinguish between different arbitrary species or roots; unless specifically stated it is not implied that the numerical sequence is the same as the alphabetical. However, $n$ and $n-1$ are always the highest species number and root number, respectively.

Double indices in separation factors, resolution distances and times, selectivity coefficients, etc., refer to pairs of species. Where double indices appear in concentrations or $H$-function roots, the first index is the species or root number and the second refers to the location (plateau zone, pulse, etc., as noted in the text). In double indices, commas are used only where ambiguity could otherwise arise. Thus, $\alpha_{ki}$, but $\alpha_{k,i+1}$.

Other index notation needed in special cases is defined in the accompanying text (see also Table of Symbols).

*Primes, Asterisks, Superscripts.* Primes, asterisks, and superscripts are normally used to designate relative locations, as defined in the accompanying text. Most commonly, superscripts zero refer to the column inlet or an influent pulse; asterisks refer to the column exit or to a plateau zone or pulse crest; primes and double primes refer to the upstream and downstream sides of a boundary, or to the influent and presaturant in systems with single abrupt influent composition changes; primes in systems with influent composition pulses refer to the base influent; single, double, triple, ... primes in systems with multiple influent composition changes refer to successive influents (in the sequence from final influent to presaturant, see p. 220).

*Series.* Series of species or variables in uninterrupted numerical order of the species number or index number are specified by their first and last members:

$$\text{species } 2, \ldots, 5 \equiv \text{species } 2, 3, 4, \text{ and } 5$$

$$h_2, \ldots, h_6 \equiv h_2, h_3, h_4, h_5, \text{ and } h_6$$

For variables, this notation is used instead of vectors where the individual members rather than the set are of interest.

*Vectors.* Where appropriate, complete sets of values of variables are given as vectors, in bold face. Indices are retained only if needed for unambiguity:

$$\mathbf{x} \equiv \{x_1, \ldots, x_n\}, \qquad \mathbf{h} \equiv \{h_1, \ldots, h_{n-1}\},$$

$$\boldsymbol{\alpha}_{1i} \equiv \{\alpha_{11}, \ldots, \alpha_{1n}\}$$

*Sums and Products.* Sums and products extend over all existing members carrying the index stated under the operator, except for specified exclusions. Thus,

$$\sum_i x_i \equiv \sum_{i=1}^{n} x_i, \qquad \prod_i h_i \equiv \prod_{i=1}^{n-1} h_i,$$

$$\prod_{i \neq j} \alpha_{ki} \equiv \prod_{i=1}^{j-1} \alpha_{ki} \prod_{i=j+1}^{n} \alpha_{ki}$$

In the text, the index follows the operator, e.g., $\Sigma_i x_i$.

## Table of Symbols

| Symbol | Quantity | Definition or first use | Dimension |
|---|---|---|---|
| $a_i$ | Coefficient in Langmuir equation | Eq. (5.24) | $cm^3/g$ |
| $A_{jk}$, $A_{jkl}$ | Constants in equation for reversal hyperplanes | Eq. (5.61) | Dimensionless |
| $b_i$ | Coefficient in Langmuir equation | Eq. (5.24) | $cm^3$/mole |
| $c$ | Total concentration in mobile phase per unit volume of mobile phase ($c \equiv \Sigma_i c_i$) | Eq. (5.13) | mole/$cm^3$ |
| $c_i$ | Mobile-phase concentration of species $i$ per unit volume of mobile phase | Eq. (2.1) | mole/$cm^3$ |
| $C$ | Total concentration in mobile phase per unit volume of column ($C \equiv \Sigma_i C_i$) | Eq. (2.3) | mole/$cm^3$ |
| $C_i$ | Mobile-phase concentration of species $i$ per unit volume of column ($C_i \equiv \varepsilon c_i$) | Eq. (2.1) | mole/$cm^3$ |
| $\bar{C}$ | Total concentration in stationary phase per unit volume of column ($\bar{C} \equiv \Sigma_i \bar{C}_i$) | Eq. (2.3) | mole/$cm^3$ |
| $\bar{C}_i$ | Stationary-phase concentration of species $i$ per unit volume of column | Eq. (2.1) | mole/$cm^3$ |
| $\tilde{C}_i$ | Overall concentration of species $i$ in both phases per unit volume of column ($\tilde{C}_i \equiv \bar{C}_i + C_i$) | Eq. (2.2) | mole/$cm^3$ |
| $D_i$ | Effective axial diffusion coefficient of species $i$ | Eq. (5.86) | $cm^2$/sec |
| $f(\ldots)$ | Function of ... | Eq. (2.12) | As required |
| $f_i(\mathbf{x})$ | Stationary-phase concentration $y_i$ in equilibrium with mobile-phase composition $\mathbf{x}$ | Eq. (5.64) | Dimensionless |
| $f_{jk}'$ | Partial derivative of isotherm function | Eq. (5.66) | Dimensionless |
| $F(h)$ | Function $F(h) \equiv \Sigma_i [x_i/(h-\alpha_{1i})]$ | page 363 | Dimensionless |
| $g_i$ | Modified $H$-function root for displacement development ($g_{i-1} \equiv \alpha_{21} h_i$) | Eq. (4.109) | Dimensionless |
| $h$ | Argument of $H$ function | Eq. (3.51) | Dimensionless |
| $h_i$ | $i$th Root of $H$ function | Eq. (3.51) | Dimensionless |
| $h(j)$ | $H$-function root attributable to trace species $j$ | page 255 | Dimensionless |
| $h_i$ | Bulk-system root in trace-component system with variable bulk composition | page 255 | Dimensionless |
| $H(q, \mathbf{r}, \mathbf{s})$ | $H$ function of arguments $q$, $\mathbf{r}$, $\mathbf{s}$ | Eq. (3.53) | Dimensionless |
| $i, j, k$, etc. | Numbers of arbitrary species | page 16 | None |
| $i \mid j$, $j \mid k$, etc. | Affinity cuts | page 57 | None |
| $i{:}j$, $j{:}k$, etc. | Selectivity reversals | page 305 | None |
| $i, j \rightarrow j, k$, etc. | Influent sequences | page 163 | None |

**Table of Symbols**—*cont.*

| Symbol | Quantity | Definition or first use | Dimension |
|---|---|---|---|
| $J_i$ | Flux of species $i$ in axial direction | Eq. (3.15) | mole cm$^{-2}$ sec$^{-1}$ |
| $k_i$ | Sorption rate coefficient of species $i$ | Eq. (5.76) | sec$^{-1}$ |
| $K_{ij}$ | Mole-fraction (or equivalent-fraction) selectivity coefficient of species $i$ and $j$ | Eq. (5.46) | Dimensionless |
| $K_{ij}{}^0$ | Practical selectivity coefficient of species $i$ and $j$ | Eq. (5.43) | $(g/cm^3)^{\nu_j-\nu_i}$ |
| $m$ | Adjusted generalized time variable | Eq. (5.4) | mole |
| $m^0$ | Generalized time variable; amount of sorbable species introduced since $t = 0$ | Eq. (5.1) | mole |
| $\bar{m}$ | Generalized distance variable; amount of sorbable species in column upstream of location | Eq. (5.1) | mole |
| $n$ | Number of sorbable species | page 10 | None |
| N | Nonsharpening boundary (for subscripts and superscripts see page 163) | Table 3.2 | None |
| $p$ | Pressure | Eq. (5.97) | atm |
| $P$ | Constant characteristic of $j \mid k-k \mid l$ common plane | Eqs. (3.83) and (3.100) | Dimensionless |
| $P_k$ | Constant in equation for velocity of $k$th boundary | Eq. (4.18) | Dimensionless |
| $\mathrm{P}_i$ | Composition point ($x_i = 1$, $y_i = 1$) | Fig. 3.9 | None |
| $\mathrm{P}_{ij}$ | Composition point of plateau zone $ij$ | Fig. 4.7 | None |
| $\mathrm{P}_{ij}$ | Distance-time point of disappearance of plateau zone $ij$ | Fig. 4.28 | None |
| $q_i$ | Stationary-phase concentration of species $i$ per unit weight of sorbent | Eq. (2.1) | mole/g |
| $R$ | Ratio of total concentrations in stationary and mobile phases ($R \equiv \Sigma_i q_i/\Sigma_i c_i$) | Eq. (5.30) | cm$^3$/g |
| $R_j$ | Enrichment factor of species $j$ | Eq. (4.61) | Dimensionless |
| $S$ | Cross-sectional area of column | Eq. (5.1) | cm$^2$ |
| S | Self-sharpening boundary (for subscripts and superscripts see page 163) | Table 3.2 | None |
| $t$ | Time | page 21 | sec |
| $T$ | Absolute temperature | Eq. (5.95) | deg |
| $u$ | Composition velocity | Eq. (3.37) | cm/sec |
| $u_i$ | Species velocity of species $i$ | page 22 | cm/sec |
| $u_0$ | Velocity of mobile-phase flow | page 21 | cm/sec |
| $u_\Delta$ | Velocity of coherent composition step | Eq. (3.39) | cm/sec |
| $u_{C_i}$, $u_{x_i}$, $u_{\Delta C_i}$, $u_{h_i}$, etc. | Velocities of entities $C_i$, $x_i$, $\Delta C_i$, $h_i$, etc. | Eq. (2.17) | cm/sec |

## Table of Symbols—*cont.*

| Symbol | Quantity | Definition or first use | Dimension |
|---|---|---|---|
| $\mathscr{u}$ | Adjusted velocity (subscript notation same as for $u$) | Eq. (2.19) | Dimensionless |
| $\dot{V}$ | Volumetric flow rate | Eq. (5.1) | $cm^3/sec$ |
| W | Watershed point | Fig. 3.9 | None |
| $W_i$, $W_i'$, etc. | Watershed points on border $x_i = 0$ in systems with more than three components | Fig. 3.10 | None |
| $x_i$ | Normalized concentration of species $i$ in mobile phase | Eq. (2.4) | Dimensionless |
| $y_i$ | Normalized concentration of species $i$ in stationary phase | Eq. (2.4) | Dimensionless |
| $z$ | Distance from column inlet | page 21 | cm |
| $z^*$ | Distance variable for constant pattern | Eq. (5.79) | cm |
| $z_i$ | Distance for disappearance of flat top of $i$th response pulse | Table 4.3 | cm |
| $z_0$ | Distance for disappearance of plateau zone having composition of influent pulse | Fig. A.5 | cm |
| $z_{ij}$ | Resolution distance of species $i$ and $j$ | Eq. (4.68) | cm |
| $z_{i/j}$ | Resolution distance of $i$th and $j$th pulses | Eq. (4.53) | cm |
| $z(h_i, \tau)$ | Trajectory of root value $h_i$ | Eq. (A.62) | cm |
| $z_{\Delta i}(\tau)$ | Trajectory of step of root $h_i$ | Fig. A.4 | cm |
| $Z$ | Column length | Eq. (5.99) | cm |
| $\alpha$ | Product of separation factors ($\alpha \equiv \Sigma_i \alpha_{1i}$) | Eq. (A.48) | Dimensionless |
| $\alpha_{ij}$ | Separation factor of species $i$ and $j$ | Eq. (2.8) | Dimensionless |
| $\varepsilon$ | Fractional interparticle void volume | Eq. (2.1) | Dimensionless |
| $\lambda$ | Eigenvalue, corresponding to reciprocal adjusted composition velocity | Eq. (5.67) | Dimensionless |
| $\lambda_i$ | Concentration parameter of species $i$ in mobile phase | Eq. (5.82) | Dimensionless |
| $\nu_i$ | Stoichiometric coefficient of species $i$ | Eq. (5.42) | Dimensionless |
| $\tau$ | Adjusted time | Eq. (2.18) | cm |
| $\tau_i$ | Adjusted time for disappearance of flat top of $i$th response pulse | Table 4.3 | cm |
| $\tau_0$ | Adjusted time for disappearance of plateau zone having composition of influent pulse | Fig. A.5 | cm |
| $\tau_{ij}$ | Adjusted resolution time of species $i$ and $j$ | Eq. (4.68) | cm |
| $\tau_{i/j}$ | Adjusted resolution time of $i$th and $j$th pulses | Eq. (4.54) | cm |
| $\omega$ | Directional coordinate of composition route | Eq. (5.72) | Dimensionless |
| $\omega_i$ | Concentration parameter of species $i$ in stationary phase | Eq. (5.82) | Dimensionless |

**Table of Symbols**—*cont.*

| Symbol | Quantity | Definition or first use | Dimension |
|---|---|---|---|
| $\Delta C_i$, $\Delta \bar{C}_i$, $\Delta x_i$, $\Delta h_i$, etc. | Differences in $C_i$, $\bar{C}_i$, $x_i$, $h_i$, across composition step | Eq. (3.19) | Dimension of respective unit |
| $\Delta z$ | Length of band in displacement development | page 239 | cm |
| $\Delta \tau$ | Duration of influent pulse, in units of adjusted time | Table 4.3 | cm |
| $\Delta \tau$ | Adjusted time required for introduction of mixture in displacement development or selective displacement | Eq. (4.88) | cm |
| $\Delta \tau'$ | Adjusted time required for introduction of first development agent in selective displacement | Fig. 4.32 | cm |
| 1, 2, 3, etc. | Species numbers (for notation $1 \mid 2$, $1:2$, $1, 2 \rightarrow k$, etc., see under $i, j, k$; for numerical sequence of species see page 16) | page 11 | None |

# AUTHOR INDEX

Numbers in parentheses are reference numbers and indicate that an author's work is referred to although his name is not cited in the text. Numbers in italics give the page on which the complete reference is listed.

# SUBJECT INDEX

## D

S